Integrated Greenhouse Systems for Mild Climates

Christian von Zabeltitz

Integrated Greenhouse Systems for Mild Climates

Climate Conditions, Design, Construction, Maintenance, Climate Control

Springer

Professor em. Dr.-Ing.
Christian von Zabeltitz
Leibniz University Hannover
Faculty of Natural Science
Institute for Biological Production System
Section Biosystems and Horticultural
Engineering
Hellwiesen 3
30900 Wedemark
Germany
zabeltitz.meitze@web.de

ISBN 978-3-642-14581-0 e-ISBN 978-3-642-14582-7
DOI 10.1007/978-3-642-14582-7
Springer Heidelberg Dordrecht London New York

Library of Congress Control Number: 2010937857

Cover design: WMXDesign GmbH, Heidelberg, Germany

Printed on acid-free paper

Springer is part of Springer Science+Business Media (www.springer.com)

Preface

Greenhouse crop production in rural areas is an increasing industry in countries with mild climate conditions, even in tropical regions with appropriate temperatures throughout the year. It provides a major source of income even for small scale growers, and fresh food with good quality for the population. Conventional open-field cultivation is either not possible in some regions (arid or tropical zones) or faces a number of problems, which protected cultivation offers the possibility of resolving. The import of highly sophisticated greenhouse technology from industrialized countries can fail because the technology is not adapted to the local conditions and climate, and due to an insufficient capital return. Plastic film greenhouses predominate in regions with mild climate. Very often, they are designed on the principle of minimum capital input. The aim should be to design low-cost systems with high technological and economical efficiency.

The contents of this book are the results of activities as researcher and teacher at the Leibniz University of Hannover (Germany), of consultancy missions in developing countries, and of the cooperation in a joint research project between the Asian Institute of Technology (AIT), the Leibniz University of Hannover and the Kasetsart University in Bangkok. The author worked for many years as consultant for appropriate greenhouse design in numerous tropical, subtropical and arid countries. The consultancy missions were carried out by order of the Food and Agriculture Organization of the United Nations, FAO, and the German Agency for Technical Cooperation, GTZ, and for private companies.

Important research results of internationally well-known scientists were taken into consideration, as well as German results and publications, written in the German language, which have not been considered in international references so far. The latest reference literature has been considered up to July 2009.

The author is thankful for the possibility to collect all these experiences and to give them back to whoever it may concern.

This book gives instructions for technical actions in practice (how to do and how not to do!), and in that way it may also stimulate engineers in the industry, teachers, students, extension service and growers to design appropriate greenhouses under varying local climate conditions. Many drawings and pictures support reading and understanding.

Wedemark, Germany Christian von Zabeltitz
June 2010

Contents

Chapter 1
Introduction

Crop production in greenhouses is an increasing industry, especially in mild climates, and is very important for the population as a source of income and clean fresh food. Greenhouses create optimal climate conditions for crop growth, and protect the crop from outside pests. At the same time, greenhouse production increases water use efficiency and provides the possibility of integrated production and protection (IPP). The environmental impact has to be considered for protected cultivation. One tool for evaluating environmental impact is, for example, the Life Cycle Assessment (LCA) (Montero et al. 2003, 2009a).

Greenhouse crop production is successful if growers produce high quality and yield.

High quality and yield depend on:

- Efficient management of production
- Knowledge and training of the growers
- Appropriate greenhouse structure
- Proper mounting, installation and maintenance of the system
- Efficient climate control during summer and winter
- Measures for integrated production and protection (IPP)

The question is:

How to produce more clean vegetables and better quality with less water, with less land and with less pesticide?

Integrated production and protection (IPP) aims at finding alternative solutions to improve yield and quality, and to reduce pesticide applications. IPP contains the following components:

- Plant materials
- Cultivation practices
- Greenhouse technology, climate control, water-use efficiency and fertigation
- Biological control and the use of bio-pesticides
- Quality norms and standards

C. von Zabeltitz, *Integrated Greenhouse Systems for Mild Climates*,
DOI 10.1007/978-3-642-14582-7_1,

A greenhouse with all measures of climate control is one important component for integrated greenhouse crop production and protection.

However, the growing period is very often limited to several months in the year, because of insufficient ventilation and cooling in summer as well as heating in winter. Structures and shapes of greenhouses are adapted insufficiently to climate conditions. Climate control and greenhouse structures have to be adapted and optimized with regard to outside climate, and growers must be trained in these adapted technologies (Baille 2001).

A greenhouse structure with light transmittance (through framework and cladding material), ventilation, heating, cooling, and protection from pest insects by screening, as well as all influences of management, irrigation, fertilisation, water quality, physical and biological plant protection, has to be considered as an integrated system. Greenhouse structures should be designed according to the climate conditions and to the general design criteria, as well as to the locally available and cost-effective materials. The import of greenhouses from countries with a different climate fails very often, because they are not adapted to the local conditions. Even an appropriate greenhouse structure can fail if mounting and continuous maintenance is not guaranteed.

Temperatures and global radiation allow open-air crop production throughout the year in many subtropical and tropical countries.

But the *problems for open air cultivation* are:

- High solar radiation intensities
- Low temperatures in some regions, in particular at night
- Damage of plants by heavy rainfall, flooding, and surface run-off
- Shortage of water in dry seasons
- Plant diseases due to rain, high humidity, and storm damage
- High weed infestation
- Erosion and decomposition of humus by floods and surface run-off
- Leaching of fertilizers
- Increasing use of pesticides, which are washed off by rain
- High evapotranspiration by global radiation
- Working conditions for people
- Climatic conditions in the open field which hamper a controlled cultivation for production deadlines and disease control

Advantages of protected cultivation in greenhouses are:

- Protection from heavy rainfall, too high global radiation and wind
- Physical plant protection by insect screens in front of the ventilators
- Possibility of biological plant protection
- Carefully directed fertigation (irrigation and fertilisation)
- Better efficiency of pesticides and fertiliser
- Lower transpiration and evaporation
- Water conservation by efficient irrigation and collecting of rainwater
- Possibility of soil-less culture

- Planned production for production deadlines, cultivating of seedlings
- More favourable working conditions

Crop production under adapted greenhouses results in higher yield and better quality, minor risks for quality and yield, extending of harvest time, and reduced water consumption.

Factors influencing the design technology are (Giacomelli et al. 2008):

1. Regional infrastructure (transportation) and market size (consumers)
2. Local climate conditions that influence construction and climate measures
3. Availability and quality of water
4. Availability and costs of fuel and water
5. Soil conditions
6. Availability of land
7. Availability and conditions of capital for investments
8. Availability and cost of labour
9. Availability of local materials for construction and equipment, including services for repair and maintenance
10. Legislation and government regulations of food safety, residuals of chemicals and emission of chemicals

The *general design criteria* for greenhouses are (see Chap. 4):

- The climate conditions in the region
- The general design requirements, including standards if available
- Measures for climate control and pest control
- The locally available cost-effective materials and life cycle
- Technical measures for integrated production and protection (IPP)

Different structures must be considered for different regions and various purposes.

(a) Small-scale farming. Small units of greenhouses for rural families to improve the productivity of the small area of their land. Low-cost but efficient constructions are required.
(b) Large-scale market gardening in rural areas around big towns and densely populated areas. Main source of income is horticulture. Larger investments for structure are possible.

Consumers want a year-round supply of high-quality products, but usually vegetable crops in particular cannot be grown during hot summertime in subtropical climates due to insufficient applied technologies (Castilla et al. 2008). The trend for crop growth management in subtropical, Mediterranean greenhouses was in the past to adapt the plants to a suboptimal environment, instead of optimizing and adapting the greenhouse design and climate control measures for a maximum of plant yield, quality, and health (Montero 2009; Castilla et al. 2008; Castilla 2002).

Investigations have been conducted to find out the most profitable and economically viable technologies. Greenhouses have been designed as simple locally made plastic-film structures or screenhouses, as well as industrial-type multispan structures with all the equipment for climate control. Properly designed and adapted locally made, low-cost greenhouse structures with limited climate control can fulfil similar cost–benefit ratios to high sophisticated greenhouse and climate control systems, if some basic rules of design and climate control are taken into consideration. This is the case in "normal" years without unusual and extreme outside climate conditions and pest infestations. For more extreme conditions, more attention should be paid to adapted and improved greenhouse systems including ventilation, cooling and pest control by mechanical and optical barriers.

Two strategies can be used to achieve year round supply with horticultural products (Giacomelli et al. 2008; Castilla and Hernandez 2007; Montero 2009):

1. High technology greenhouses at one site with all necessary design components and climate control measures.
2. Adapted greenhouses at different sites with complementary climate conditions and harvesting periods, for example greenhouses in coastal areas during winter and spring and screenhouses in highlands during hot summer season.

The greenhouse area in the world increases yearly. Most of the greenhouses are built in mild climate areas, and more than 90% of the greenhouses are plastic film greenhouses. Therefore, this book focuses in particular on plastic-film greenhouses. The estimated greenhouse areas in the world are (Giacomelli et al. 2008):

	Plastic film greenhouses and large plastic film tunnels (ha)	Glasshouses (ha)
Western Europe	140,000	29,000
Eastern Europe	25,000	1,800
Africa	27,000	600
Middle East	28,000	13,000
North America	9,800	1,350
Central/South America	12,500	0
Asia/Oceania	450,000	2,500

Some existing books describe more the theoretical bases for greenhouse design and climate control (Castilla 2005; Bakker et al. 1995; von Zabeltitz 1986; Tantau 1983).

This book describes the climatic conditions in various regions in the first part, and deals then with the corresponding demands on the structures for protected cultivation and materials as well as on climate control measures. It gives instructions for technical actions, and shows by many examples of greenhouse construction, mounting, ventilation, heating, cooling, irrigation, screening, and rainwater collection "*how to do and how not to do*".

Chapter 2
Climate Conditions and Classification

2.1 The Climate Elements

The climatic elements to be considered for open-air and protected cultivation of plants are (von Zabeltitz and Baudoin 1999):

- Solar radiation
- Temperature
- Precipitation
- Humidity
- Evaporation and evapotranspiration
- Wind velocity

The relationship between precipitation and evapotranspiration plays an important role for open-field cultivation. Crops under protected cultivation receive all their water by irrigation systems. If rainwater storage for irrigation is planned, the relationship precipitation to evapotranspiration inside the greenhouse has to be taken into consideration.

2.1.1 Solar Radiation

The solar radiation at the edge of the earth's atmosphere, called solar constant, is 1,349 kW/m^2. Solar radiation is reduced in the earth's atmosphere by reflection, absorption and scattering, so that only part of it reaches the earth's surface. The solar radiation on the earth's surface changes with latitude, season, and time of day, as well as by the various radiation losses in the earth's atmosphere, e.g., clouds and haze.

Figure 2.1 shows the *annual mean of daily solar radiation* energy (kWh/m^2 day) at the earth's surface in relation to the latitude (von Zabeltitz and Baudoin 1999). Average solar radiation increases from very low mean values at the poles up to 20°

C. von Zabeltitz, *Integrated Greenhouse Systems for Mild Climates*,
DOI 10.1007/978-3-642-14582-7_2, © Springer-Verlag Berlin Heidelberg 2011

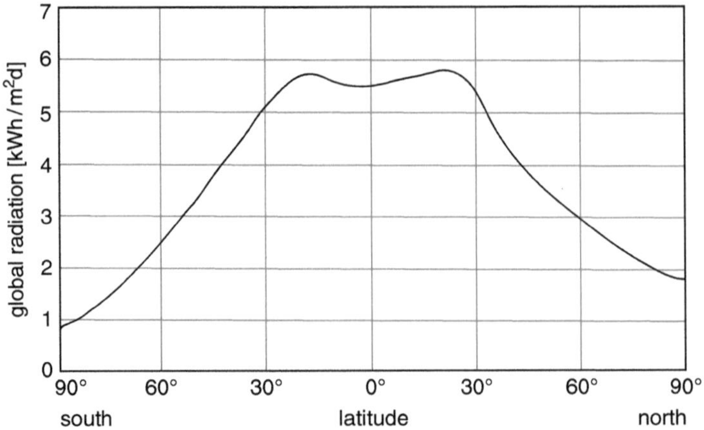

Fig. 2.1 Annual means of global radiation depending on the latitude

Fig. 2.2 Mean daily sum of global radiation for different latitudes. (*1*) Equator zone, (*2*) 25°–30° north, (*3*) 35°–40° north, (*4*) 45°–55° north (von Zabeltitz and Baudoin 1999)

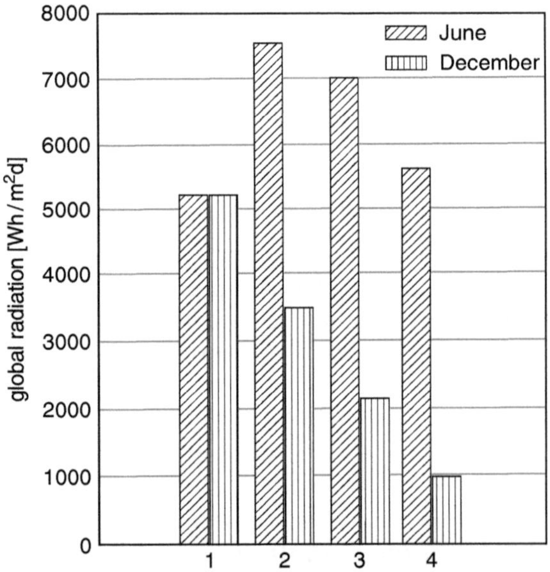

latitude. In the equatorial zone, the mean solar radiation remains almost constant. A high amount of vapor in the atmosphere results here in extremely high radiation losses.

For the production of plants, it is important to know the *monthly mean* of daily solar radiation energy. Figure 2.2 shows the mean daily sum of solar radiation for different latitudes of the northern hemisphere in the months of June and December. The daily solar radiation in the equatorial zone (1) is the same in summer and in winter, while the other zones show considerable differences between summer and winter. Daily solar radiation decreases in summer and winter evenly with growing latitudes. The same is true for the southern hemisphere.

Figure 2.3 indicates the annual course of the daily solar radiation for different locations. Variations are small in the equatorial zone in the course of the year. Some locations show maxima in March/April and in September/October (Mogadishu, Bongabo). Higher latitudes show strong annual amplitude. Solar radiation sums for Mediterranean countries in summer, i.e., in latitudes between 30° and 40°, are often higher than at the equator. This also depends on the different day lengths and on the mean sum of daily hours of sunshine (Fig. 2.4).

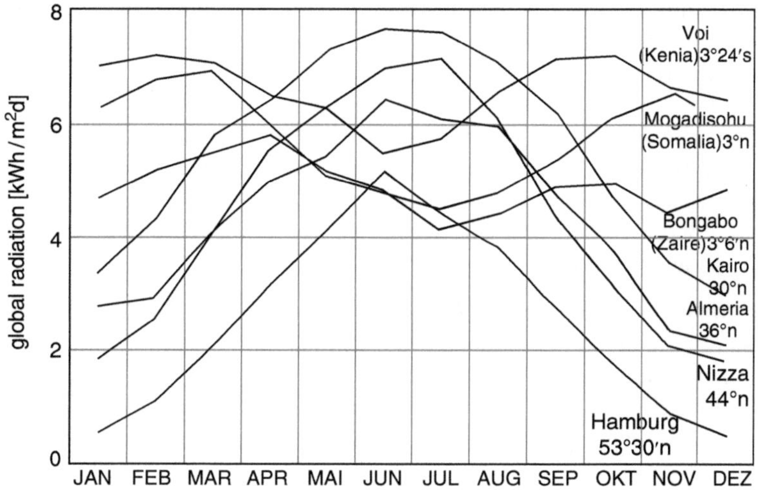

Fig. 2.3 Mean daily sum of global radiation in different months for different locations (von Zabeltitz and Baudoin 1999)

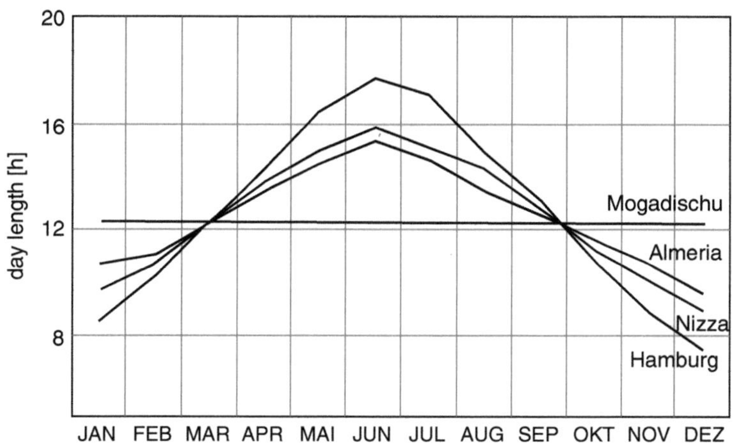

Fig. 2.4 Day lengths of different locations

The production of plant dry matter decreases almost linearly with the radiation at low solar radiation values. The growth comes to a halt at the compensation point, which is at 14–30 W/m^2 light power for plants (0.1 kWh/m^2 day) (Krug et al. 2002). An efficient production of dry matter cannot be expected in higher northern and southern latitudes during winter without artificial lighting.

The minimum amount of irradiation necessary to ensure sufficient growth and flowering corresponds to a daily global radiation of 2.0–2.3 kWh/m^2 day (Nisen et al. 1984). The solar radiation values in the equatorial zone do not drop below the minimum during the year, and do not fall below the minimum even in winter in higher latitudes up to 40°. Vegetable varieties are adapted to the day length. For tropical regions, appropriate varieties must be chosen with regard to the shorter day length during summer. The light loss in greenhouses should be considered.

Most important for photosynthesis, i.e., the growth of plants, is solar radiation power (W/m^2). Figure 2.5 shows the mean solar radiation (W/m^2) in relation to time of day at various locations. The maximum solar radiation at lower latitudes at noon can be very high (Voi, Kenya) because of the relatively short day length in summer. Shading may be necessary if the radiation power becomes too high.

Cloud conditions and altitude above sea level produce significant deviations from the mean radiation distribution.

- Due to *heavy cloud and high precipitation*, radiation in equatorial zones is constantly reduced. In regions with distinct rainy seasons, the decrease in radiation is limited to these seasons.
- With *increasing altitude*, radiation is intensified. Corresponding climatic conditions can be found, e.g., in the tropic plateaus of Ecuador, Colombia and Kenya near the equator.

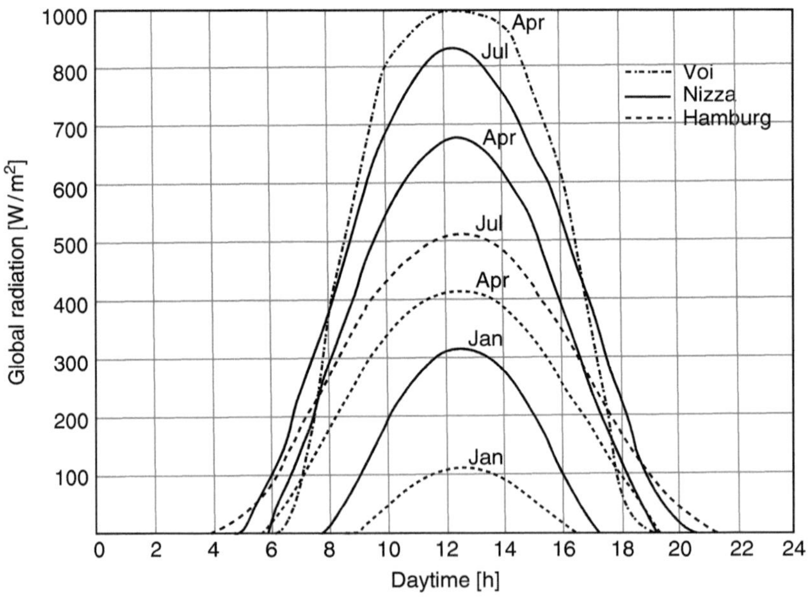

Fig. 2.5 Mean hourly global radiation at different locations (von Zabeltitz and Baudoin 1999)

The mean daily sum of global radiation in the months of the year can be found for many stations all over the world in Müller (1996) and from FAO, Rome (http://www.fao.org/ag/AGLN/climwat.stne).

2.1.2 Temperature

The course of temperature for a location depends on radiation, season and altitude above sea level, distance to seas, wind conditions and cloud conditions. Therefore, it is difficult to make general statements.

The conditions shown in Figs. 2.6 and 2.7 result from considering only mean temperatures up to an altitude of 500 m above sea level (von Zabeltitz and Baudoin 1999). Figure 2.6 shows the mean daily maximum temperatures for the warmest (t_{maxw}) and the coldest (t_{maxc}) month in relation to latitudes of the northern hemisphere. The same is true for southern latitudes.

The mean of maximum temperatures in the warmest months is 33°C at the equator. It rises to 37°C at the 27th latitude. Then it drops to 23°C at the 50th latitude. Scattering around the mean value increases from the equator to the 39th latitude and then decreases. The mean of the maximum temperatures in the coldest month hardly differs from the warmest month at the equator. It only starts sinking considerably from the 18th latitude on. Scattering around the mean grows with increasing latitudes.

Correspondingly, Fig. 2.7 shows the mean daily minimum temperatures in the warmest (t_{minw}) and in the coldest (t_{minc}) month. The course of the minimum

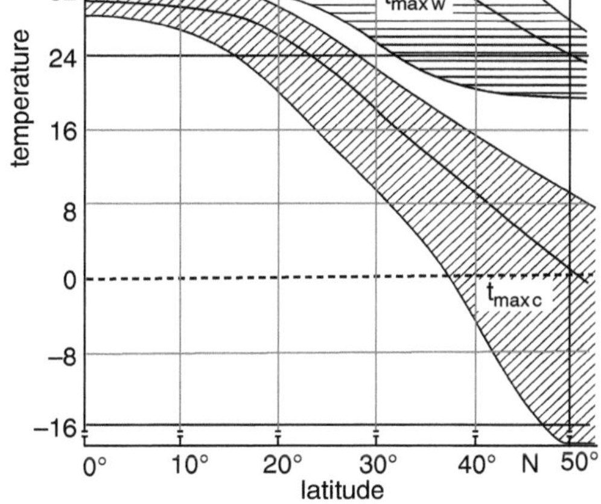

Fig. 2.6 Mean daily maximum temperatures for the warmest (t_{maxw}) and coldest (t_{maxc}) month of the year and extreme maximum temperatures depending on northern latitudes

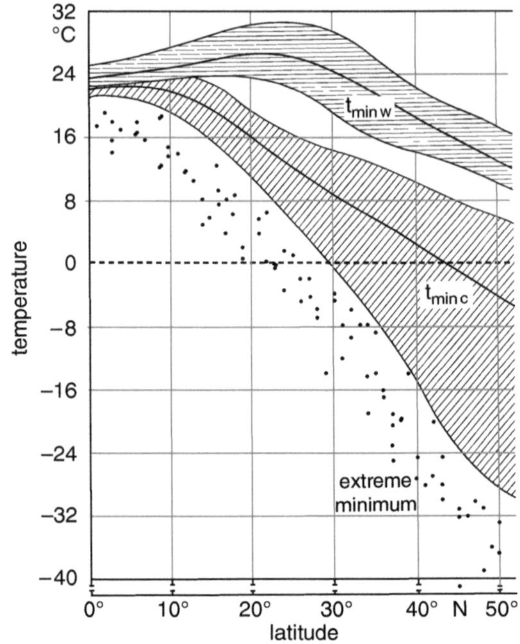

Fig. 2.7 Mean daily minimum temperatures for the warmest (t_{minw}) and coldest (t_{minc}) month of the year and extreme minimum temperatures depending on northern latitudes

temperatures is similar to that for the maximum temperatures; only the temperature scales are shifted.

A survey of the mean temperature variation between the warmest and coldest months is given in Fig. 2.8. The differences of the means of Figs. 2.6 and 2.7 are as follows:

- Latitudes 0°–10°:

The mean temperature differences between the warmest and coldest months are smaller than 5°C.

- Latitudes 0°–23°:

The mean temperature differences are smaller than 13°C.

- Latitudes 23°–48°:

The mean maximum temperature differences between the warmest and coldest months rise continuously up to 24°C at the 48th latitude. The mean minimum temperature differences rise up to ~17°C and then stagnate. The temperature differences between the extreme values are much bigger.

At temperatures between 0°C and 8°C, the growth rate of vegetables is small, and fruit may even be destroyed. Absolute minimum temperatures should be above 0°C. Optimal growth rates are guaranteed at temperatures between 20°C and 30°C.

Compared to those temperature values for vegetable growth, good conditions for vegetable cultivation are given up to the 23th latitude. Open-air cultivation throughout the year is possible.

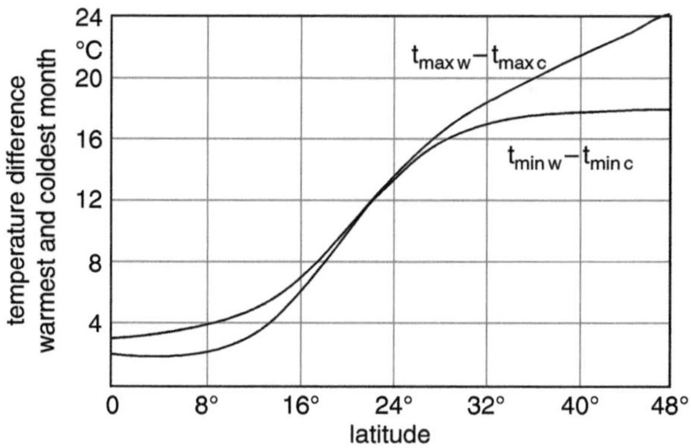

Fig. 2.8 Temperature differences between warmest and coldest month depending on latitude

In the climatic zones between the 23th and 40th latitudes, problems resulting from extreme excess temperatures and the danger of frost have to be considered. A year-round plant production in greenhouses without cooling and heating encounters difficulties.

For more detailed information, the local microclimate of the country or location in question has to be examined, as temperatures also depend on altitude above sea level, distance to the seas, wind, and cloud conditions. Data can be found in Müller (1996) and FAO (http://www.fao.org/ag/AGLN/climwat.stne). More examples and figures for crop growth requirements are given in Chap. 3.

Some main conclusions are:

- Temperatures sink with increasing altitude above sea level. In tropic plateaus and mountains, temperature extremes do not change between summer and winter (Fig. 2.9), but every 12 h between day and night. In spite of high day temperatures, near-zero temperatures can be reached at night in highlands. This cannot be seen from the monthly means.
- The temperature amplitude grows with increasing distance to the seas (continental climate).
- The effects of cold air can advance further down than to the 30th latitude.
- The cloud conditions influence the sun radiation and the long-wave radiation of the earth. During the rainy seasons, temperatures change because of clouds. Lower radiation during the day reduces the warming up; reduced long-wave radiation during the night reduces the cooling. The course of temperatures is more even.

Figure 2.9 shows the extreme maximum and minimum temperatures for the Seychelles at sea level and for Bogota, Colombia, at an elevation of 2,556 m. Even temperatures suitable for vegetable production throughout the year are given at regions near sea level. Extreme minimum temperatures can range below the biological optimum in highlands.

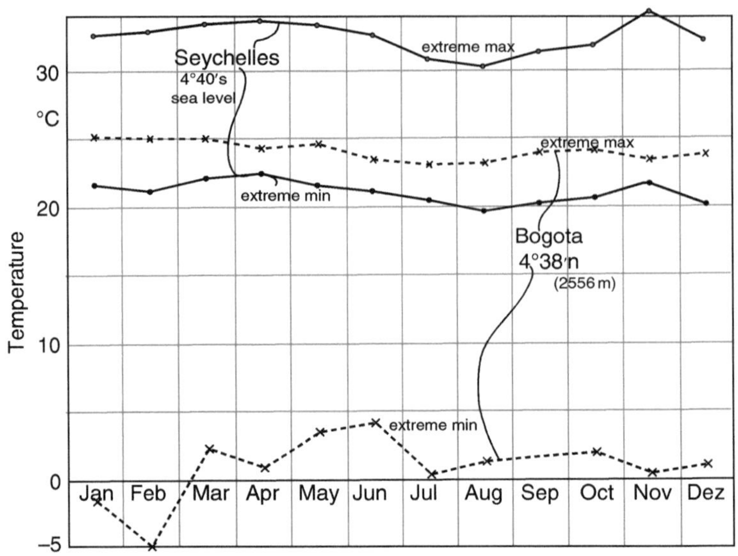

Fig. 2.9 Extreme maximum and minimum temperatures for the Seychelles (sea level) and Bogotá (highland), Colombia

2.1.3 Precipitation

Water is a vital element for plant growth. In order to evaluate protected cultivation in a region, it is necessary to have information about the most important characteristics of rainfall, e.g., total quantity, seasonal distribution, intensity, and frequency. Variability from year to year is also an important factor.

Greenhouses must have gutters in regions with heavy rainfall to drain off the water and to avoid penetration into the greenhouse.

Together with precipitation, evapotranspiration is an important element, especially for the water supply (water balance) in open-air cultivation. Under protected cultivation, the total water consumption of the plants is covered by irrigation systems, but rainwater can be collected for irrigation if enough rainfall is available for economical design of storage facilities (see Chaps.13 and 14). Evapotranspiration in the greenhouse is important in order to estimate water consumption and necessary water quantity for irrigation. A comparison of the crop water requirement and the possible amount of storable rainwater makes it possible to calculate the volume of rainwater storage.

Figures 2.10–2.12 (von Zabeltitz and Baudoin 1999) show the mean annual precipitation in different countries of the earth (Trewartha et al. 1980) and the coefficients of variation as standard deviation in percent of the average (Duckham and Hasefield 1971). Variation in precipitation may change over the years due to global climate change, but the given coefficients give an idea of the possible rainfall.

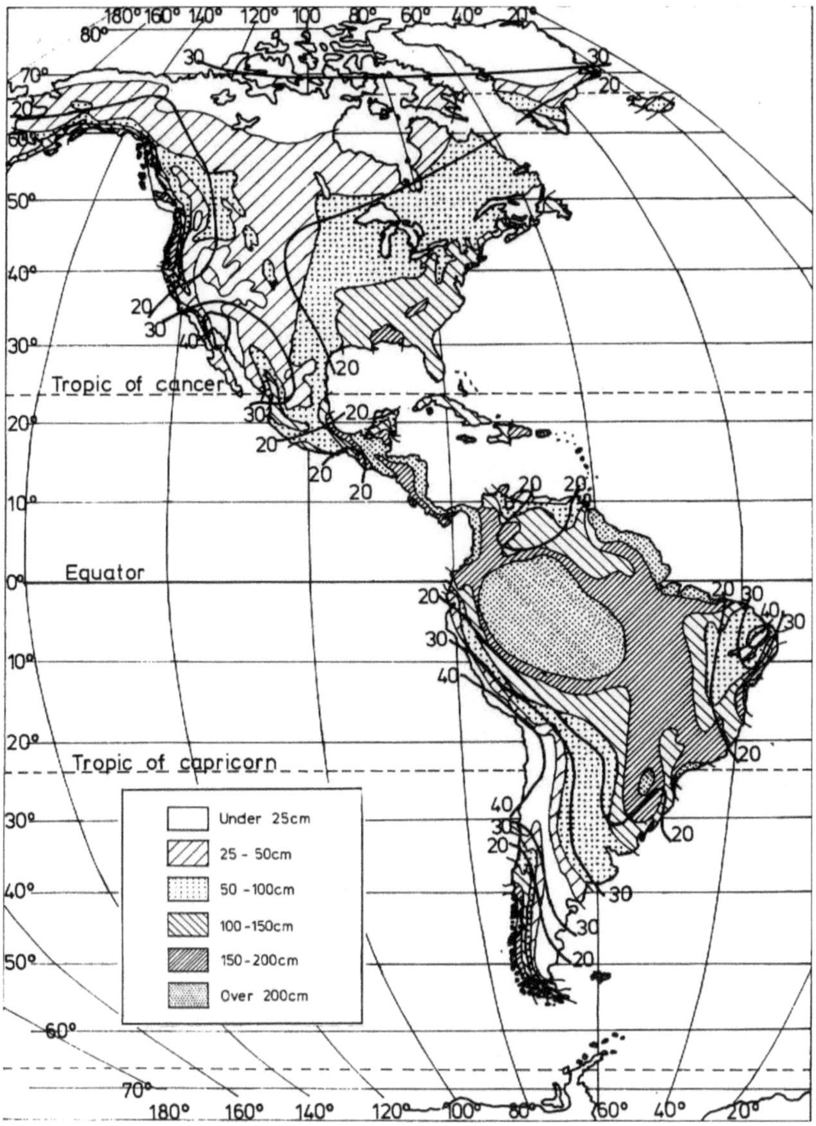

Fig. 2.10 Mean annual precipitation and coefficient of precipitation variation (%) for America

Regions of precipitation are determined by:

- Belts of ascending air at the equator and at polar fronts.
- The direction and kind of moisture-bearing winds, and their relation to ocean currents.
- The distance to the main source of water.

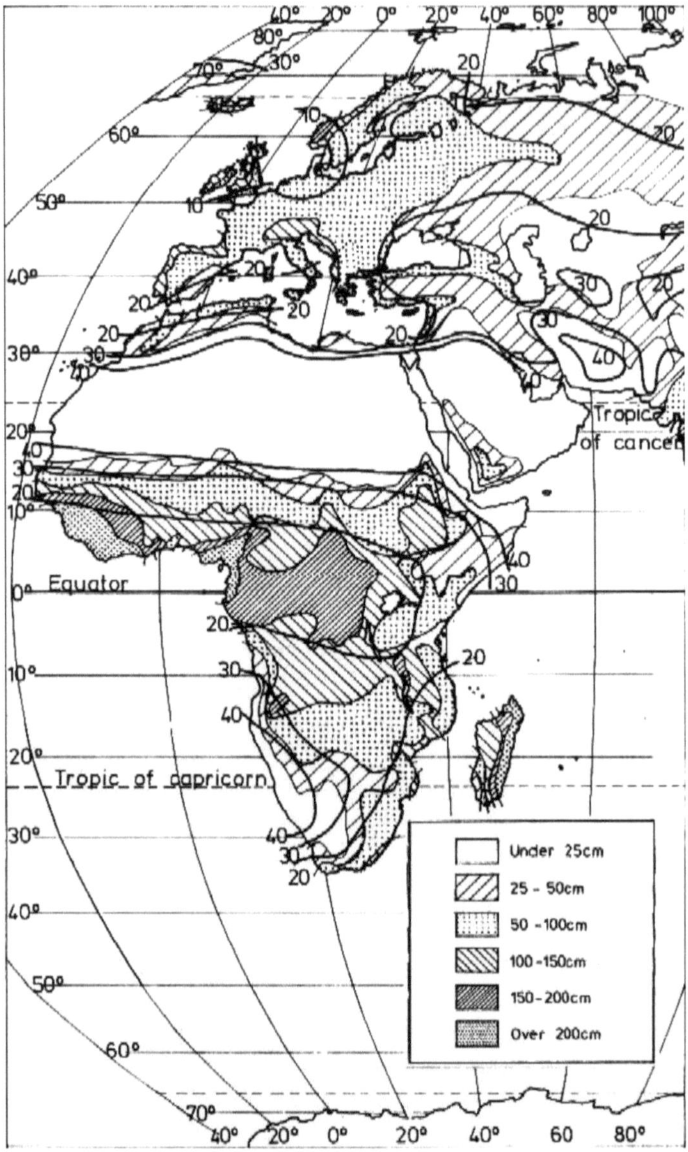

Fig. 2.11 Mean annual precipitation and coefficient of precipitation variation (%) for Europe and Africa

Rainfall in tropical regions tends to occur in high intensities together with storms. Excess rainfall, which the soil can not absorb, flows away as surface run=off. It is not available to the plants, and causes strong soil erosion. Generally speaking, rain intensities larger than 10 mm/h are dangerous with regard to erosion. These intensities, however, occur frequently in tropical regions. Therefore,

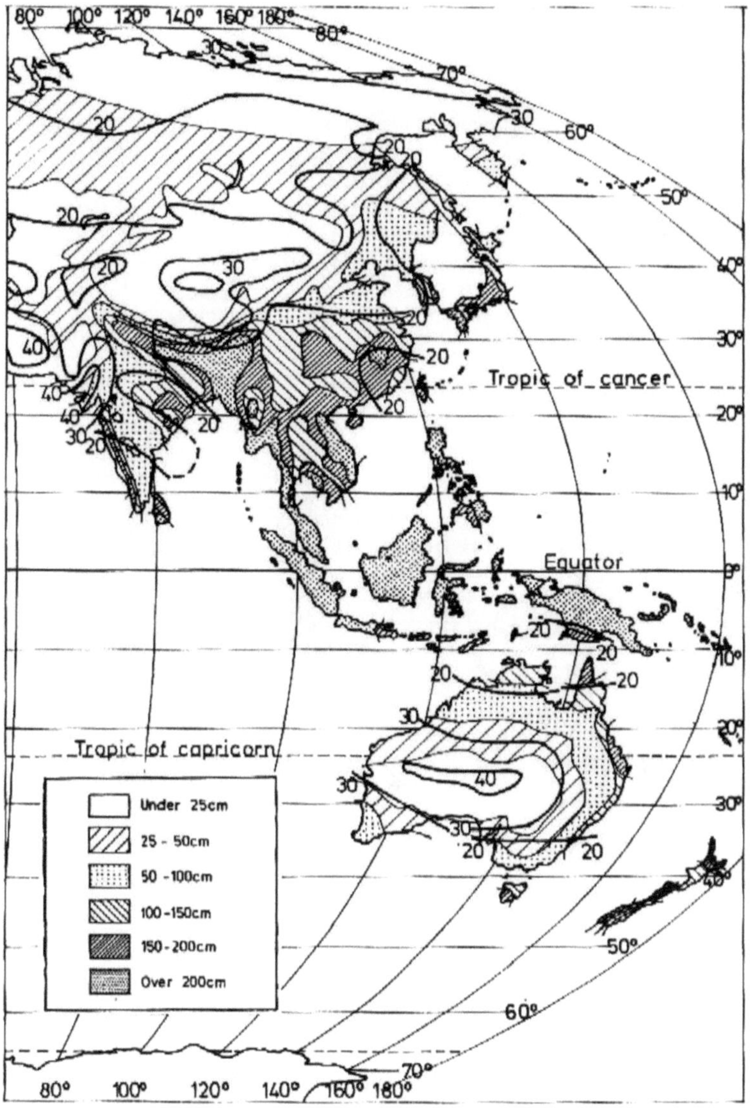

Fig. 2.12 Mean annual precipitation and coefficient of precipitation variation (%) for Asia and Australia

protected cultivation is advantageous even in the tropics, because crop and soil are protected from high rain intensities. The rain can be collected for irrigation in seasons with little rainfall.

Tropical humid climates suffer from intensive rainfall, whereas subtropical climates have large areas with only very little rainfall. Usually, rainfall shows the biggest variation where the total quantity of rain is smallest.

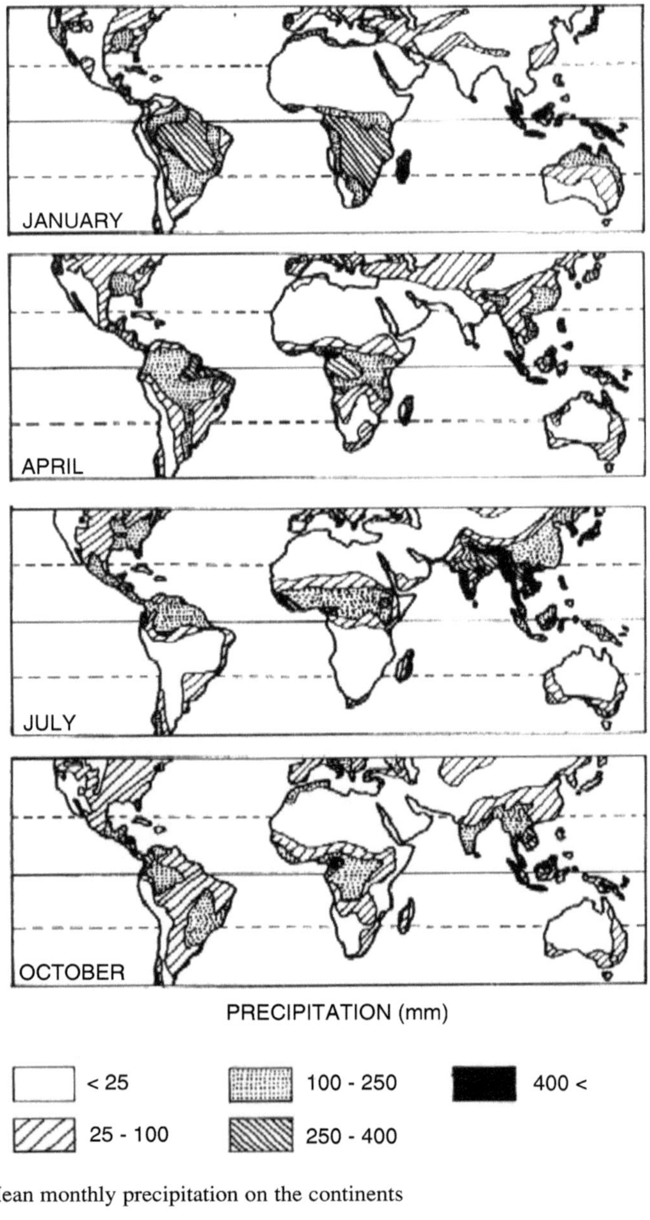

Fig. 2.13 Mean monthly precipitation on the continents

Figure 2.13 shows mean precipitation in January, April, July, and October (Jackson 1977).

- In January, a zone of heavier rain can be found south of the equator in South America and Africa. In Asia including northern Australia, these zones can be found on both sides of the equator.
- In April, the rain zones shift north, especially in Southeast Asia.

- These shift northward progresses in the month of July. In Southeast Asia, the association with the monsoon is noticeable.
- In October, the rain zones move back south.

The seasonal distribution of rainfall differs considerably within the global regions at different locations.

Mean monthly rainfall as well as maximum and minimum precipitation for many stations in the world can be found in Müller (1996) and FAO http://www.fao.org/AGLN/climwat.stne

2.2 Classification of Climates

The climatic conditions of the different regions are very important for the protected cultivation of vegetables and ornamental plants in greenhouses or shading halls. A classification of climate has therefore to be considered. The most important factors of influence are temperature, precipitation, and global radiation. In addition, evapotranspiration is an important factor for crop growth in open field and greenhouses.

This study is based on the classification of Trewartha et al. (1980), which is based on the classification of Köppen. Temperature and precipitation are the main factors of influence in most of the classifications.

Climate classification is divided into six main climatic groups; five of them are based on the five great thermic zones, and the sixth group is the dry group which overlaps with four of the thermic groups.

Figures 2.14–2.16 show climatic maps of the earth with the climatic zones (von Zabeltitz and Baudoin 1999). The five thermic zones A, C, D, E, and F have a zonal orientation, and are based on temperature boundaries. Group B is based on precipitation criteria. The definitions of the climatic symbols and boundaries are (according to Trewartha et al. 1980):

2.2.1 A. Tropical Humid Zones

Killing frost is absent; in marine areas, temperature of the coolest month is above 18°C.

- r (rainy) = 10–12 months wet; 0–2 months dry
- w = winter (low-sun period) dry; more than 2 months dry.
- S = summer (high-sun period) dry

2.2.2 B. Dry Zones

Potential evaporation exceeds precipitation.

- W = desert or arid
- S = steppe or semiarid

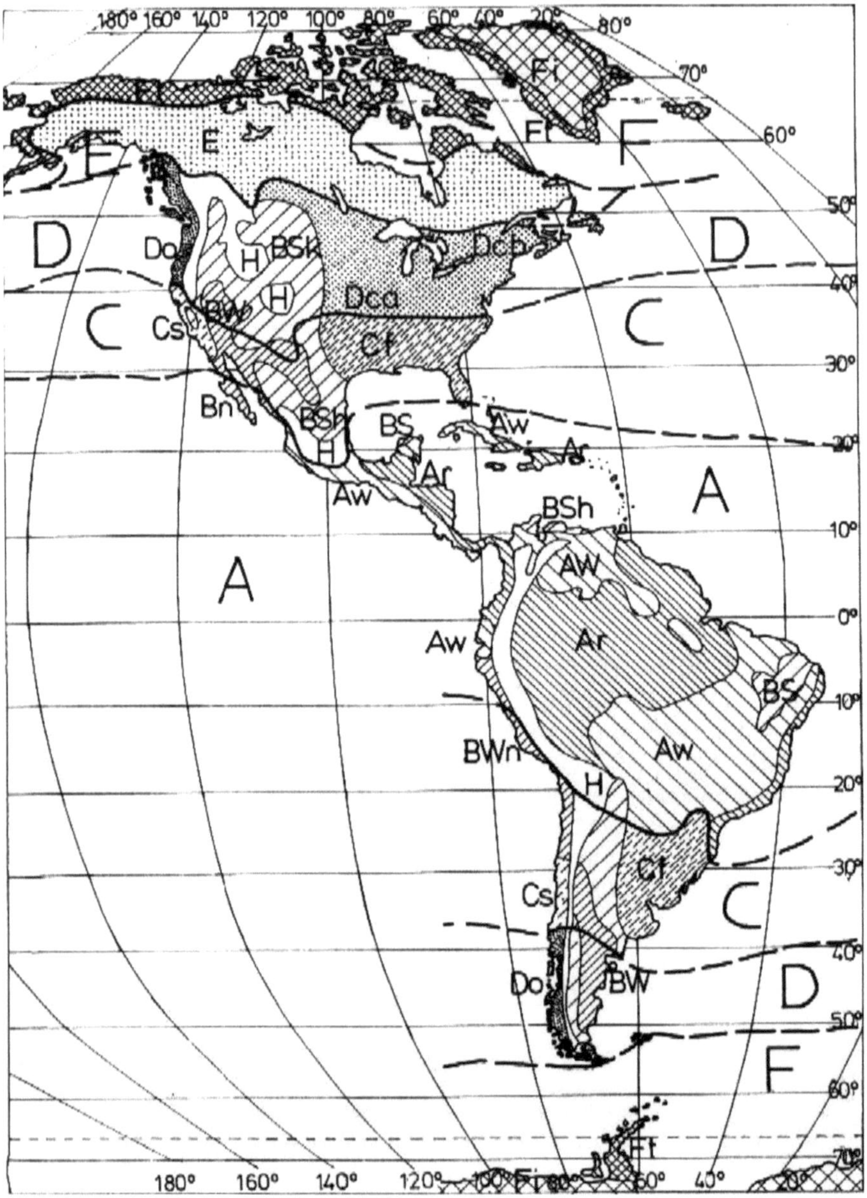

Fig. 2.14 Climatic map of America

- h = hot; 8 months or more with mean temperature above 10°C
- k = cold; less than 8 months with mean temperature above 10°C
- s = summer dry
- w = winter dry
- n = frequent fog

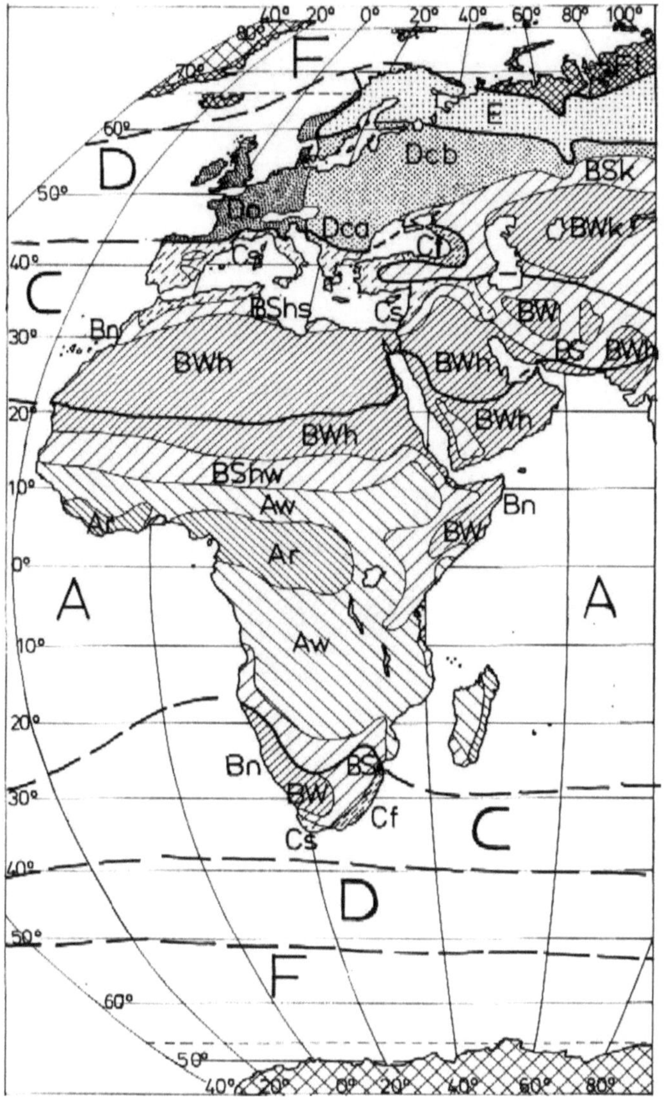

Fig. 2.15 Climatic map of Africa and Europe

2.2.3 C. Subtropical Zones

For 8–12 months temperatures are above 10°C; coolest month below 18°C.

- a = hot summer; warmest month above 22°C
- b = cool summer; warmest month below 22°C
- f = no dry season; difference between driest and wettest months less than required for s and w; driest month of summer more than 3 cm rainfall

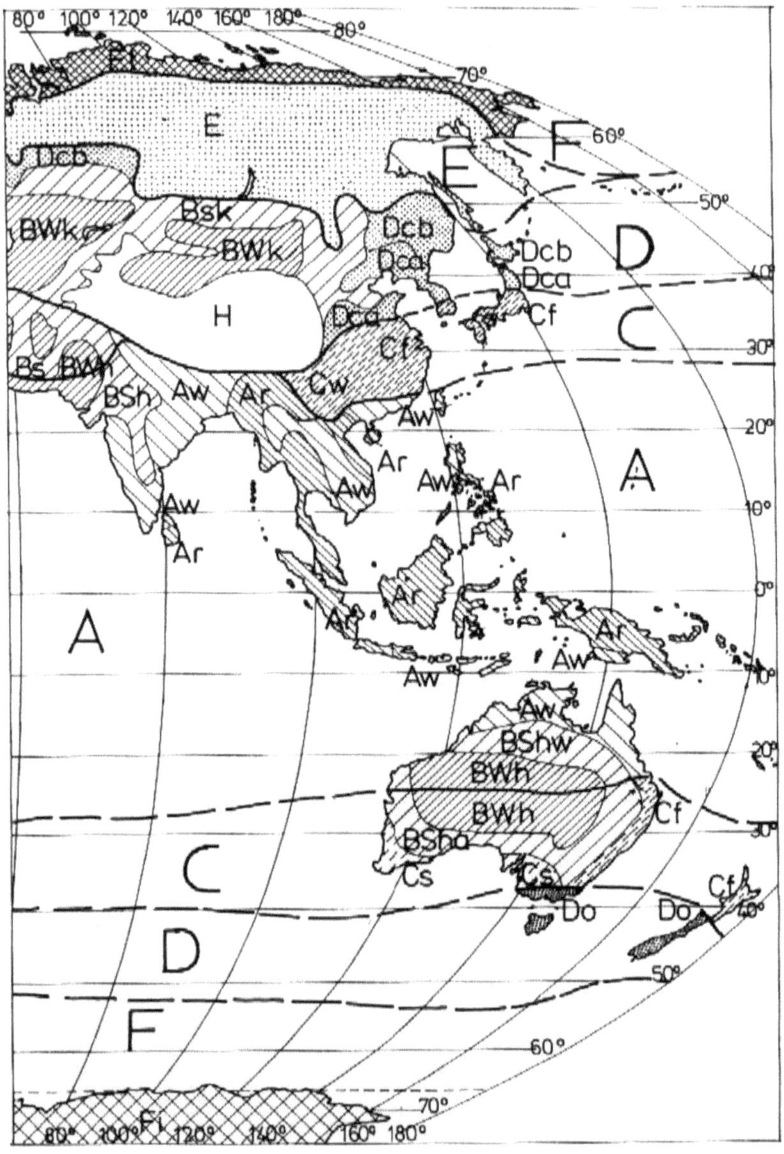

Fig. 2.16 Climatic map of Asia and Australia

- s = summer dry; at least three times as much rain in winter half of year than in summer half of year; driest summer month less than 3 cm; annual total under 89 cm rainfall
- w = winter dry; at least ten times as much rain in summer half of year as in winter half of year

2.2.4 D. Temperate Zones

4–7 months over 10°C

- o = oceanic or marine; cold month over 0°C
- c = continental; cold month under 0°C
- a, b, f, s, w: same definition as for C.

2.2.5 E. Boreal

- 1–3 months over 10°C

2.2.6 F. Polar

All months below 10°C

- t = tundra; warmest month between 0°C and 10°C
- i = ice cap; all months below 0°C

The Boundaries in Figs. 2.14–2.16:

- A/C boundary = equatorial limits of freeze. In marine locations, the isotherm of 18°C for the coolest month
- C/D boundary = 8 months 10°C
- D/E boundary = 4 months 10°C
- E/F boundary = 10°C for warmest month
- BA, B/C, B/D, B/E boundary = potential evaporation equals precipitation.

The following climatic areas in particular have to be distinguished for the construction of greenhouses and shading halls, without taking into consideration regionally limited microclimatic conditions. The details for greenhouse construction are described in Chap. 5.

2.2.7 Tropical and Equatorial Zone

- A_r = tropical wet, not more than 2 dry months,
- A_w = tropical wet and dry, high sun wet, low sun dry,
- BW and BS = arid and semiarid, mostly dry and short rainy season.

Within latitudes 30° North and 30° South, also (Cf) subtropical humid.

The following characteristics of tropical climates are important for protected cultivation.

A climate: tropical humid

- The humid regions have a constantly warm climate. It is frost-free. In many locations, the temperature differences between day and night are larger than the temperature difference of the mean values between the warmest and coldest days.
- Usually, the climatic conditions reach up to an altitude of about 600 m.
- The amount of rain is high and usually occurs in large quantities, together with storms and thunderstorms. The annual and seasonal distribution of rainfall differs considerably.

Ar climate – tropical wet

- Characteristics for those climatic zones are even monthly temperatures with mean values from 25–27°C, with very small monthly temperature variations only; the daily temperature variations are higher and reach 6–14°C.
- There is heavy rainfall throughout the year, and there are no dry seasons, or only 2 dry months at most. The annual rain quantities are 1,800–2,500 mm.
- Solar radiation and humidity are high.
- The typical course of temperature and precipitation in a tropical-wet Ar climate is shown for Singapore in Fig. 2.17.

Aw climate – tropical wet and dry

- The annual rainfall is distributed less evenly over the year than for Ar climate. The wet season is shorter, the dry season at low-sun time longer and there are also severe droughts.

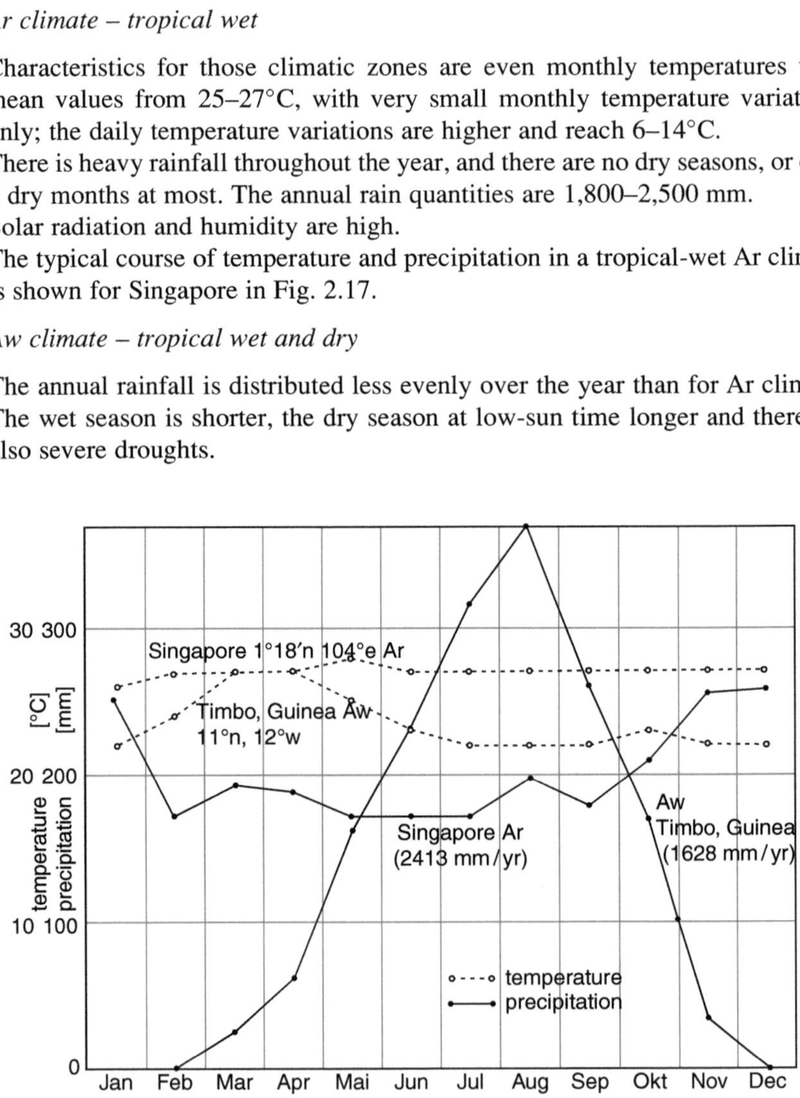

Fig. 2.17 Typical tropical wet, A_r and tropical wet and dry climate, A_w

- Temperature courses are similar to those in Ar climates. With almost vertical position of the sun and same day and night length throughout the year, the monthly variations of mean temperatures are small and range from 3 to 8°C.

  Figure 2.17 shows a typical tropical wet and dry climate for Timbo, Guinea.

2.2.8 Subtropical Climate: C

The subtropical climates stand out due to strong seasonal rhythms of temperature between summer and winter (Fig. 2.18).

Cs climate: Subtropical dry summer (Mediterranean)

The Cs climates are on the west side of the continents between the latitudes 30° and 40°. In simplified terms, they show three features:

1. Most of the rain falls during the winter. Summers are very dry
2. Summers are very hot, winters mild
3. Solar radiation is very intensive, especially in summer; only little clouding

- The mean temperatures in winter are between 4° and 13°C, and in summer between 21° and 28°C. The annual variations of the monthly mean temperatures range from 11° to 17°C. During 3 months in winter, there are occasional frost nights.
- The annual rainfall of 380–760 mm is usually not enough for plant production.

A typical example is shown for Haifa, Israel, in Fig. 2.18.

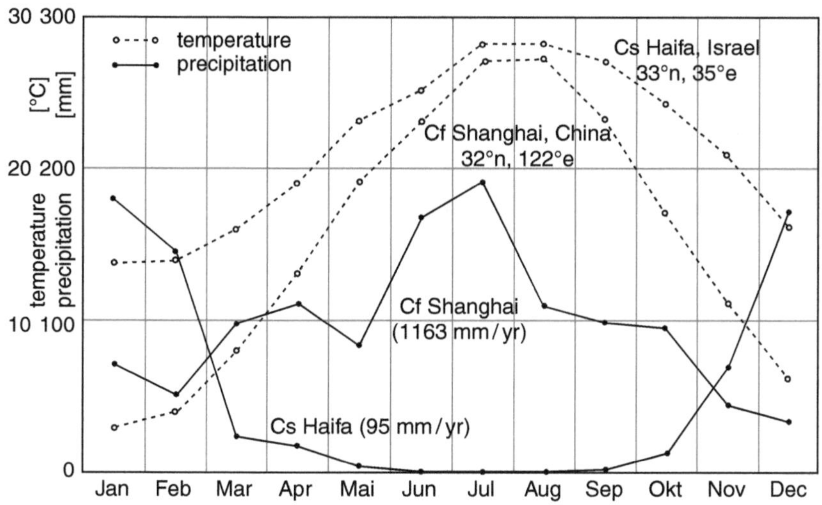

Fig. 2.18 Typical subtropical climates for Haifa (Cs) and Shanghai (Cf)

Cf climate: subtropical humid

Subtropical humid climates are found between latitudes 25° and 40° and differ from subtropical dry-summer Cs climates in three points, Fig. 2.18:

1. They are on the east side of the continents.
2. The annual rainfall is higher.
3. The distribution of rainfall is extended over the whole year, with a maximum in the warm months. There is no summer drought.

- The summer months are very hot along with a high humidity. Winters are relatively mild with mean temperatures of 4°–13°C. Frost can occur occasionally in some winter months during the night.
- The total annual quantity of rain comes to 760–1,700 mm.

2.2.9 Dry Climates: B

Dry climates are not only defined by temperature boundaries but also by the quantity of rain.

- In dry climates, the annual potential evapotranspiration is larger than the precipitation. The boundaries of the dry climates are determined by the fact that the annual potential evapotranspiration equals the precipitation.
- Daily temperature variations are very high, especially in winter. With a mostly cloudless sky, there is a high solar radiation that makes temperatures rise during the day. The relative humidity during the day is low at 12–30%. With a (usually) cloudless sky, thermal radiation from the ground is very high during the night; consequently, temperatures drop correspondingly.

2.2.10 Highlands: H

The climate in highlands varies considerably, and is determined by many local influences. A description is therefore difficult.

- With growing altitude, irradiation increases and temperature drops.
- Rainfall can vary considerably according to location.
- In equatorial highlands, such as in Kenya, Colombia and Malawi, very good conditions for the production of plants are given

Figure 2.19 shows a section of Colombia in the area of 4°N.

Figure 2.20 illustrates the climatic conditions of the highlands of Bogotá, at an altitude of ~2,600 m. There are about 1,500 ha of greenhouses for ornamental plants in this highland around Bogotá. In comparison to Bogotá, the climate of Medellin (Colombia) is shown at an altitude of 1,550 m. Here, average temperatures and rain quantities are higher.

The mean maximum and minimum temperatures, as well as solar radiation, are important for the evaluation of climate conditions for plant growth.

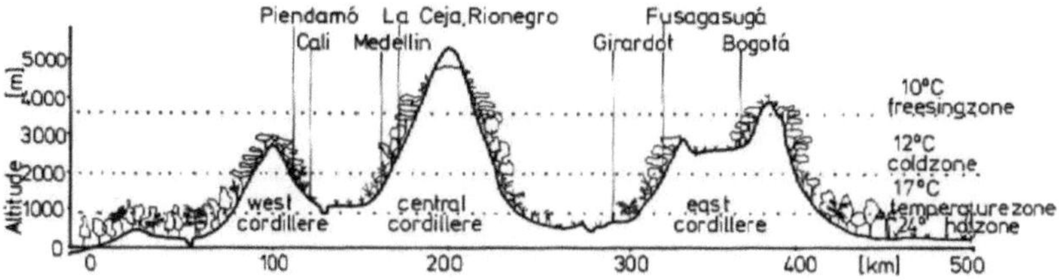

Fig. 2.19 Section of Colombia

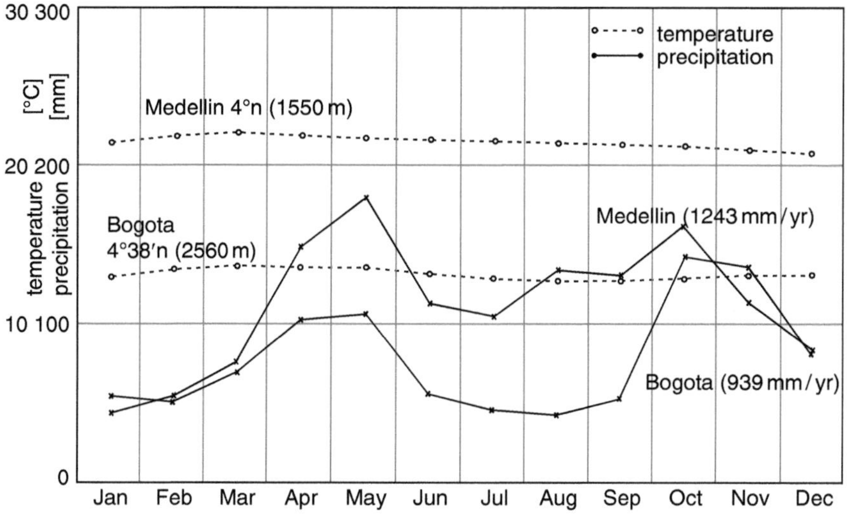

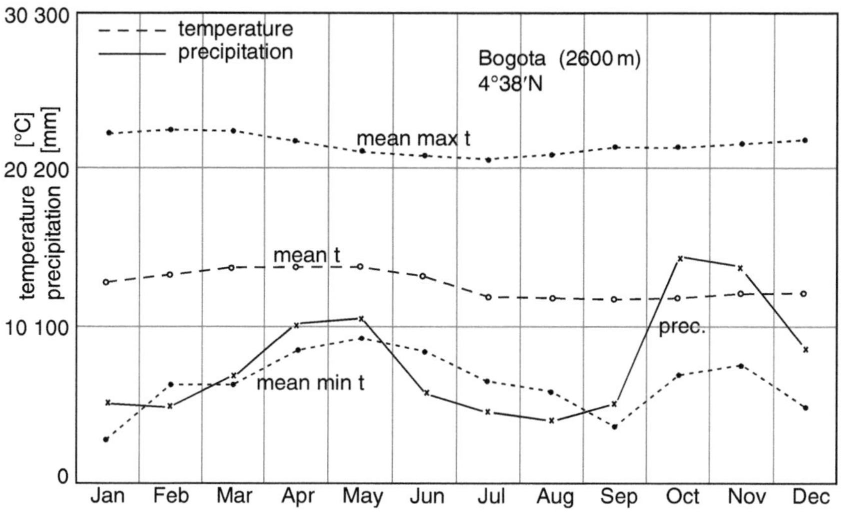

Fig. 2.20 Climates of highlands in Colombia

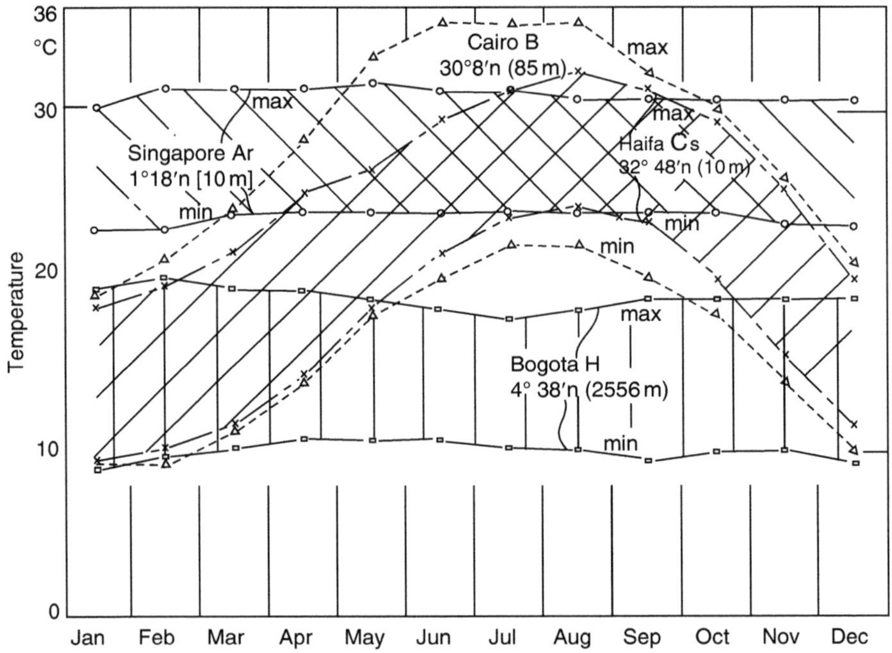

Fig. 2.21 Mean maximum and minimum temperatures for different climate regions

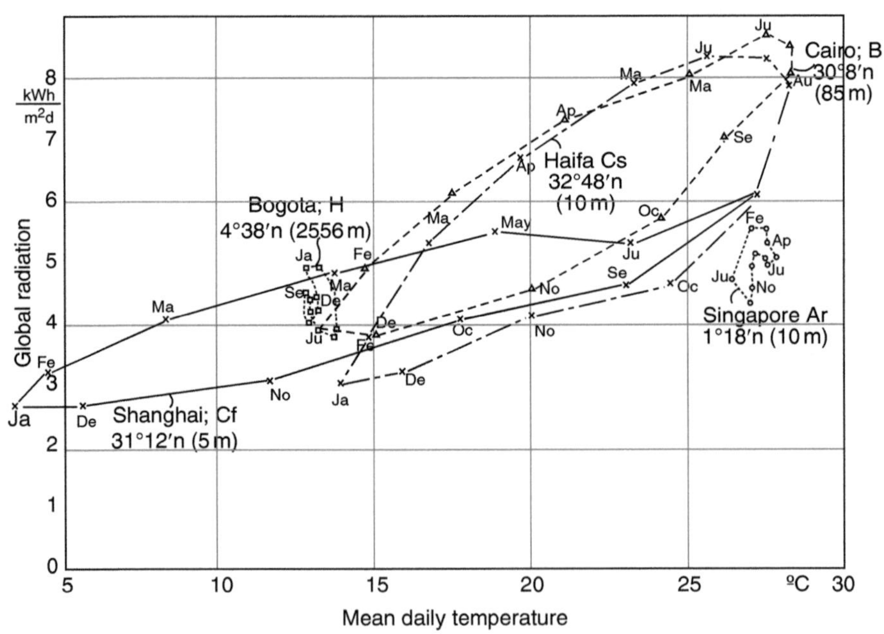

Fig. 2.22 Mean daily global radiation depending on mean daily temperature

Figure 2.21 shows the range between mean maximum and minimum temperatures for some typical climates throughout the year.

In the tropical wet climate Ar of Singapore, the mean maximum temperature does not exceed 31°C. In the dry climate B of Cairo and the subtropical dry summer climate Cs of Haifa, the mean maximum temperatures are higher than 32°C in summer. There are many days with much higher temperatures at daytime.

The mean minimum temperatures do not fall below 9°C, even in the Highland H of Bogotá at an elevation of 2,556 m.

The suitability of the climatic conditions of different regions can also be compared in Fig. 2.22, where the mean daily global radiation is plotted against the mean daily temperatures for the 12 months of the year for different climate regions. The mean temperatures are close together in a narrow range in tropical lowland (Singapore), and the mean daily global radiation does not exceed 5.5 kWh/m^2 day. The temperatures vary very much from winter to summer in dry and subtropical climatic regions, and the daily global radiation exceeds 6–8 kWh/m^2 day during summer (Cairo, Haifa). The mean daily temperatures are also close together in tropical highlands (Bogotá) but they are significantly lower than in tropical lowlands.

Chapter 3
Crop Growth Requirements and Climate Control

To check the suitability of a region for protected cultivation, the climate data should be compared with those for other regions and with the main requirements for plants.

3.1 Climate Requirements

Many plants grown in greenhouses are warm-season species. Vegetables and flowers have different climate requirements. The main climate requirements of plant growth can be defined and summarized as follows (Baudoin et al. 1991; Krug et al. 2002; Nisen et al. 1984; Nisen et al. 1990):

1. Plants can be killed by frost. The absolute minimum temperature in the greenhouse has to be above 0°C. The risk of subzero temperatures can be ignored when the daily minimum outside temperature exceeds 7°C.
2. The growth, yield and quality of greenhouse fruit vegetables are affected when temperatures are below 12°C or exceed 30°C (Castilla and Hernandez 2007). Plants grown under protected cultivation are particularly adapted to average temperatures ranging from 17 to 27°C. Optimal temperatures range between 22 and 28°C in the daytime and 15–20°C at night (Castilla and Hernandez 2007). Taking into account the warming-up effect in greenhouses caused by solar radiation, one can define the climatic limits of suitability to be between 12 and 22°C average daily outside temperature if the greenhouses are not heated.
3. The mean absolute maximum temperature for plants should not be higher than 35–40°C. The maximum temperature for a tomato crop should not be higher than 35°C (Verlodt 1999).
4. A minimum of 500–550 h of sunshine for the 3 winter months (November, December and January in the Northern Hemisphere) is desirable. This corresponds to a daily insulation of about 2.3 kWh/m^2 day. The limit for effective production is 1.0 kWh/m^2 day (Krug et al. 2002). Artificial lighting may be used for intensive production.

C. von Zabeltitz, *Integrated Greenhouse Systems for Mild Climates*,
DOI 10.1007/978-3-642-14582-7_3,

5. The minimum threshold for soil temperature is 15°C for heat-requiring plants.
6. Nisen et al. (1990) suggest a threshold of the average night temperature to be 15–18.5°C for heat-requiring plants such as tomato, pepper, cucumber, melon, and beans. The minimum night temperatures depend on the species. Verlodt (1999) gives the following figures
   12°C for hot pepper
   14°C for melon, sweet pepper and eggplant
   7–8°C for strawberry
   The minimum day temperature for a tomato crop should not be lower than 15°C.

   Relative humidity of 70–90% can be regarded being within a safe range.

3.2 Consequences for Greenhouse Construction

The consequences for greenhouse construction and climate control are:

1. If the mean outside temperature is above 7°C, there is no risk of frost. Heating is absolutely necessary if the temperatures fall below 0°C.
2. If the mean minimum outside temperatures are below 10°C in the coldest month, and if the crop needs higher temperatures (cucumber 18°C), heating will be necessary for the improvement of quality and yield.
3. If the mean minimum outside temperature is <12°C (absolute minimum temperature <9–10°C), ventilation openings (ventilators) of greenhouses have to be closed to keep the temperature higher inside. Heating may possibly be necessary.
4. If the mean minimum temperature is >12°C, ventilators can be permanently open. Ventilation becomes necessary if the inside temperature reaches >18–20°C.
5. If the mean maximum temperature is >27°C, ridge or roof ventilation is necessary in a hot and humid climate and for multi-span greenhouses. Evaporative cooling may be recommendable in a dry climate.
6. If the mean maximum outside temperature is >36°C, artificial evaporative cooling is necessary if outside humidity is low enough.
7. If relative humidity at day time is <55–60%, inside humidity should be increased by fog systems or evaporative cooling.

3.3 Measures for Climate Control

The measures for climate control depend on the outside climate conditions. The relationship can be shown by special diagrams showing the influencing factors of global radiation, mean minimum temperature, mean maximum temperature, and humidity.

3.4 Tropical Regions

As the temperatures depend on altitude as well as on latitude, different greenhouse constructions have to be designed for tropical lowlands and tropical highlands (see Chap. 5).

The main characteristic design criterion is the mean minimum outside temperature, depending on altitude and latitude.

Greenhouses for tropical highlands with mean minimum temperatures below 12°C have to have ventilation openings that can be closed.

Greenhouses for tropical lowlands with mean minimum temperatures above 12°C can be permanently open at side walls and ridge ventilators throughout the year.

The climate conditions can be very different even in the same country, so different greenhouse constructions have to be considered. That means the local climate conditions should be considered for successful site selection.

The following figures show the dependence of mean minimum versus mean maximum temperatures in tropical regions:

Karonga (Malawi) has mean minimum temperatures above 15°C and mean maximum temperatures lower than 33°C. Greenhouse constructions with permanently open ventilators are suitable. Lilongwe (Malawi) has mean minimum temperatures below 12°C. Greenhouses have to have vents that can be closed at night, and very good ventilation efficiency by ridge ventilation in the daytime for crop production throughout the year. Almeria in the Mediterranean region (subtropics) has similar conditions, but in many cases there is no production in July and August (Figs. 3.1 and 3.2).

The temperatures in Bangalore (India) at an altitude of 921 m are higher than in Pune (India) at an altitude of 560 m. The reason is the lower latitude of Bangalore. Greenhouses in Pune should have closable ventilators, while ventilators of greenhouses in Bangalore can be permanently open. That means both the altitude and latitude have to be considered for decisions on greenhouse construction (Fig. 3.2).

The mean daily global radiation is above 4 kWh/m^2 day throughout the year in Lilongwe (highlands) and Karonga (lakeshore plain, Malawi) as well as in Bangalore and Delhi (India) (Fig. 3.4). Mean maximum daily global radiation is not much higher than 7–7.5 kWh/m^2 day in tropical regions. Solar radiation may be too high around noon in some months, as the day length is relatively short. Shading may be necessary for some crops. The construction and cladding material cause a reduction of incoming light by about 20%, and the shading effect is increased by dirt deposition on the plastic film (Fig. 3.5).

The mean humidity is higher than 80% for some months in Karonga and Lilongwe when precipitation is highest (Fig. 3.6). These are months with mean outside temperatures below 30°C. The humidity is relatively low in months with higher temperatures. Evaporative cooling as well as fog systems may possibly be necessary for flower crops.

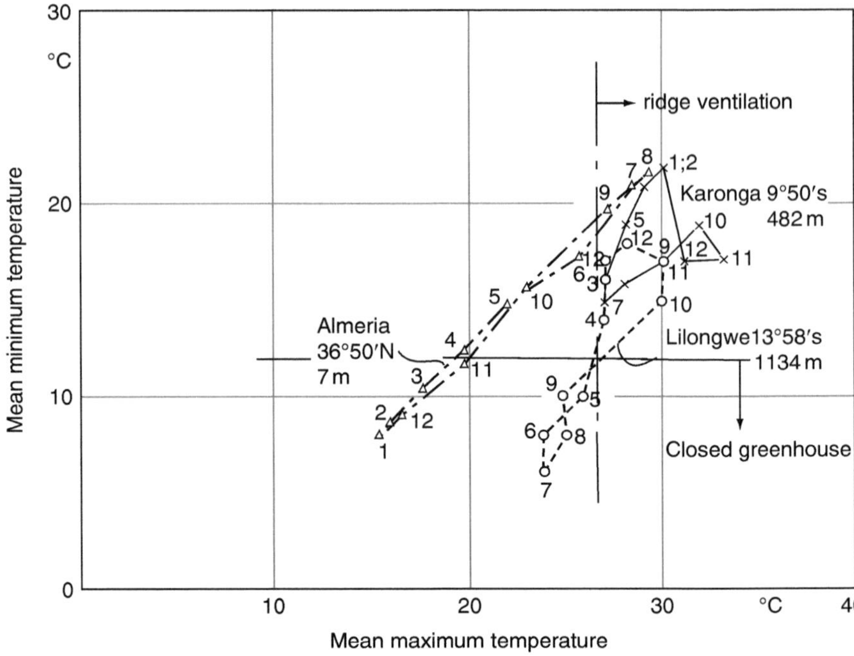

Fig. 3.1 The mean minimum temperatures versus the mean maximum temperatures for Karonga (lakeshore plain) and Lilongwe (Highland) in Malawi from January (1) to December (12), in comparison to the subtropical region of Almeria (Spain)

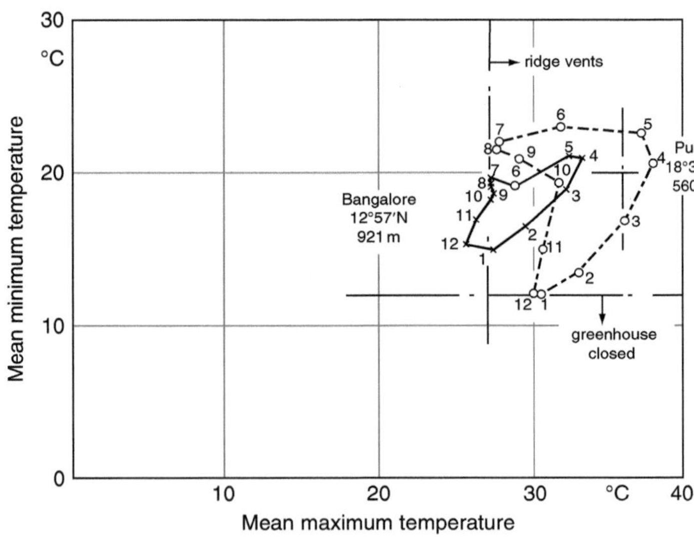

Fig. 3.2 Mean minimum and maximum temperatures for Bangalore and Pune (India)

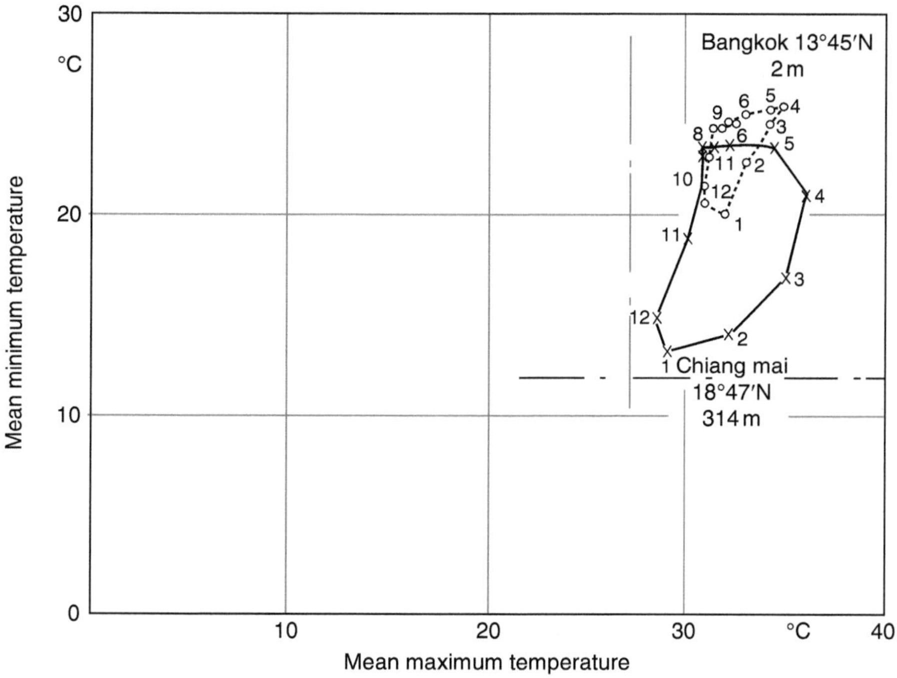

Fig. 3.3 Mean minimum and maximum temperatures for Bangkok and Chiangmai (Thailand). Both locations have suitable temperatures for crop production in greenhouses if ridge vents are installed

3.5 Subtropical Regions

The maximum temperatures in subtropical regions (Mediterranean) are high in summer and the minimum temperatures below 10°C in winter (Fig. 3.7).

Heating may be necessary for successful crop production in both regions for some months in winter. It becomes very hot in summer in Antalya, so ridge vents are necessary in multi-span greenhouses. Possibly evaporative cooling may be necessary. Greenhouses for vegetables are normally out of production in summer months.

Figure 3.8 shows the precipitation for Antalya and Almeria, in addition to the temperatures. Almeria is a very dry area with low precipitation in winter and summer. Antalya has high precipitation, in particular in winter. Gutters are necessary for greenhouses in Antalya to lead off the rainwater and to protect crops from water penetrating through the side walls. Collecting of rainwater for irrigation is recommendable and profitable in Antalya. Greenhouses in Antalya need more ventilation efficiency in spring, summer and autumn.

A typical greenhouse type for Almeria, the Parral type, is not transferable without problems to Antalya or other Mediterranean regions.

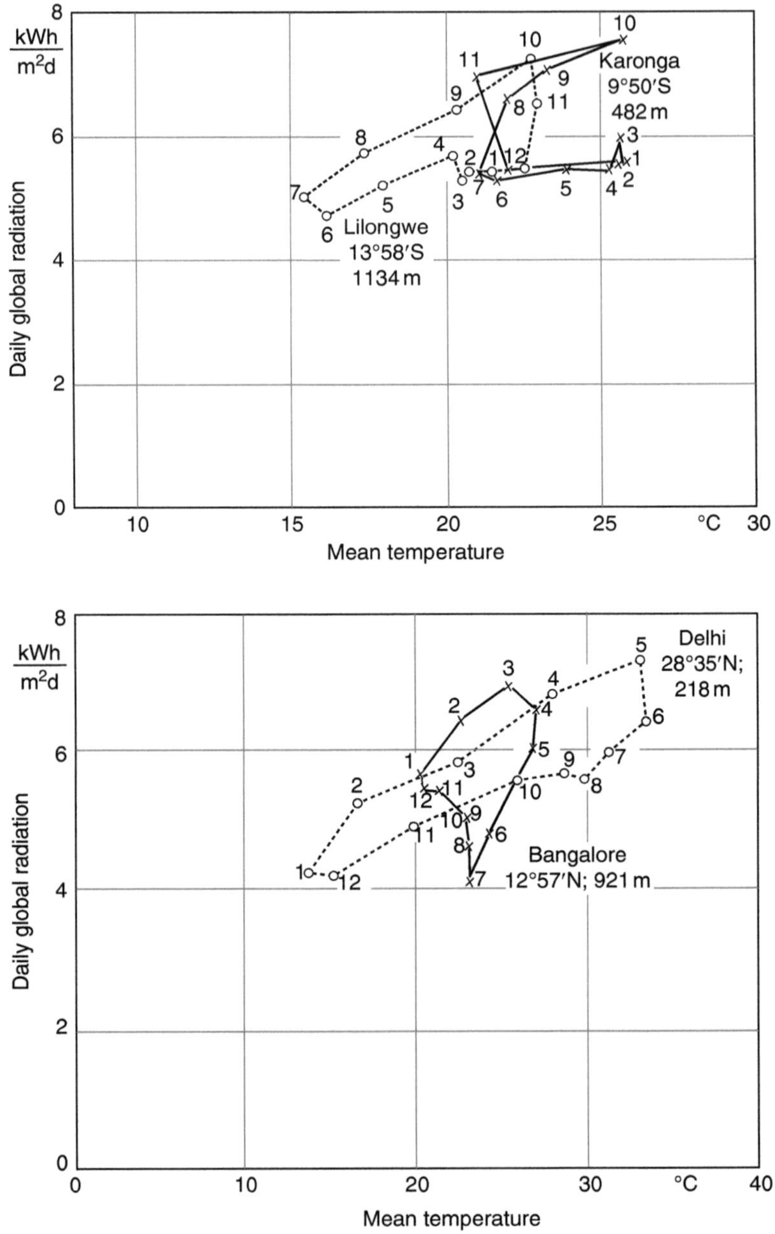

Fig. 3.4 Mean daily global radiation versus mean temperature in some tropical locations

The subtropical climate near the Mediterranean Sea is much more suitable for crop production in greenhouses than the more continental climate in Delhi. Greenhouses have to be heated in the highlands of Kashmir during winter and very well-ventilated in summer (Fig. 3.9).

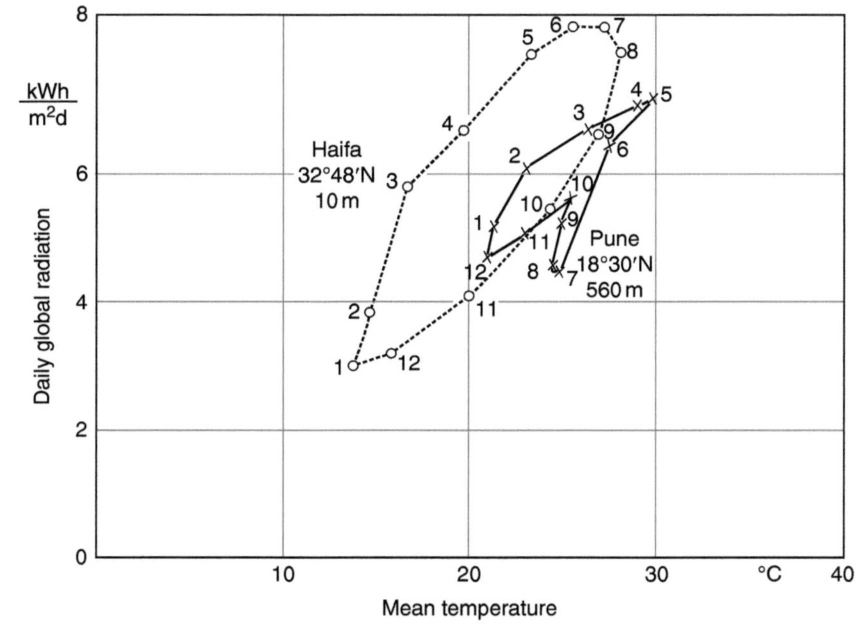

Fig. 3.5 Mean daily global radiation versus mean temperature for Haifa (subtropics in comparison to Pune (tropics). Haifa (subtropics) has, in comparison to Pune (tropics), very high global radiation in summer and values below 4 kWh/m^2 day in winter

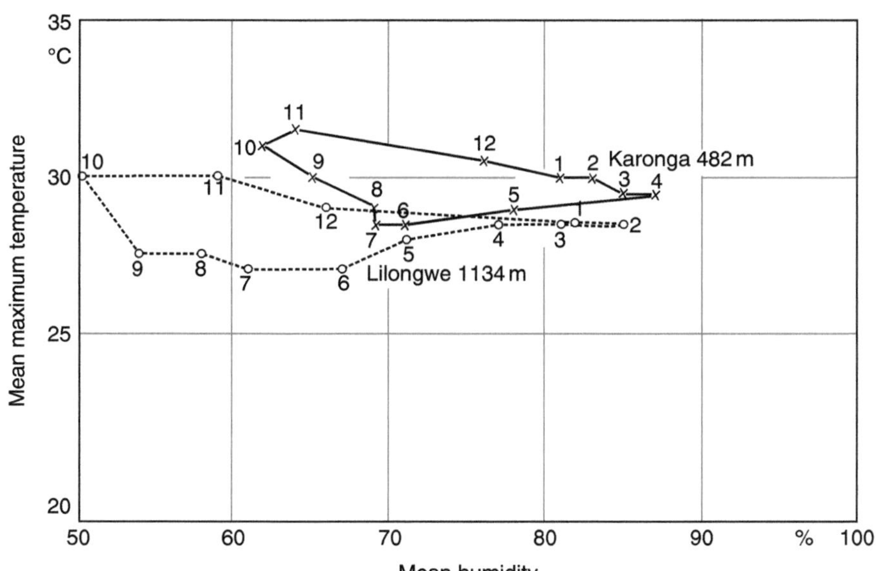

Fig. 3.6 Mean maximum temperature versus mean relative humidity for tropical locations in Malawi

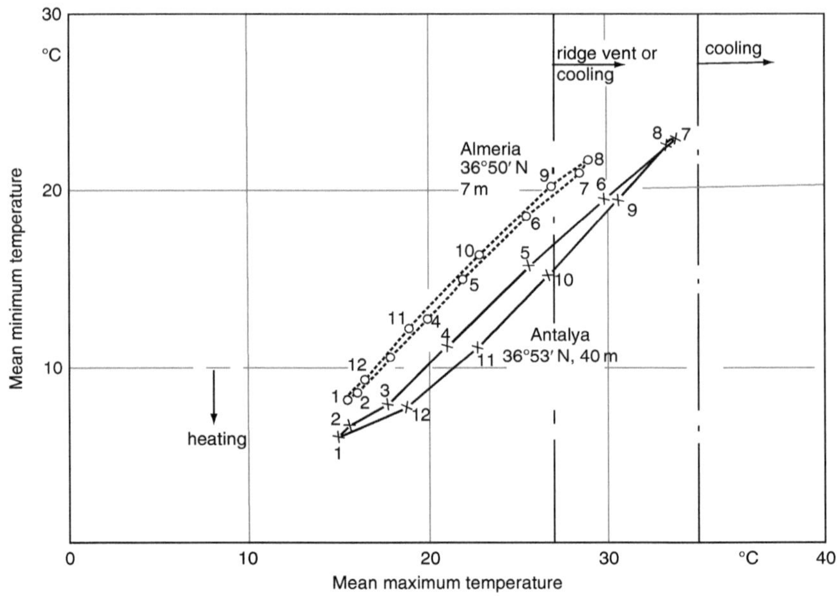

Fig. 3.7 Mean maximum and minimum temperatures for subtropical sites in Almeria (Spain) and Antalya (Turkey), both on the Mediterranean, where large areas of greenhouses are located on the Northern Mediterranean sea coast

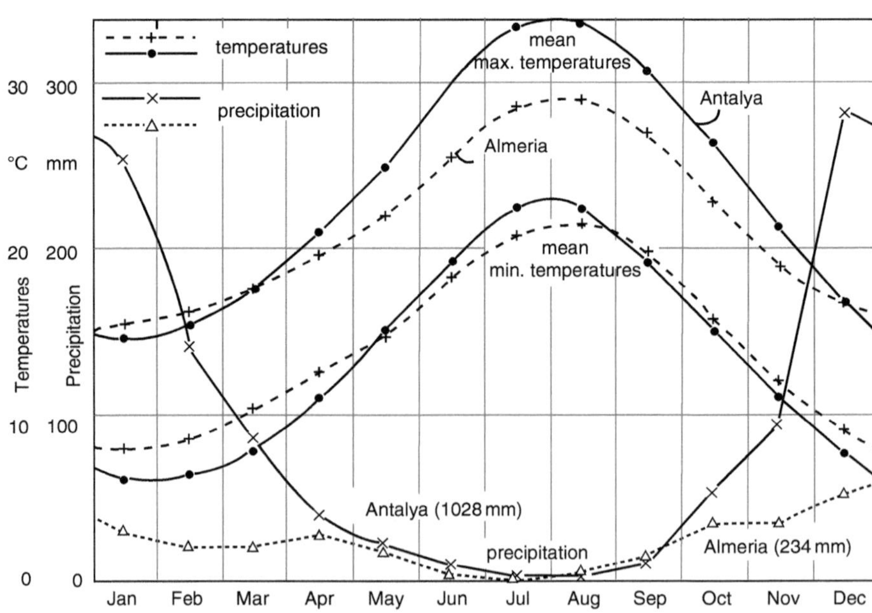

Fig. 3.8 Precipitation and temperatures for Almeria and Antalya

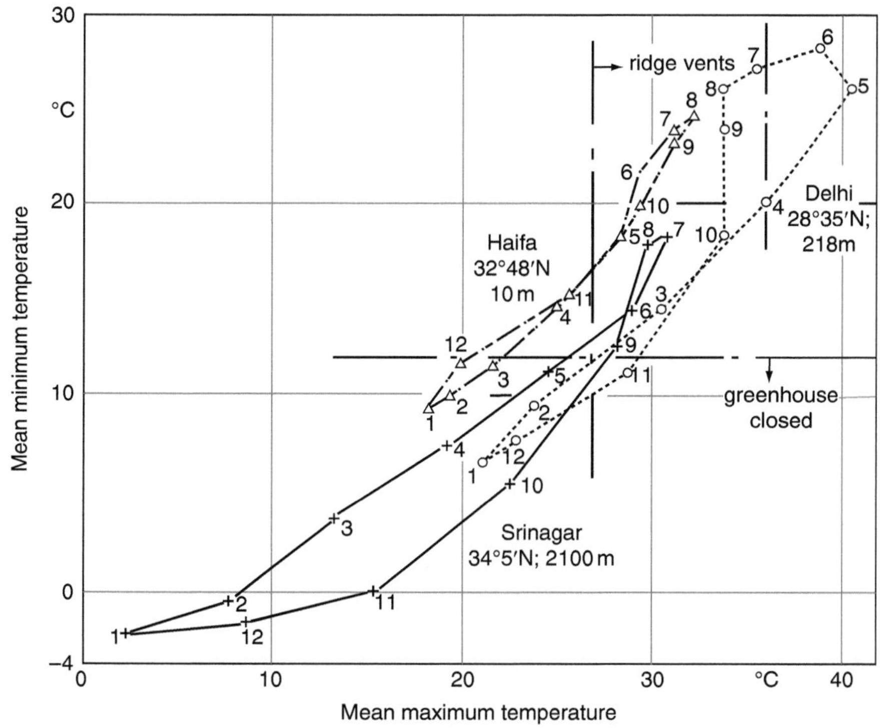

Fig. 3.9 Minimum and maximum temperatures for Haifa (Mediterranean) in comparison to Delhi (India) and Srinagar (highlands of Kashmir, India)

The coastal climate in Lebanon (Beirut) also fits the conditions for protected cultivation more than in Aleppo (Syria) with more arid climate (Fig. A.1 in Annex 1). Cooling and heating is necessary for year-round cultivation in Aleppo.

Large greenhouse areas are located in the region of Agadir (Morocco), where the climate is suitable for protected cultivation. Some heating may be necessary in winter. Cairo (Egypt) also has some cold winter days and very hot summers (Fig. A.2). Jixi County in central China, with a continental climate, has unsuitable temperatures compared to Cairo at the same latitude, both in winter and in summer (Fig. A.3).

Figure A.4 shows the climates of Famagusta (Cyprus) and Catania (Sicily). Both islands are well-known for protected cultivation.

Figure 3.10 shows the mean daily global radiation versus the mean daily temperature for Almeria and Antalya (subtropics) in comparison to a temperate climate in De Bilt (Netherlands). If the mean daily global radiation is considered to be 2.3 kWh/m^2 day, the Antalya region has not enough light in December and January. There is lower global radiation but higher mean temperatures in Antalya than in summer in Almeria. In both centres, crop production is difficult

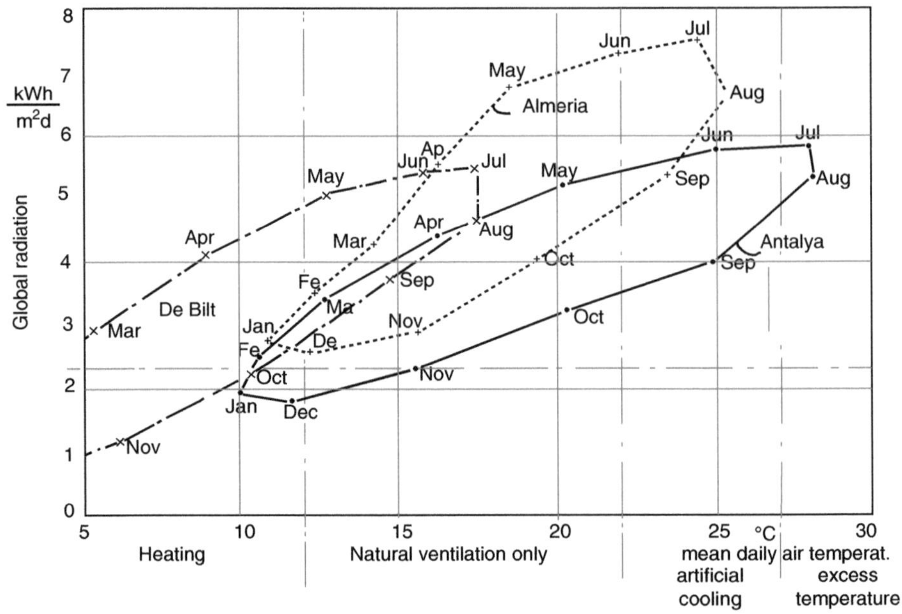

Fig. 3.10 Mean daily global radiation versus mean temperature for Mediterranean sites in comparison to temperate climate in Netherlands

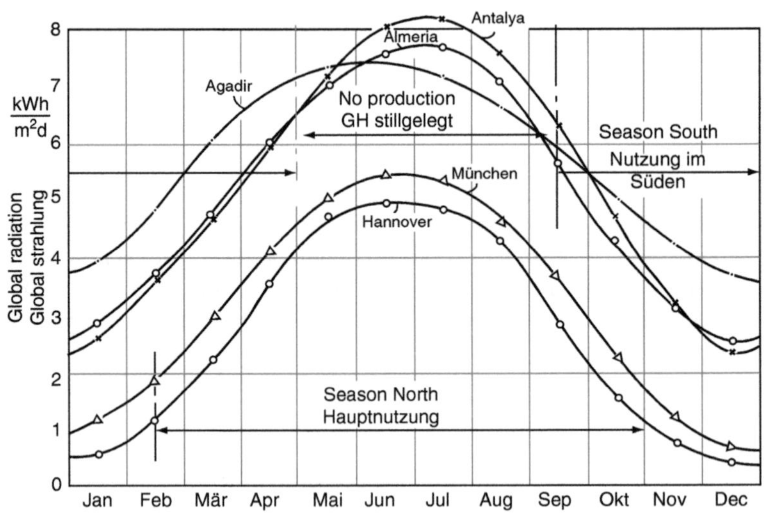

Fig. 3.11 Mean global radiation in subtropical (Mediterranean) sites compared to temperate climates (Germany)

without artificial cooling in summer, but most of the greenhouses are out of production during this period.

Figure A.5 shows the global radiation for Agadir (Morocco) and Gafsa (South Tunisia) compared to Almeria (Spain).

The mean daily global radiation in subtropical climates (Agadir, Almeria and Antalya) is higher than in the temperate zones (Hannover, Munich, Germany) (Fig. 3.11). But one has to take into consideration the main cropping seasons, for vegetables in particular. Transplanting starts in September/October in the Mediterranean, and the plants grow and develop with decreasing light intensity up to the middle

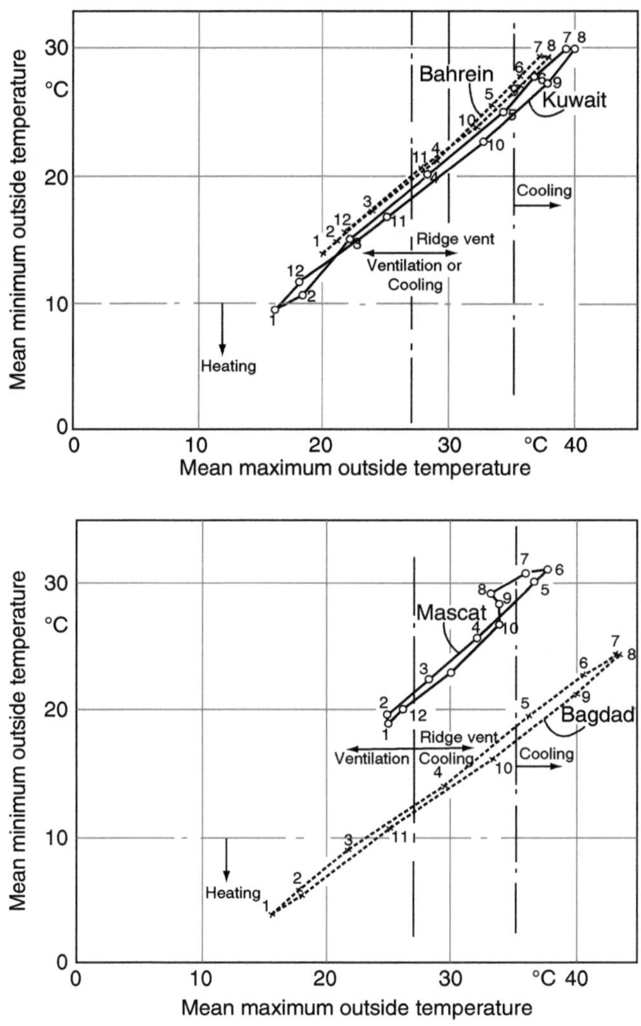

Fig. 3.12 Mean minimum outside temperature versus mean maximum temperature for some sites in arid regions

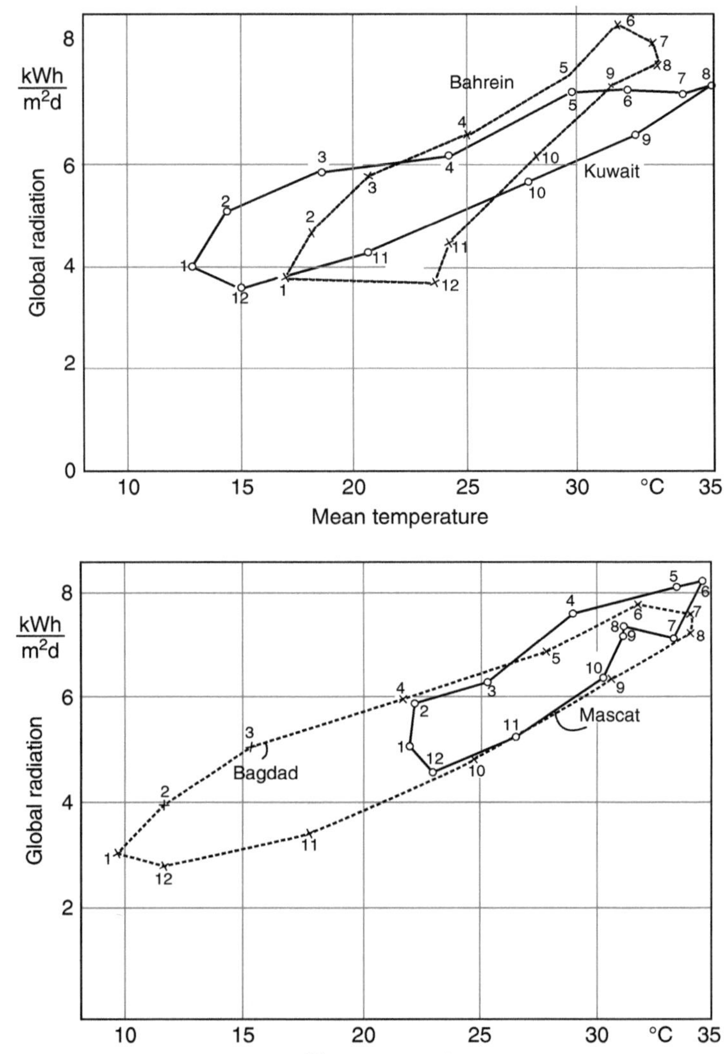

Fig. 3.13 Mean daily global radiation versus mean temperature for arid regions

of the cropping season, when light intensity is at a minimum. Transplanting in the Northern countries (temperate climate) starts in February, and crops grow with increasing light intensity by a maximum in the middle of the season. As light is a limiting factor during the growing season in southern subtropical countries, one has to increase light transmittance of greenhouses as much as possible for vegetable crops.

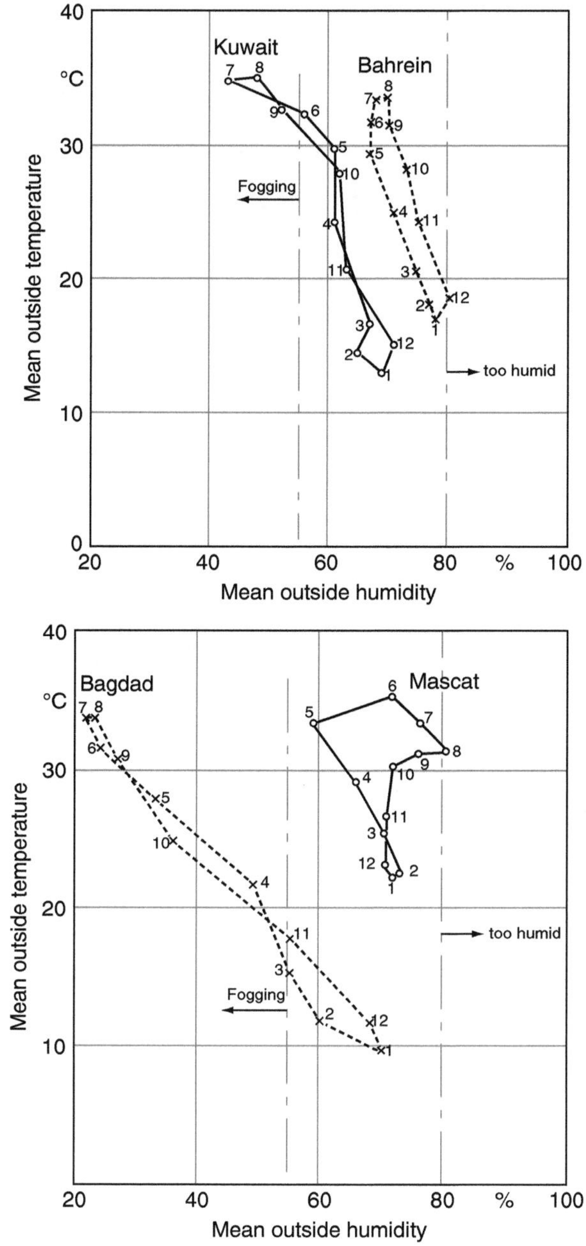

Fig. 3.14 Mean outside temperature versus mean outside humidity for arid regions

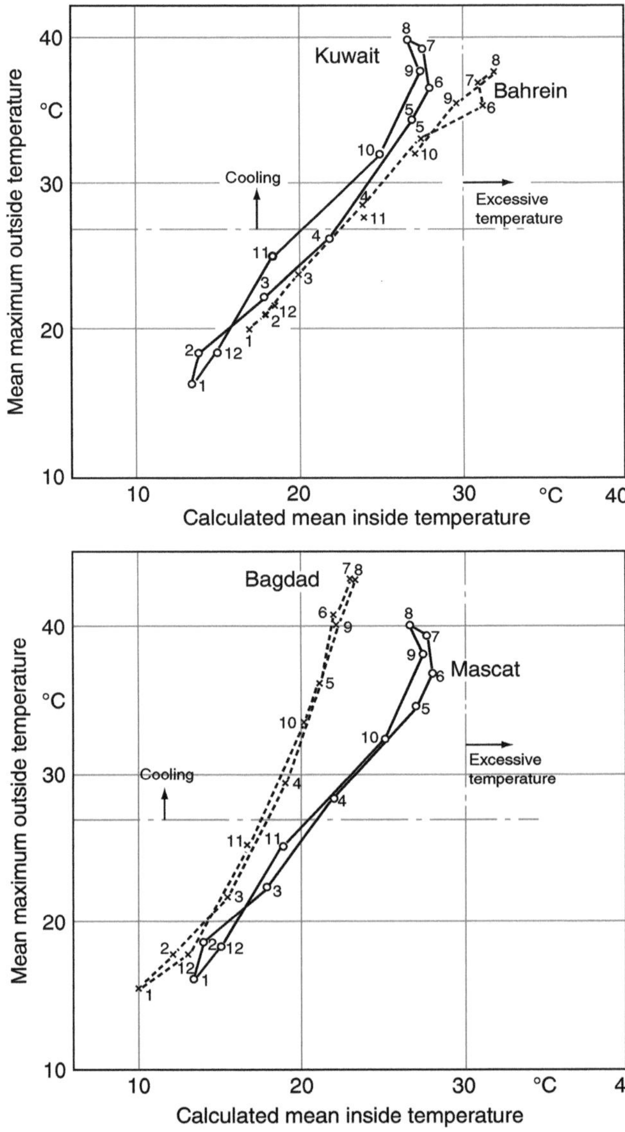

Fig. 3.15 Mean maximum outside temperature versus the calculated mean inside temperature for arid regions that can be achieved by evaporative cooling

3.6 Arid Regions

Figure 3.12 shows the mean minimum outside temperature plotted against the mean maximum outside temperature for different places in arid regions (Bahrein, Kuwait City, Bagdad and Mascat).

If evaporative cooling is necessary with average maximum outside temperatures of more than 27°C, and if heating at night is necessary with average minimum temperatures less than 10°C, evaporative cooling is necessary from April to November in Kuwait, Bahrein and Bagdad and from March to December in Mascat. Multi-span greenhouses should have forced or roof ventilation.

Heating at night may be necessary in Kuwait and Bagdad. Mascat and Bahrein need no heating, perhaps only emergency heating for some cold nights.

Figure 3.13 shows mean daily global radiation plotted against mean daily temperature:

- The minimum global radiation is higher than 2.5 kWh/m^2 day at all places, which is enough for plant production in winter.
- Global radiation is very high during summer months; shading may be necessary.
- The average temperature exceeds 30°C in all countries.

Figure 3.14 shows the mean outside temperature plotted against the mean outside humidity. If evaporative cooling is necessary during summer from April to November, the outside humidity has to be low enough for the cooling.

The average humidity in Kuwait is below 70% throughout the year. Bahrein has an average humidity below 75% from July to September, and evaporative cooling may be inefficient. Cooling in Bagdad is no problem in summer. Mascat has high humidity during summer, and cooling is difficult.

Figure 3.15 shows the mean maximum outside temperature plotted against the *calculated* mean minimum inside temperature which can be achieved with evaporative cooling.

The calculation was done with the average outside temperature, the average outside humidity and the average maximum outside temperature in Mollier's h–x diagram. The assumption was a constant water content with rising temperature from average temperature to average maximum temperature. If the average minimum inside temperature is assumed to be not higher than 30°C, evaporative cooling may be difficult from June to August in Bahrein. Greenhouses in the other climate zones can be cooled during summer.

Chapter 4
Design Criteria for Greenhouses

Greenhouses have to provide optimal climate conditions for plants inside. They should be built less in accordance with national traditions but more with regard to general cropping needs and conditions of different climate zones (von Zabeltitz 1999, von Zabeltitz and Baudoin 1999). Greenhouses should withstand minimum action forces by wind, rain, snow, and crop loads. Minimum standards and technical specifications should be observed.

Most of the greenhouses in the world are plastic film greenhouses. Glass greenhouses can be found in particular in temperate climates of Northern Europe and relatively often in Turkey. Therefore, plastic-film greenhouses in particular are dealt with in this book.

The analysis of storm damage after a severe storm in the Netherlands 1990 gave the following results (Waaijenberg and Denkov 1992):

1. The profiles used in the bent trusses of plastic film greenhouses were too light, and the arches were deformed by wind.
2. The plastic film sometimes appeared to be stronger than the structure.
3. The plastic film tearing started at sharp edges of construction components which were in contact with the film.
4. The foundations were often not sufficiently secured against uplift forces.
5. Damage to plastic film greenhouses often started from the ventilation openings.
6. The connections of arch, purlin and ridge tubes were often not sufficient, especially clamp connections.
7. Too few bracings were fixed between structure components to give sufficient stability.

Experiences in other countries show that those weak points occur frequently.

C. von Zabeltitz, *Integrated Greenhouse Systems for Mild Climates*,
DOI 10.1007/978-3-642-14582-7_4, © Springer-Verlag Berlin Heidelberg 2011

4.1 Greenhouse Types

Greenhouse types have been developed for the following climates:

- Tropical wet climates, lowlands with mean minimum temperature in the coldest month >12°C
- Subtropical climates and tropical highlands. Mean minimum temperature in the coldest month <12°C
- Dry and arid climates
- Temperate climates

In addition to the climate criteria, one has to design greenhouse types depending on the user (small-scale farmer or industrialised horticultural production), the material for construction, life cycle and the cost-effective availability of materials in the country. Greenhouses for a shorter life cycle, which can be cheaper, are wooden structures with untreated poles that can be corroded by rotting or by termites. The life cycle is 4–6 years. Greenhouses for a longer life cycle have galvanised steel tube or treated timber structures, with appropriate foundations. Special greenhouse designs are necessary for nurseries, and for larger plants and trees like bananas.

4.2 Site Selection

The following criteria should be considered with regard to site selection:

- The microclimate conditions
- The water and electricity supply
- Labour availability
- The distance to the markets and transportation costs
- The orientation of the greenhouses depends on the main wind and rain direction. Penetration of rain through ventilator openings has to be avoided. Fans for evaporation cooling should face the main wind direction.
- Natural circumstances such as high trees and buildings have to be considered with regard to shade effects and windbreaking.
- Windbreaks by natural planting or by artificial windbreak structure should be considered in the planning of the greenhouse if wind velocities are high.
- The soil at the greenhouse location should have adequate drainage provision.
- The soil topography should have a slight slope (about 0.5–1%) in the longitudinal direction to drain off rainwater.
- The Indian standard for greenhouse structures ISI 4462–1997 recommends a slightly southern-facing orientation for greenhouses in areas above 40° longitude, to present maximum winter sunlight to the crop.

4.3 General Design Criteria

The general design criteria for greenhouses are:

- The climate conditions in the regions
- The general design requirements for glass and plastic film greenhouses, including standards for different loads (European standard EN 13031-1, December 2001, ASAE EP 460 Dec 01, Indian standard for greenhouse structures ISI4462-1997).
- Measures for climate and pest control
- The most cost-effective locally available materials, and the expected duration of life.
- Technical measures for integrated production and protection (IPP).

4.4 Loads for Greenhouse Structures

Greenhouses should be low-cost but cost-effective structures. Nevertheless, they have to withstand outside weather stresses such as storm, rain, hail, and snow if snowfall occurs. Damage to greenhouses is very often caused by too weak structures (Fig. 4.1). To avoid severe damage, the greenhouse structure should be calculated and designed in accordance with available standards or technical specifications that give figures for different loads or actions to withstand influencing forces.

Fig. 4.1 Too weak tunnel structure damaged by wind

The European standard EN 13031-1 (December 2001) "*gives rules for structural design and construction of greenhouse structures for the professional production of plants and crops*".

According to the standard, "*greenhouses shall be designed by verifying that no relevant limit state is exceeded. The relevant limit states to be considered depend on the class of the greenhouse*".

"Greenhouses shall be classified in accordance with a minimum design working life for the structure and the tolerance to frame displacements of the cladding systems".

Ggreenhouse structures are divided into two classes, A and B, depending on tolerance to possible frame displacements, and into three groups, depending on the minimum design working life of 5, 10, and 15 years respectively. Glass-covered greenhouses have to be designed according to A15 (Class A, 15 years minimum design working life), sophisticated plastic film covered multi-span greenhouses according to B15, and simple plastic film tunnels and shade houses according to B10 and B5 (Waaijenberg 2006; EN 13031-1, 2001).

The main loads or actions to be considered are:

- *Dead load* or permanent load "self-weight of structural and non-structural elements, excluding the installations even if they are permanently present".
- *Wind loads* are "actions imposed on the structure by wind".
- *Snow loads* have to be considered in regions with snowfall.
- *Crop loads* have to be considered where structures support crops. Where crops are suspended on separate horizontal wires the horizontal tensile forces transmitted to the structure have to be taken into assessment.

The different actions shall be considered in combinations which are given in the standard. The European standard is valid for the EU. It can be taken as an example also for other regions outside the EU.

4.5 General Requirements

The climatic conditions are the essential basis for the protected cultivation of plants and for the construction of protective structures. Chapter 2 describes the most important climatic zones (Figs. 2.14–2.16). That results in the following additional requirements for greenhouses and shade houses for the climatic zones (von Zabeltitz and Baudoin 1999).

4.5.1 Ar: Tropical Wet Climates

4.5.1.1 Regions

Amazon basin of South America, equatorial West Africa, insular and peninsular area of Southeast Asia.

4.5.1.2 Climatic Conditions

- High monthly precipitation quantities and high mean humidity throughout the year.
- Only small variations of temperature and solar radiation in the course of the year.
- Day and night temperatures above the biological minimum for plant production throughout the year.
- Irradiation which can be too high.

4.5.1.3 Greenhouse Structures

A protection from precipitation and too high solar radiation is necessary throughout the year. Only covering of the roof area is needed for protection from rain, solar radiation, and wind. The cladding material has to reduce solar radiation and shade the cultures. As outdoor temperatures are constantly high, sidewalls and gables can remain open. This results in good ventilation. Temperatures and humidity do not rise much higher than in the open air. If necessary, ventilation openings have to be equipped with insect screens in order to keep out insects, but the reduction of ventilation efficiency has to be considered. Rainwater should not penetrate the greenhouse, but should be led off by gutters or ditches.

4.5.2 Aw: Tropical Wet and Dry Climates

4.5.2.1 Regions

North and south of the equator on all continents between latitudes 5° and 20°.

4.5.2.2 Climatic Conditions

- Separation of rainy and dry seasons
- High monthly precipitation and high mean humidity during the rainy season
- Slightly higher temperatures and higher average irradiation sums during the dry period
- Relatively even course of temperatures with small variations throughout the year
- Day and night temperatures above the biological minimum throughout the year

4.5.2.3 Greenhouse Structures

The crop has to be protected from high solar radiation throughout the year. Protection from rain is only necessary during the rainy season. In case of serious droughts during the dry period, rainwater should be collected for irrigation purposes during the dry period. The greenhouses must have gutters, and rainwater reservoirs have to be built. In the dry season, humidity can drop considerably during the day. Temperature conditions do not demand closable ventilators. They might, however, become necessary in the case of extremely low humidity. In the case of closed ventilators, humidity in the greenhouse actually rises due to evapotranspiration, but temperature also rises. If humidity is too low, evaporation cooling or fog systems are possibilities to create a favourable climate.

4.5.3 Cf: Subtropical Humid

4.5.3.1 Regions

Uruguay, coastal regions in the east of China, South Japan, southeast coast of Australia, South-East regions of the USA

4.5.3.2 Climatic Conditions

- High monthly precipitation quantities and high average humidity throughout the year
- Strong seasonal differences in temperature and solar radiation values between summer and winter
- Night frosts which can occur during winter

4.5.3.3 Greenhouse Structures

Precipitation is also a problem in subtropical humid climates throughout the year. Greenhouses serve to protect the cultures from too heavy rain. Gutters are necessary to discharge the rainwater. Solar radiation is a problem only during summer months. In winter, the daily solar radiation sum drops to values 50–70% lower than those in summer. It ranges from approximately 2–3.5 kWh/m^2 day. Greenhouses should therefore let through as much light as possible. Daily temperatures are relatively high; greenhouses must have good ventilation efficiency. During the night, temperatures can sink considerably; therefore, greenhouses must have ventilators that can be closed overnight. In addition, the covering material should have a transmittance as low as possible for long-wave radiation, in order to reduce thermal radiation of the plants.

In summer, temperatures and daily sum of solar radiation are higher than in the tropical zones. Therefore, shading and very good ventilation are required. Shading facilities should be movable, in order to let through as much light as possible into the greenhouse in winter.

4.5.4 Cs: Subtropical Dry Summer. Mediterranean

4.5.4.1 Regions

Borderlands of Mediterranean Sea, lowland California, central Chile, southern part of South Africa, parts of southern Australia.

4.5.4.2 Climatic Conditions

- Distinct seasons with very hot dry summers and mild winters
- Sometimes, temperatures below 5°C at night ,and possible occurrence of frost
- Rainfall concentrated on wintertime

4.5.4.3 Greenhouse Structures

Greenhouses with lockable ventilators have to be tight, in order to protect crops from too low temperatures in winter. Often, it is necessary to heat the greenhouse. The covering material should be impervious to long-wave radiation to minimize thermal radiation out of the greenhouses. Gutters are advantageous for the collection of rainwater for irrigation purposes. During the hot summer months with high temperatures, vegetables are often not cultivated in greenhouses.

4.5.5 H: Tropical Highlands

4.5.5.1 Regions

Colombia, Kenya, Malawi.

4.5.5.2 Climatic Conditions

- Mean temperatures depend on altitude. They are very even throughout the year, but with considerable temperature variations between night and day.
- Precipitation distributed over the year, with stronger variations from year to year
- High even solar radiation throughout the year.

4.5.5.3 Greenhouse Structures

According to altitude and crop, greenhouses with or without lockable ventilators are necessary. Additional requirements are large ventilation openings, covering material impervious to rainwater, and gutters for the collection of rainwater.

4.5.6 B: Dry Climates

4.5.6.1 Regions

Arid and semiarid regions.

4.5.6.2 Climatic Conditions

- Evapotranspiration is higher than precipitation
- Not enough water
- Brackish water sometimes available in the underground
- Strong temperature variations between day and night. Frost can occur at night
- Low humidity and high solar radiation during the day
- Sand storms

4.5.6.3 Greenhouse Structures

Greenhouses have to protect crops from excessively high irradiation, wind, sand storms and too low humidity. They must have efficient ventilators that can be closed at night. In the case of low humidity and high temperatures, evaporation cooling can be installed that operates with brackish water if available. Brackish water can, if there is a lack of irrigation water, be desalinated by solar energy.

4.6 General Design Requirements

The general design requirements for plastic-film greenhouses are:

4.6.1 A. Structure

1. Sufficient stability against wind and crop loads. That means sufficient dimensions of the construction components, and installation of wind braces (Fig. 4.2: also see Chap. 8).
2. Mounting and installation according to technical specifications (Chap. 8).
3. The connections and connectors between the different construction components must not move or slide by load forces (Fig. 4.3) (Sect. 8.2).
4. The foundations under the stanchions have to endure pressure and suction forces by wind (Sect. 8.1).

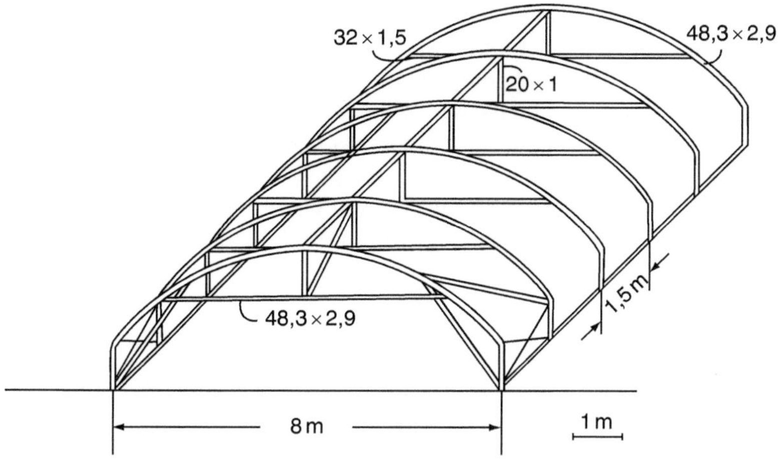

Fig. 4.2 Round arched tunnel calculated according to Dutch Greenhouse standard NEN 3859 (Waaijenberg 1990; Waaijenberg and Denkov 1992)

Fig. 4.3 Connectors and clamps: how to do and how not to do

5. The plastic film must not flutter by wind forces. It has to be stretched and fixed tightly on the structure (Sect. 8.3).
6. The use of simple and detachable fastening devices for changing the film. The nailing of the film on the structure is not to be recommended (Sect. 8.3).

7. Insulation of Steel Components
   The plastic film must not touch directly steel components that are heated up by solar radiation. The steel components should be insulated by plastic strips, or the plastic film should be painted white where it touches the steel components (Fig. 4.4). Unprotected steel components can be heated up to 70°C by solar radiation and the film will be destroyed quickly.
8. Drops of condensed water should not fall down from the inner surface of the film onto the plants, but they have to run off at the film.

Fig. 4.4 Insulation of steel components or protection by white painting of the plastic film

Fig. 4.5 Drops at the inner surface of the roof fall down and reduce light transmittance

Droplets on the inner film surface reduce the light transmittance of the greenhouse by 10–15% (Fig. 4.5). No horizontal roof areas but sufficient roof slope of more than 10° and the use of plastic film with No-Drop additives can avoid those disadvantages. Wires below the plastic film hinder the run-off of droplets. They fall down at the wires. The installation of wires should be avoided. Condensation itself is desirable to avoid too high humidity. Condensation should appear as film condensation (Sect. 7.4).

9. Gutters or deep ditches are necessary to drain off and collect the rainwater. No water penetration from side wall, gable and roof into the greenhouse (Fig. 4.6).
10. Durable materials which are available and cost-effective in the country should be used for the greenhouse design and construction. The maximum width of the greenhouse unit depends on the maximum available width of plastic film, if the film is stretched in longitudinal direction of the greenhouse.
11. Windbreaks should be installed if the wind velocity is too high.

The most important factors for the stability of the greenhouses are:

- The dimensions of the construction components
- The connection of the components by clamps
- The foundation
- The fastening and stretching of the plastic film

Fig. 4.6 If there is no gutter or ditch, the consequence is heavy rainwater penetration inside the greenhouse. Ditches of dug plastic film filled with gravel are to be recommended

4.6.2 B. Climate Control

Appropriate climate conditions are prerequisite for plant growth and quality. That means in particular appropriate temperature on hot days and in cold nights, appropriate humidity level, no CO_2 deficit, and enough light. The main requirements are:

1. Sufficient ventilation efficiency (Chap. 9).
   The ratio of ventilation opening to greenhouse floor area should be more than 20–25%, if no insect screens are installed. Multi-span greenhouses have to have ridge or roof ventilation, if the mean maximum temperature is above 27°C and if the greenhouses have more than three spans.
2. Lockable ventilators should be installed in regions where mean minimum temperatures are lower than 12°C.
3. Larger greenhouse volume is favourable for climate control on hot days.
   Side walls should be 3 m high or more. This also provides a buffer for CO_2 supply if the vents are closed on colder days and the light conditions are sufficient.
4. The plastic film should have a low transmittance for long-wave radiation, to reduce heat transfer by radiation from plants and soil to the outside atmosphere. Materials with high transmittance for long-wave radiation increase the heat transfer by radiation from the crop through the cover, with the consequences of lower air temperature, lower plant temperature and temperature inversion (inside temperature lower than outside). If the plant and air temperature sink below the dew-point temperature, condensation and fog may occur, and the danger of diseases increases. Thermic film with IR absorber or co-extruded PE and EVA film should be used for greenhouses, to keep the temperature as high as possible in unheated greenhouses during cold nights (Chap. 7).
5. The structure has to be tight.
   It is very important to avoid leaks, even in unheated greenhouses, to keep the warm air inside as long as possible during night. Main sources of leaks are the ventilators and the doors. If a temperature inversion occurs, or if the inside humidity is too high, it is more efficient to ventilate the greenhouse, in particular in the morning (Sect. 8.4).
6. Heating becomes necessary for heat-requiring crops in some winter months.
   Cheap fossil fuel heating systems, simple solar heating equipment or geothermal energy should be used if available (Chap. 12).
7. Energy-saving measures in heated as well as in non-heated greenhouses.
8. Irrigation systems with high water use efficiency and low evaporation should be used. These systems are drip irrigation with fertilizer distribution.

4.6.3 C. Integrated Plant Production and Protection (IPP)

Integrated production and protection (IPP) aims in finding alternative solutions to reduce pesticide application.

Insect screens are used in front of the ventilation openings and doors to keep useful insects inside, and to prevent pest insects from penetrating the greenhouse (Chap. 10).

Insect screens with different mesh openings are used because the insects are of various sizes.

Criteria for the choice of insect screens are:

- The species of insects to be screened out
- The influence on the greenhouse climate
- UV stability and mechanical durability (thickness of threads)
- The cost in comparison to the economic value of the crop

Insect screens reduce the ventilation efficiency remarkably. Sufficient ventilation efficiency has to be guaranteed, even in screened greenhouses. To guarantee climate conditions similar to those in unscreened houses, enlarged ventilation openings should be designed. If the ratio of vent opening to greenhouse floor area is 20–25% in unscreened houses, the vent opening should be enlarged in screened houses by a factor of about 2 for a insect screen against white fly.

Different principles of greenhouse structure, cladding materials and climate control measures are available to meet all these requirements (Chap. 10).

Chapter 5
Greenhouse Structures

Greenhouses have to provide optimal climate conditions for the plants growing inside. They have to protect plants against too low temperatures, wind, rain, hail, birds, and insects (Waaijenberg 2006). The following components are very important for successful greenhouse design: shape, orientation, structure, cladding material, foundation, ventilation, and technical equipment for climate control.

Structures for protected cultivation can be classified into greenhouses and net or shade houses mostly used in tropical regions.

Shade and net houses have water-permeable cladding nets. They shade the plants and protect them from incoming insects if roof and side walls are completely covered by nets, and if the mesh is small enough. They reduce too high radiation, wind speed and the impact of heavy rain, but they do not protect the plants from being wet by precipitation. Fertilizer will be washed out much more easily, and controlled fertigation is not possible. They have no positive effect on water-use efficiency.

Greenhouses have cladding material that is impermeable to water and has high transmittance for natural light. Greenhouses protect the crops from rain and other climate factors described in Chaps. 2 and 4. Usually, the cladding material is plastic film in mild climates and glass or rigid plastic in temperate climates.

The width of one span depends on the available width of plastic film if the roof is covered by the film in a longitudinal direction, which is the recommended approach. The width of available cladding material is limited in many countries by the manufacturing process. The necessary width of plastic film for the roof on multi-span structures, depending on the width of the span, is:

Width of span (m)	Minimum width of plastic film (m)
5	6
6	8
8	10

The shape of the construction, the height to the eave and ridge (the whole volume), the cladding material, and the number of spans influence internal climate conditions such as temperature, humidity, light transmittance, and CO_2 buffer.

C. von Zabeltitz, *Integrated Greenhouse Systems for Mild Climates*,
DOI 10.1007/978-3-642-14582-7_5, © Springer-Verlag Berlin Heidelberg 2011

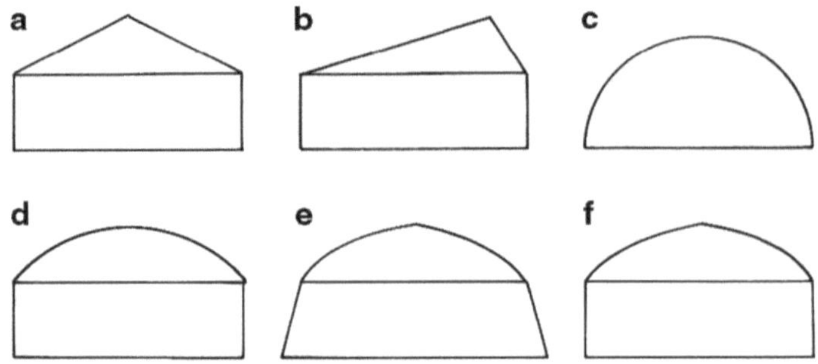

Fig. 5.1 Most frequent shapes of greenhouses: see text

The most frequent shapes for greenhouses (see Fig. 5.1) are saddle roof (*a*), saw tooth or shed roof (*b*), round arched tunnel (*c*), round arch with vertical side wall (*d*), pointed arch with sloping side wall (*e*) and pointed arch with vertical side wall (*f*). Preferable for plastic film greenhouses are designs (*e*) and (*f*).

Single-span and multi-span greenhouse can be distinguished.

Ventilation efficiency by natural ventilation, which is proportional to pressure differences (Bot 1983), depends on the height of the greenhouse if ventilation openings are positioned at the ridge and the side wall. The higher the ridge and the bigger the distance between ventilators at ridge and side wall, the higher are the pressure differences. On the other hand, wind loads and the strength of the structural components depend on the greenhouse height, just as the heat requirement depends on surface area. High greenhouses with large volumes provide better climatic conditions, but also increase the heat requirement if heating is needed.

The design of the greenhouse has to be chosen according to the climate conditions and the general design requirements, and not to national traditions.

Simple low-cost greenhouse structures (Fig. 5.2) allow the economic development of regions in development countries, but usually they have more unfavourable climate conditions, with higher temperature and humidity followed by lower yield, higher pest infestation and a higher amount of non-marketable crop.

Simple plastic film greenhouses predominate in warmer subtropical countries (De Pascale and Maggio 2005). Crops cannot be grown all year round, and the yield does not fulfil quality standards completely, because of inadequate climate conditions inside the simple greenhouse structures (Baille 1999b, 2001). The structure and the shape are often not adapted to the climate conditions of the region in development countries. Low-cost plastic film greenhouses in particular are designed on the principle of minimum capital and technological input as well as low running costs. Sophisticated greenhouses often are too expensive. The aim is to find a compromise between suitable greenhouse technology, increasing costs, and economical output for the grower.

Greenhouse crop production in warmer climates requires a specific design of greenhouse structures and climate control equipment. There is a need for higher

Fig. 5.2 Cheap low-cost structures

technological standards to face the increasing competition by products and marketing from other countries (Pardossi et al. 2004).

Cost-effective greenhouse structures with higher investment in height of the structure, volume and better ventilation efficiency need less expense for spraying and provide better quality, healthier crop and better yield. They may have a better cost–benefit ratio and permit environmentally friendly production (Fig. 5.3).

Plastic film greenhouses and glass greenhouses are being built side by side in Turkey. Growers and scientists are still discussing the advantages and disadvantages of glass and plastic film greenhouses for Mediterranean climates. Many growers are still building glass greenhouses, supposed to have less maintenance and better light transmittance. The light transmittance of plastic film greenhouses should be higher because of less construction components if the greenhouses are cleaned regularly (see Chap. 6).

One important point of crop production in greenhouses should be "the sustainable development that meets the needs of the current generation without undermining the ability for future generations to meet their needs" (De Pascale et al. 2005).

The environmental compatibility of plastic film greenhouses is higher than that for glass greenhouses. This can be investigated by the life cycle assessment method (Sect. 5.8).

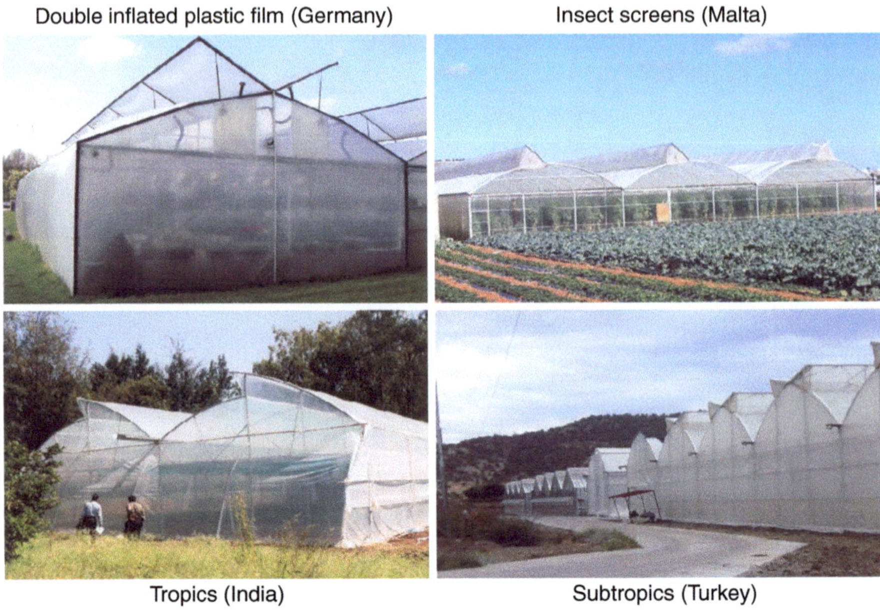

Fig. 5.3 More cost-effective greenhouses

5.1 Round Arched Tunnel Greenhouses

The *round arched tunnel* greenhouses are still the most used greenhouses in many developing countries with mild climates (Fig. 5.4).

The *advantage* of the single-span tunnels is the relatively simple construction system and the wind resistance if they have foundations and if the steel tubes are not too weak, but the disadvantages are not negligible with reference to yield and quality.

The *disadvantages* of the round arched tunnel greenhouses are:

- The net greenhouse floor area fit for plant cultivation is small compared to the ground occupied by the tunnels. The space left between two tunnels is 1–3 m (Fig. 5.4).
- The plastic-film consumption is higher per net floor area.
- The surface area, and as a result the heat consumption, is higher in heated greenhouses.
- The greenhouse volume is too small for appropriate climate control.
- There is a relatively wide nearly horizontal zone, 1–2 m wide, at the top where condensation drops fall down from the covering material.
- 6–30 wires are stretched in longitudinal direction. The plastic film can be damaged by the wires, which are fastened to the bent pipes. Water droplets cannot run off along the inner surface of the film even if No-Drop films are used

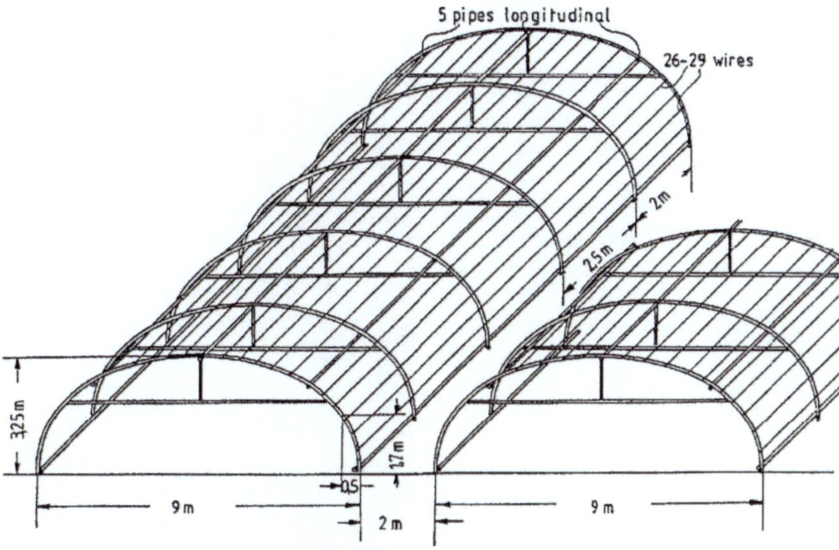

Fig. 5.4 Round-arched tunnel greenhouse

Fig. 5.5 Drop condensation at wires

(Fig. 5.5). The droplets drop down where the wires touch the film. Thus, wires should be avoided in greenhouse construction.

- When tall plants (tomatoes, cucumbers, and roses) are grown, the arched design of the wall causes about 1 m of ground to be not usable between the sidewall and the first row on each side (Fig. 5.6).

Fig. 5.6 Free space at bent side wall

Fig. 5.7 The border effect

- The plants near the sidewall ventilation and near the gable grow less than those in the middle of the tunnels. This border effect (Fig. 5.7) is caused by lower temperatures, by lower humidity next to the vents and gables, and by wind effects.
- Very often, round arched tunnels are built without any foundation or the foundations are not sufficiently secured against uplift wind forces. They will be destroyed by uplift wind force (Fig. 5.8).
- *Ventilation* efficiency is not sufficient if only the overlapping plastic film and the doors at the gables are "opened". This kind of ventilation cannot be operated mechanically, but each opening has to be opened manually. This takes a long time, and the greenhouses can be overheated in the morning (Fig. 5.9).

Fig. 5.8 Greenhouses without foundations are secured against uplift forces by sand sacks (Jordan). Point foundations are preferable

The simple ventilation of round-arched tunnels by opening the overlapping film sheets is not sufficient. If the film has a width of 6.5 m, there is an opening every 6 m at the sidewall. The ratio of vent area to greenhouse floor area becomes less than 10%. For optimum ventilation, the ratio should be more than 20% (Chap. 9).

Even if the vent openings are opened by parting the overlapping film sheets over the whole round-arched surface, the ventilation efficiency is not always sufficient (Fig. 5.10).

If round arched tunnel greenhouses are supposed to be built, then only with *through ventilation openings* at both side walls — but the border effect has to be taken into consideration (Fig. 5.11).

5.1.1 Improvements for Tunnel Greenhouses

The disadvantages of single-span tunnel greenhouses can be reduced by some improvements such as

- Sufficient stability resulting from the dimensions of the construction, sufficient fixed connections of the components, and using foundations.
- Pointed arched construction to avoid drop fall at the top of the roof.
- Installation of additional straight side walls with through ventilation openings.

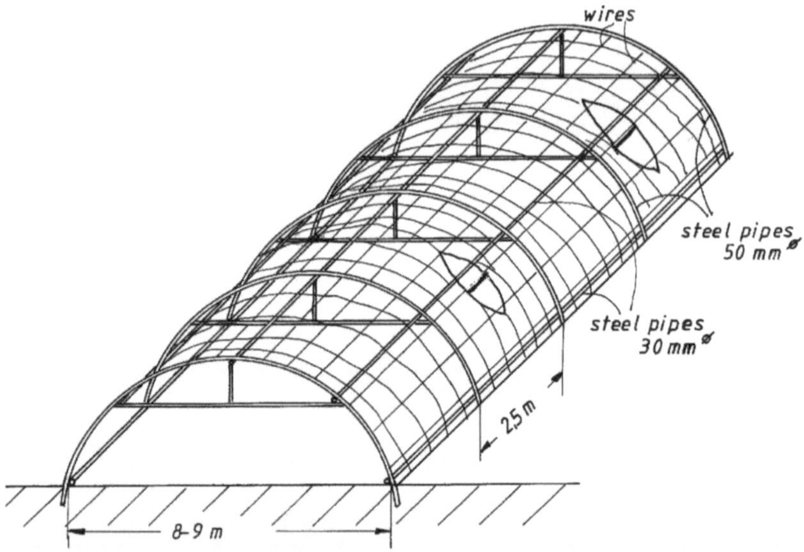

Fig. 5.9 Inefficient ventilation

5.1.2 Dimensions

The dimensions of the construction are responsible for the stability of the greenhouse. Figure 5.12 shows some dimensions of steel tubes for a simple single-span tunnel greenhouse from experience sufficient for normal wind loads in mild climates. But point foundations have to be installed. These dimensions can be used if no standards are available.

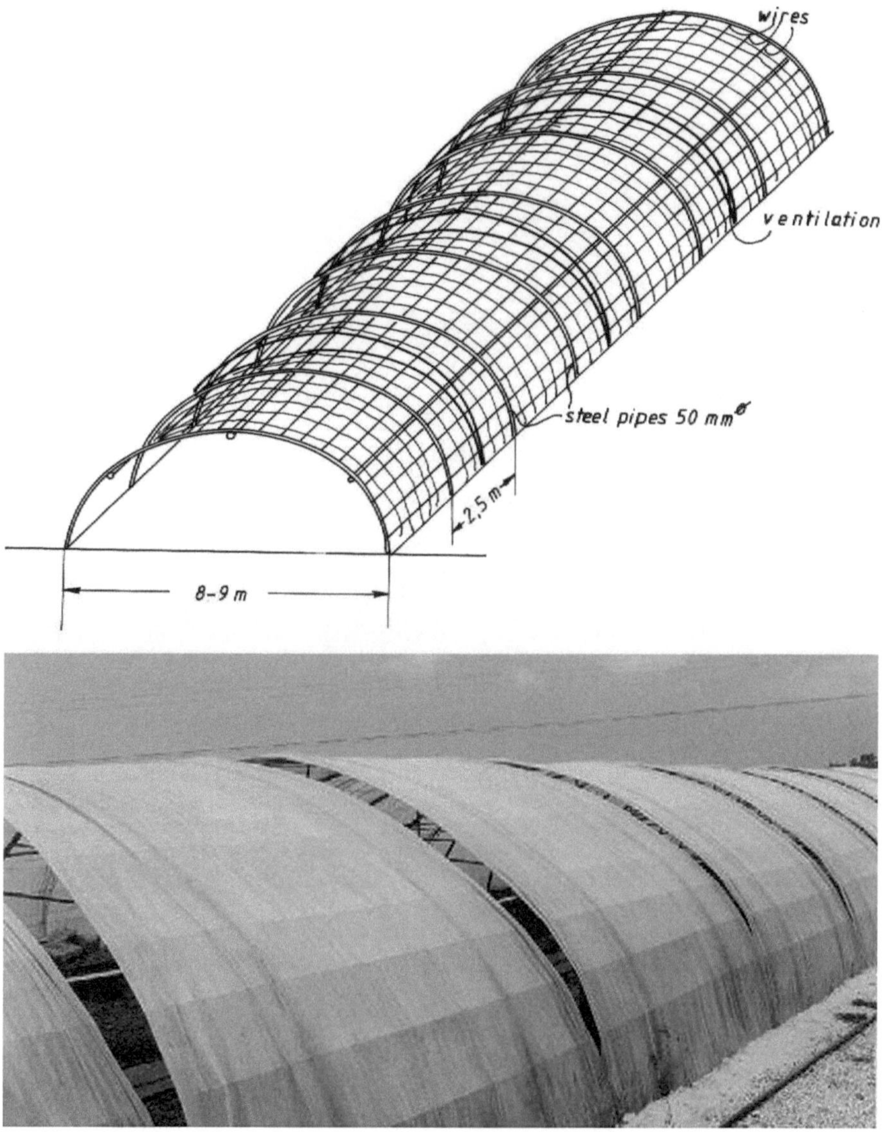

Fig. 5.10 Ventilation by parting the overlapping film over the whole arch is not sufficient

Waaijenberg (1990), Waaijenberg and Denkov (1992) has made calculations in accordance with the climate conditions of Tunisia and with a crop load of 0.15 kN/m^2. Figure 5.13 shows a structure with horizontal trellis girders and another one with extra bracings (Von Zabeltitz 1999). The arch tubes, combined with trellis girders and extra bracings, can be designed using tubes with dimensions

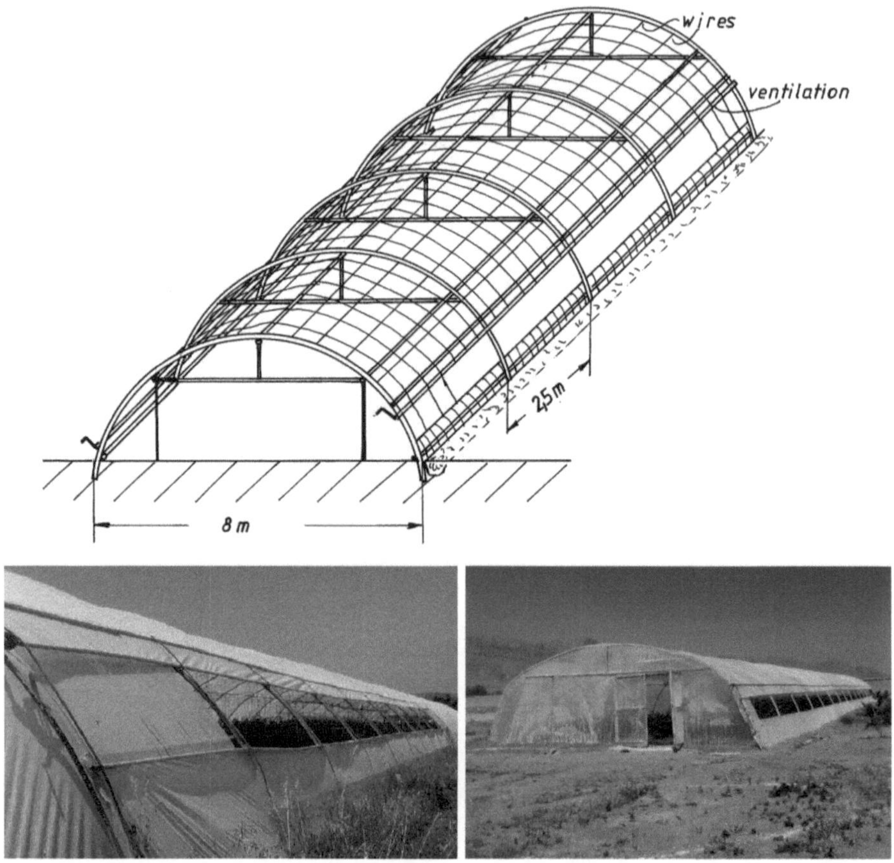

Fig. 5.11 Through ventilation at both sides.

48.3 × 2.9 mm even if the arch distance is 2.0 m. The same figure shows the dimensions of a bitunnel, calculated also with crop load.

5.1.3 Pointed Arched Construction

The round arched tunnel normally has a wider more or less horizontal zone at the ridge. Water drops fall down even when using special No-Drop film. The steel tube arches of the tunnel greenhouses are frequently made of four parts (1–4 in Fig. 5.14a). If the upper bends *2* and *3* are turned and fixed at the ridge by a new connector, the result is a more pointed arched roof built with the same construction components (Fig. 5.14b). If a more distinct pointed arch is desired, the lower bends *1* and *4* can be kept and new bends *5* and *6* can be installed at the ridge with a corresponding connector (Fig. 5.14c).

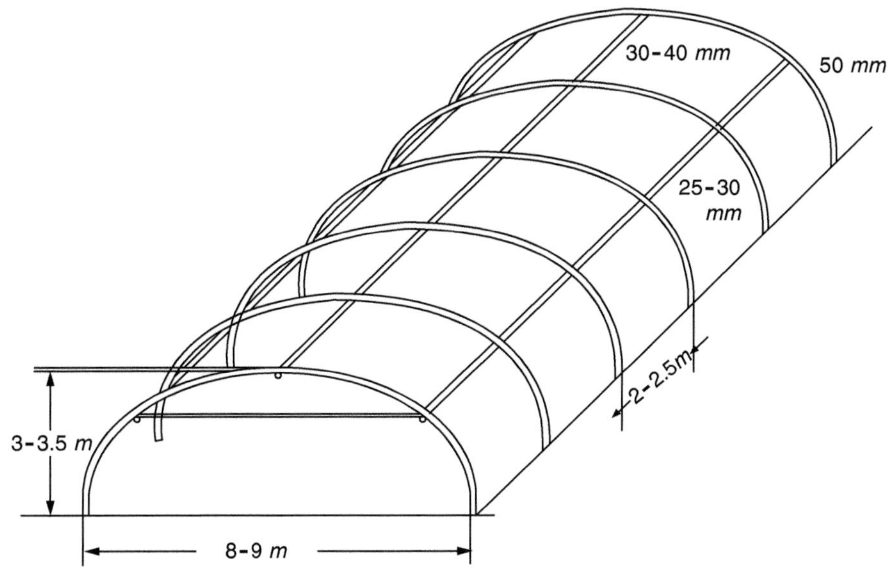

Fig. 5.12 Some dimensions of a single-span round-arched tunnel.

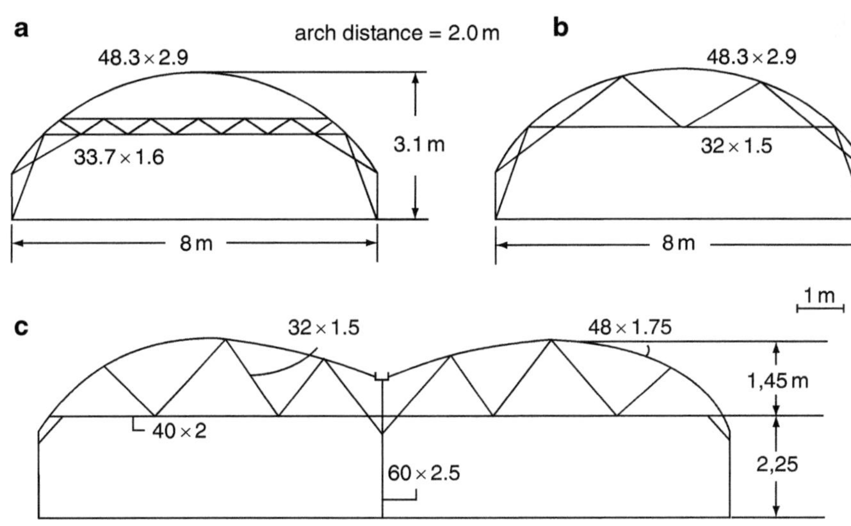

Fig. 5.13 Single-span and Bitunnel designed according to European standard EN 13031-1 (2001).

5.1.4 Enlarged Tunnel Greenhouse

If farmers can not remove the traditional round arched greenhouses and build appropriate ones, they can improve the existing greenhouses as shown in Fig. 5.15 for better climate control. The height of the greenhouse can be enlarged

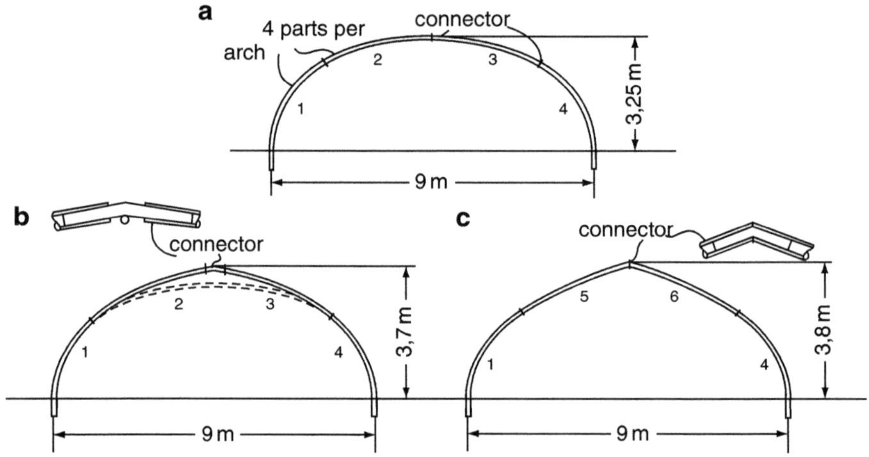

Fig. 5.14 Modification of round-arch to pointed-arch tunnel.

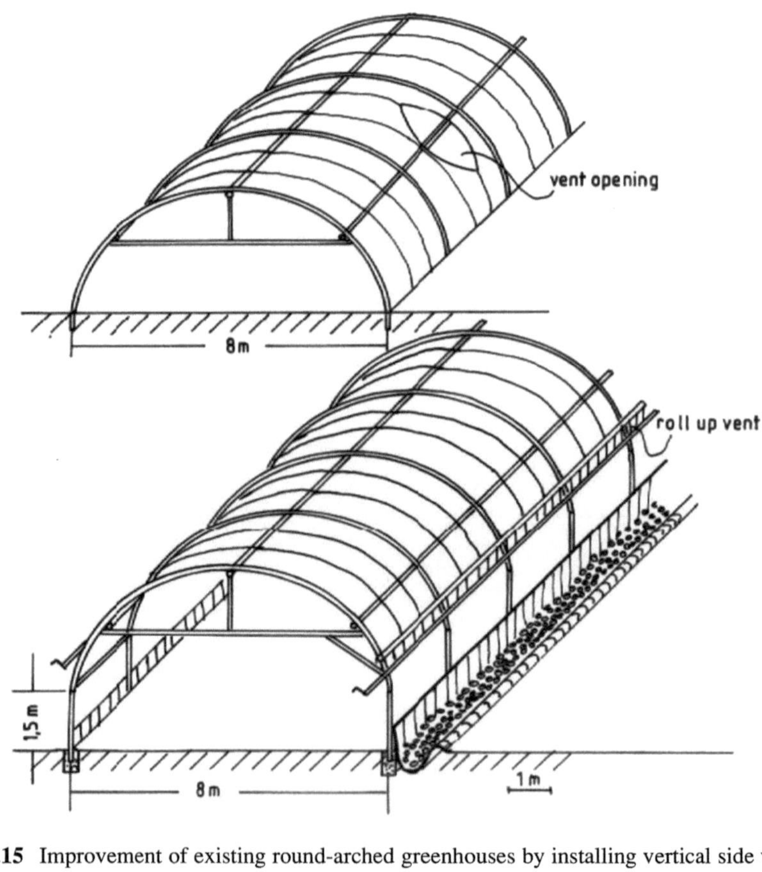

Fig. 5.15 Improvement of existing round-arched greenhouses by installing vertical side walls

by connecting pipes of about 1.5 m length, fixed to the ends of the original truss tubes. In this case, concrete point foundations are absolutely necessary. Through side-wall ventilation has to be installed. The film below the ventilation openings is buried into the soil and filled with gravel. In this way, it works as a gutter to drain off rainwater, and it gives additional stability to the whole construction against uplift wind forces.

5.2 Multi-Span Greenhouses

Multi-span gutter-connected plastic-film greenhouses fulfil most of the design criteria. They have advantages as follows:

- The greenhouse volume is larger, and the climatic conditions are better during day and night. The sidewalls should be as high as possible. A sidewall height of 3 m is favourable, but wind resistance has to be guaranteed.
- Ventilation with sidewall and gable ventilators can prove sufficiently efficient, if the total width of the multi span unit is limited to about 18 m.
- Ventilators can be operated mechanically.
- Crop density is higher, and the border effect is less influential. Vertical sidewalls avoid losses of space along the sidewalls and allow the use of machines to work inside the greenhouse.
- The usable greenhouse area per ground is higher.
- Pointed-arched roofs can be built in order to reduce dripping.

According to the prevailing climate conditions, multi-span greenhouses should have roof or ridge ventilation, if the mean maximum outside temperature exceeds 27°C.

But one has to take into consideration that roof or ridge ventilation is very costly. The investment for these ventilation systems represents 25–30% of the whole investment for the greenhouses. The following investment costs for greenhouse structures including plastic film are given in €/m^2 floor area (Castilla and Hernandez 2007):

	Without roof vents	With roof vents
Sicily	10–16	27
Greece	11	14.8

Small units with two to three spans only can therefore be built with side-wall and gable ventilation only if the gutter height is 3 m or more. Roof or ridge ventilation has to be built for multi-span greenhouses with four or more spans. Many greenhouses in the subtropics do not have ridge vents, but one has to take into consideration that the vent openings have to be enlarged if insect screens ought to be used (see Chap. 10).

Figure 5.16 shows a pointed arched steel tube construction with 5 m width of span and gutter height of 2 m only. Pointed-arched or gothic-arched structures have

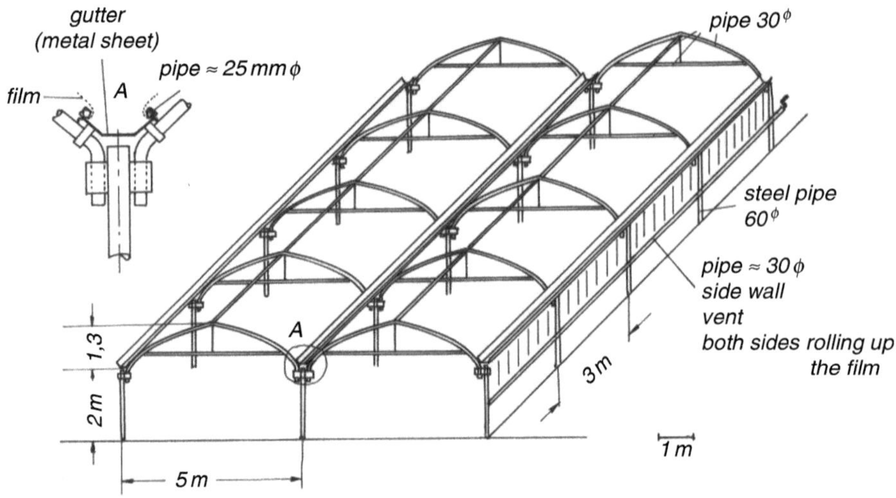

Fig. 5.16 Pointed-arched greenhouse

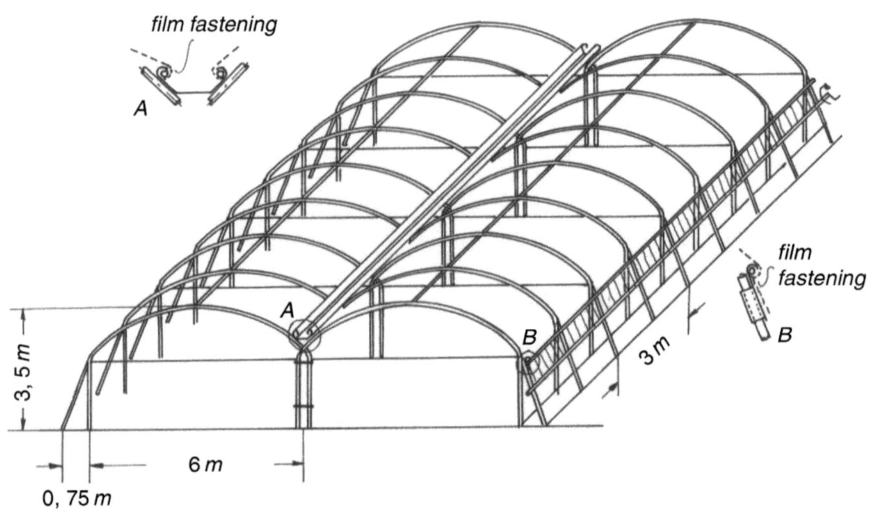

Fig. 5.17 Round-arched greenhouse with sloped side walls

advantages over round-arched structures, because condensation water can flow off better on the inner side of the film, and therefore little water drops down on the plants (see Sect. 7.4). The side wall should be 3 m or more. The plastic film at the gutter is fixed and stretched by rolling it up on a steel tube in the gutter. The roof tubes are fixed at the stanchions by clamps in which they are inserted and screwed. The width of the single span depends on the maximum width of the plastic film available in the country.

Figure 5.17 shows a round-arched gutter-connected construction with a sloped side wall to stabilise the construction against wind forces. The plastic film is fixed and stretched by rolling it up on a tube in the gutter profile as shown in *A*. The fastening and stretching of the film at the side wall without gutter can be done also by rolling it up on a tube. Both films the roof and the side wall cladding are rolled on one tube as shown in section *B*. The film is rolled up on the sloping side wall for ventilation. Round-arched constructions may be easier to design, but pointed-arched constructions are preferable for plastic film.

Sloped side walls have advantages if insect screens are installed, because ventilators with insect screens need bigger opening areas for sufficient ventilation efficiency (Fig. 5.18) (also see Chap. 10).

Figure 5.19 shows the possibility of fastening the plastic film by rolling it up on a steel tube. Both the plastic film at the side-wall ventilation as well as the plastic film of the roof can be fixed by rolling up on one tube.

Figs. 5.20–5.22 is show some more examples for multi-span greenhouses.

Various roof or ridge ventilations are possible if the greenhouse units have more than two spans and if the total width is more than 18–20 m (Figs. 5.23–5.25) (see Chap. 9).

Figure 5.26 shows two kinds of roof ventilation. Both open one half of the roof. Design *A* is linked at the gutter and opens at the ridge. This ventilation has to have good resistance against perpendicular wind forces, and has to be installed tightly. Design *B* opens at the gutter and is linked at the ridge. Both ventilations should open a minimum of 1 m. This means that a sufficiently long and strong rack-and-pinion drive have to be installed.

All those multi-span greenhouses can have a pointed-arched roof, which has advantages for the run-off of condensation water.

At the side wall, there is a plastic film fastened to the posts and dug into the soil on the other edge. The film in soil is filled with gravel, and thus acts as a gutter. In addition to that, it stabilises the whole structure against uplift forces

Fig. 5.18 Sloped side walls covered with insect screens and rolling-up ventilation in Jordan

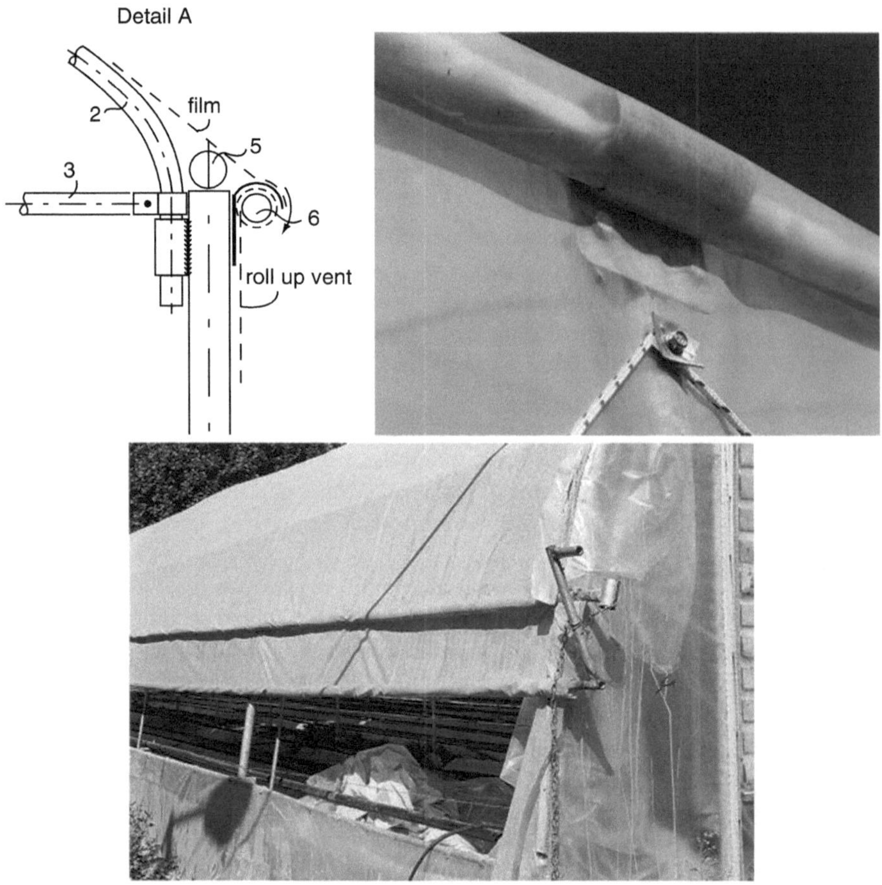

Fig. 5.19 Fixing and stretching both the plastic film of the roof and the ventilation opening on one tube

If greenhouses do not have a gutter at the eave of the side wall, they have to have ditches deep enough to drain off the rainwater. The plastic film at the side wall has to be dug a minimum of 20–30 cm into the soil to prevent water from penetrating the greenhouse from the side wall.

A recommendable solution is shown in Fig. 5.27. The plastic film of the side wall forms a ditch of 30–40 cm depth. The ditch is filled with gravel of 10–30 mm diameter. As a result of this, rainwater can drain off; the side wall is tight against penetrating water, and the dug plastic film gives additional stability to the construction.

Greenhouses have to have a sufficient slope in longitudinal direction to drain off rainwater by gutters or ditches.

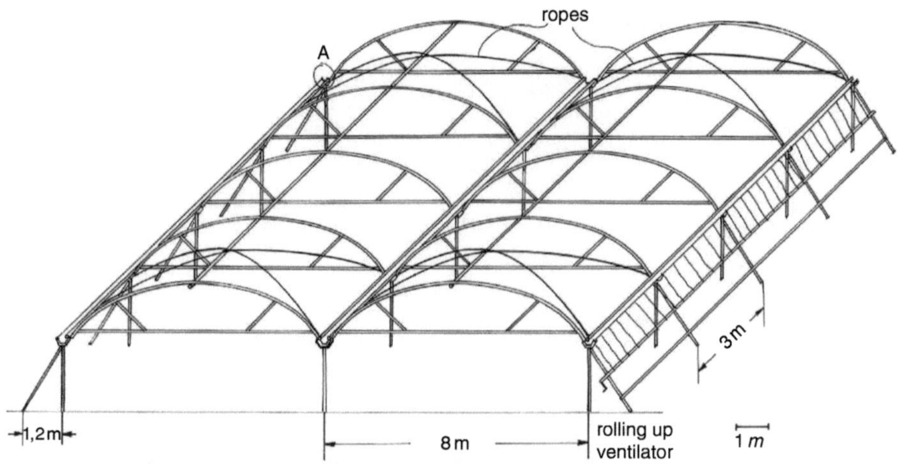

Fig. 5.20 The additional stretching of ropes across the roof at the gable ends may have advantages in areas with high wind speed (Iran)

If the farmer has chosen a greenhouse construction, he has to observe the *mounting and installation* very carefully. Even if the construction principally fulfils the requirements and specifications, mistakes during the mounting can weaken the whole construction and can make the investment useless.

After the construction is ready for operation, the farmer has to take care of *continuous maintenance* so that the greenhouse can be operated as long as possible.

Stable connection of the steel components by clamps is very important for the wind resistance of the greenhouse construction (see Sect. 8.2).

The clamps must not slide on the tubes but have to be tightened firmly. After the mounting has been finished, one has to check whether all clamps, screws, and bolts are screwed and firmly fixed.

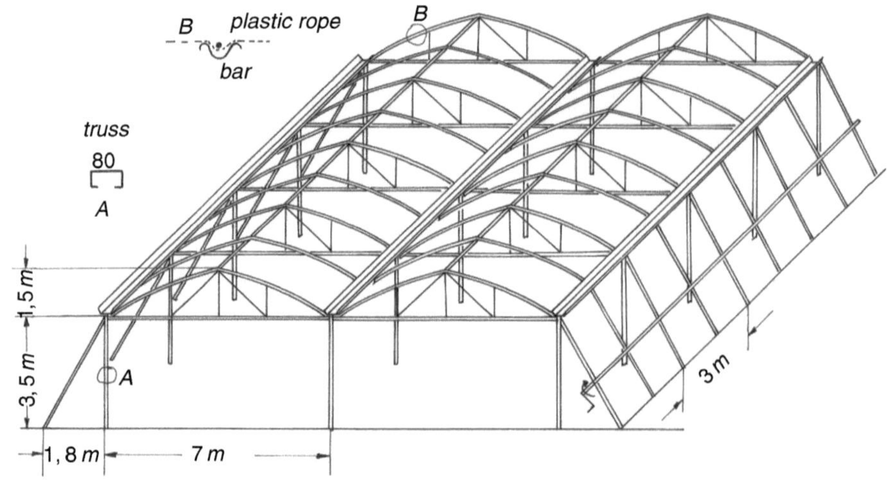

Fig. 5.21 Pointed-arched multi-span greenhouse with sloped side wall and gutter height of 3.5 m (Israel)

5.3 Greenhouses for Tropical Lowlands

Greenhouses are very important for tropical regions because they avoid the disadvantages of open-air production and increase the yield and quality remarkably (Figs. 5.28–5.30).

Average yields for tomatoes in tropical countries (Malawi, Seychelles, Thailand) are:

Open field in dry season	6 kg/m^2
Open field in wet season	2–3 kg/m^2
Protected cultivation	10–15 kg/m^2

Average commercial yields of tomatoes (long cycle from September to May) in Almeria, Spain, are 14 kg/m^2 (Castilla and Hernandez 2005), and the commercial yields of cherry truss tomatoes in South Spain are 8.5–11.9 kg/m^2 for one season (Hita et al. 2007).

The available cost-effective construction materials and the available width of the plastic film vary in different countries. Design materials for greenhouses are wooden poles, timber profiles and steel tubes depending on the available materials in the country. The dimensions of the greenhouse spans are restricted by the available dimensions of materials and plastic film.

The design also varies from simple structures for small-scale farmers to more industrialised multi-span structures for large-scale farmers. Small-scale farms are in the majority in many developing countries. Therefore, appropriate cost-effective structures have to be designed.

Fig. 5.22 Multi-span pointed-arched greenhouses with flap ventilation in Turkey. Thermal screen and hydroponic system inside

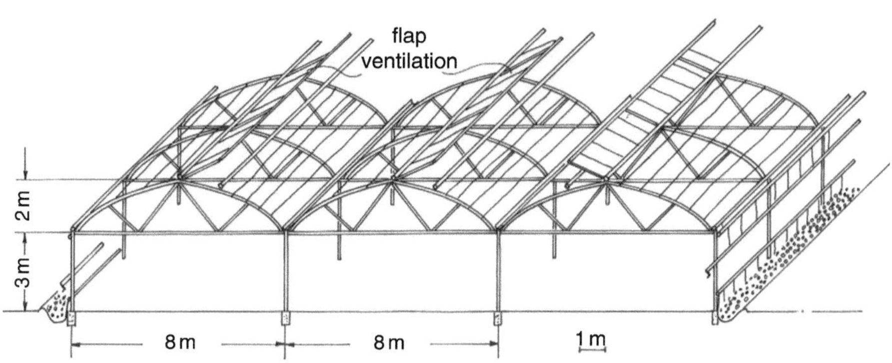

Fig. 5.23 Multi-span greenhouse with flap ventilation at ridge and roll-up ventilation at side wall

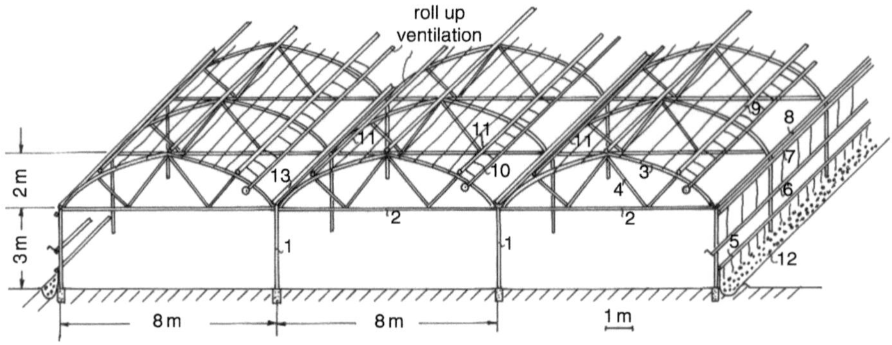

Fig. 5.24 Multi-span greenhouse with roll-up ventilation at roof and gutter. A good holder is necessary at the gutter to close the vent opening tightly and to prevent fluttering by wind forces

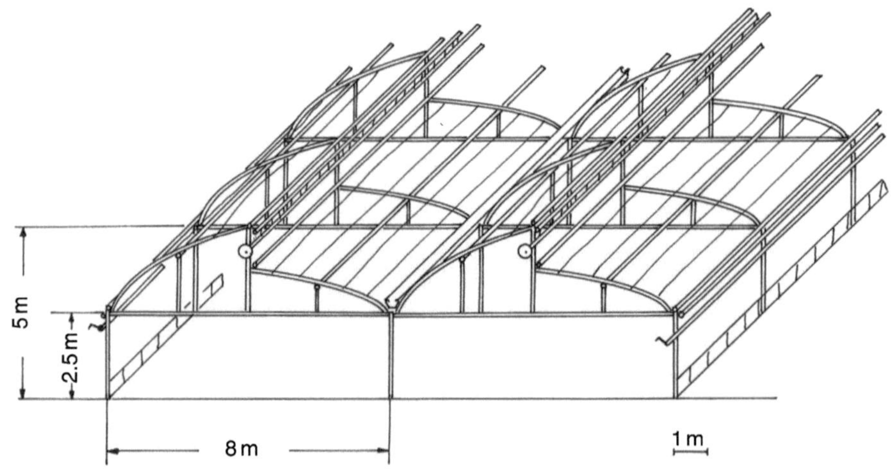

Fig. 5.25 Vertical roll-up ventilation at the ridge, which is very effective in hot climates

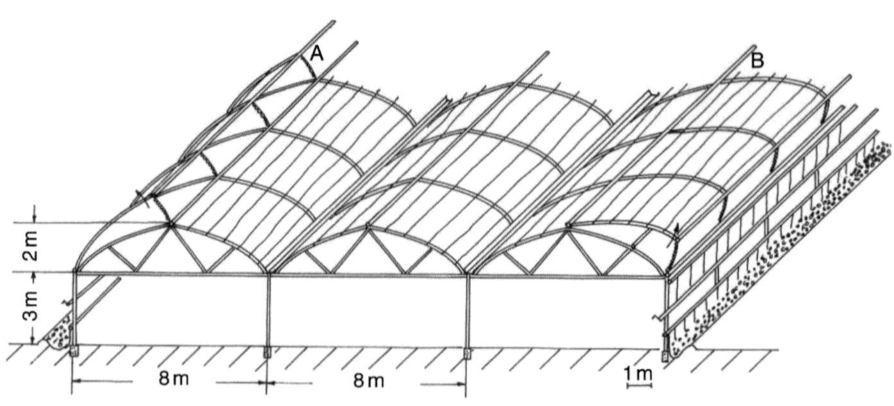

Fig. 5.26 Multi-span greenhouse with roof ventilation linked to gutter and ridge

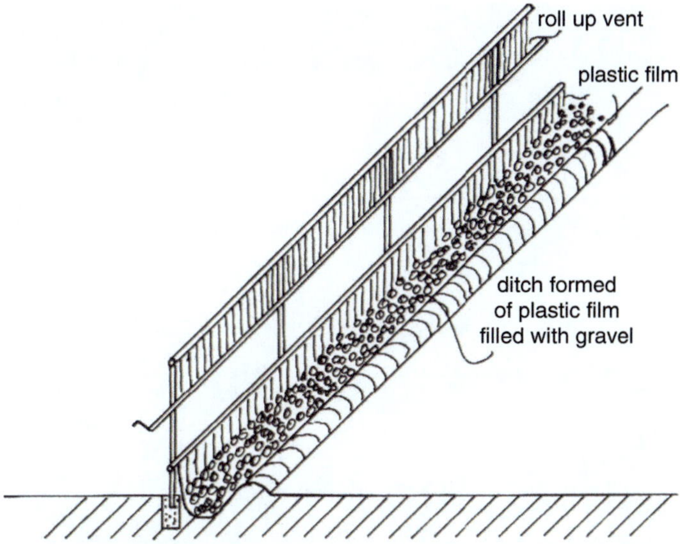

Fig. 5.27 Ditch as gutter

Fig. 5.28 An open-air tomato crop destroyed by a heavy tropical rain shower (Seychelles)

One can find different greenhouse types:

- For short duration of life, structures designed of wood, with untreated poles and without foundations. The duration of life is only 3–5 years.
- For longer duration of life, structures designed of treated timber profiles or steel tubes. Those structures need more investment for the structure itself, but they are more cost-effective and economical in the long term.

Fig. 5.29 Comparison of a pepper crop in the open air and under a shelter structure, both planted on the same day in the Seychelles

The general requirements for greenhouse structures, depending on the climate conditions, are described in Chap. 4.

Greenhouses for tropical lowlands should have the following characteristics (Von Zabeltitz 2000):

- Crop protection from rain, wind, and too high global radiation. Cladding material impermeable to rain. UV-stabilised plastic film with long duration of life is sufficient for most of the tropical lowland greenhouses.
- Crop protection from birds and insects.
- Ventilation openings should be equipped with insect screens if necessary, but ventilation efficiency should not be influenced too much (Chap. 10).
- Very efficient ventilation. Ventilation openings at side walls and the ridge. Ridge ventilation is absolutely necessary.
- The relation of greenhouse volume to ground floor area should be as large as possible. The gutter height should be about 3 m minimum.

The higher the structure, with ridge vents, the higher the ventilation efficiency by the chimney effect.

- Gutters are necessary to drain off the rainwater and to prevent the rainwater from penetrating the greenhouse.
- The roof at the gutter should overlap the side wall to avoid the penetrating of sloping rainfall.
- Resistance of the construction for wind and crop loads.
- Foundation construction should guarantee wind resistance and prevent stanchions from rottenness.

Net houses with screens as roof cover are not suitable for sustainable production.

The ridge ventilation systems of the structures (*a*) to (*d*) in Fig. 5.31 have got better ventilation efficiency than structure (*e*). The wind blows straight through the two openings at the ridge in structure (*e*). One vertical opening at the ridge causes a better air exchange through the whole structure by suction and pressure forces (chimney effect). Growers in India are going to build vents with one opening at the ridge like (*d*), but with alternate directions of the openings to improve the ventilation efficiency (see Chap. 9). If the roof construction is made of steel tubes, a

Fig. 5.30 A tomato crop in the open air and under a very simple shelter structure in Malawi. The protected crop has higher yield and better quality, even when the simple structure does not fulfil the mentioned characteristics

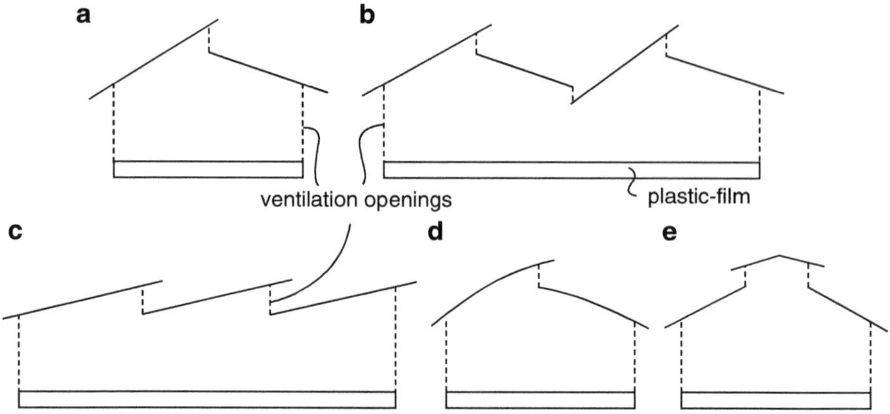

Fig. 5.31 Some principal shapes for tropical greenhouses. All of them have ridge or roof ventilation. The side walls and gables are open in warm and humid tropical climates. They have to be closed at night in tropical highlands

round- or pointed-arched structure like (*d*) is recommendable, because the plastic film can be stretched better over curved constructions.

A 50–60 cm-high plastic film should be fixed near the soil at side walls and gables. That way, rainwater and small animals can not penetrate into the greenhouse. The plastic film can be dug into the soil and fixed to the structure using stretched wires, timber profiles or steel tubes.

A simulated comparison of greenhouse structures for conditions in Indonesia similar to types (*d*) and (*e*) in Fig. 5.31 gave no significant differences in mean air temperatures inside the greenhouse (Hemming et al. 2006a).

But our own measurements in single-span greenhouses in the Seychelles indicated significantly higher temperatures in the daytime in a greenhouse type like (*e*) compared to type (*d*). The temperature difference at a global radiation of 4 kWh/m^2 was 3.5°C (Von Zabeltitz 1994) Figs. 5.32–5.42 show examples and discussions for tropical greenhouse.

Gutters of plastic film only, which are sometimes used, are not recommendable because water pockets occur very easily (Figs. 5.34–5.41, Tables 5.2 and 5.3).

Figure 5.42 shows a comparison of two trusses manufactured in India. The round-arched commercial truss is connected together by clamps, and designed for high wind speeds. The width of the span is 9 m and the truss distance is 3.84 m. The second truss with straight side wall is a welded structure designed by a farmer in India. The dimensions of the components are given in Tables 5.4 and 5.5.

The material costs of the bent truss were 6,140 Indian Rupees for galvanised steel and 4,418 Indian Rupees for mild steel.

The material costs for the welded truss was 1,358 Indian Rupees for mild steel. That means only 30% of the cost of the bent truss, but more than 20 welds were necessary to connect the truss.

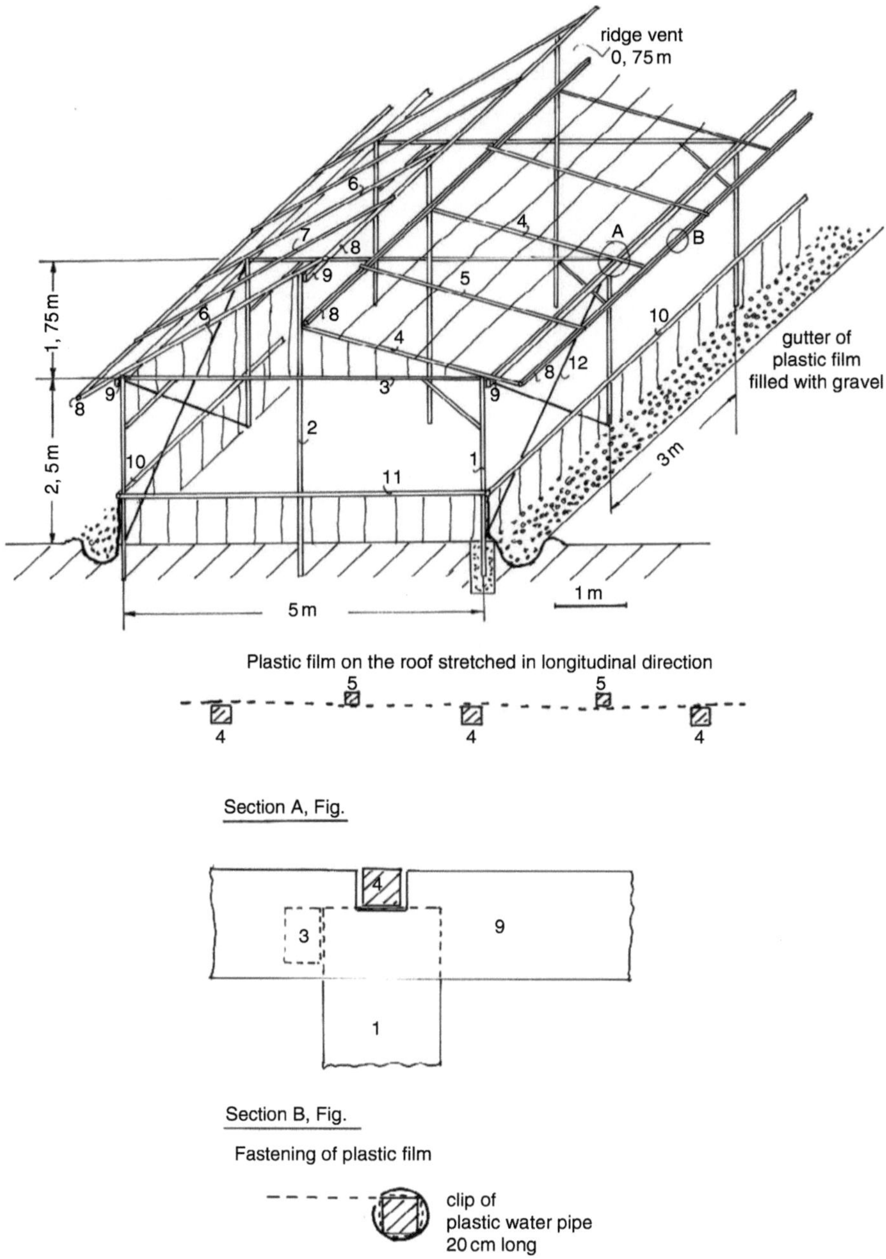

Fig. 5.32 A design for small-scale farmers similar to that called Kenya type in Africa (Von Zabeltitz 1998; Pauer 2002). The vertical stanchions (*1*) and (*2*) are round poles which are dug directly into the soil or which should be put into concrete foundations for the extension of duration of life. All other components are rectangular timber profiles. The numbers in the drawing signify the profiles, and their dimensions are given in Table 5.1. The width of the span is limited by the width of available plastic film. The details show how the plastic film is fixed

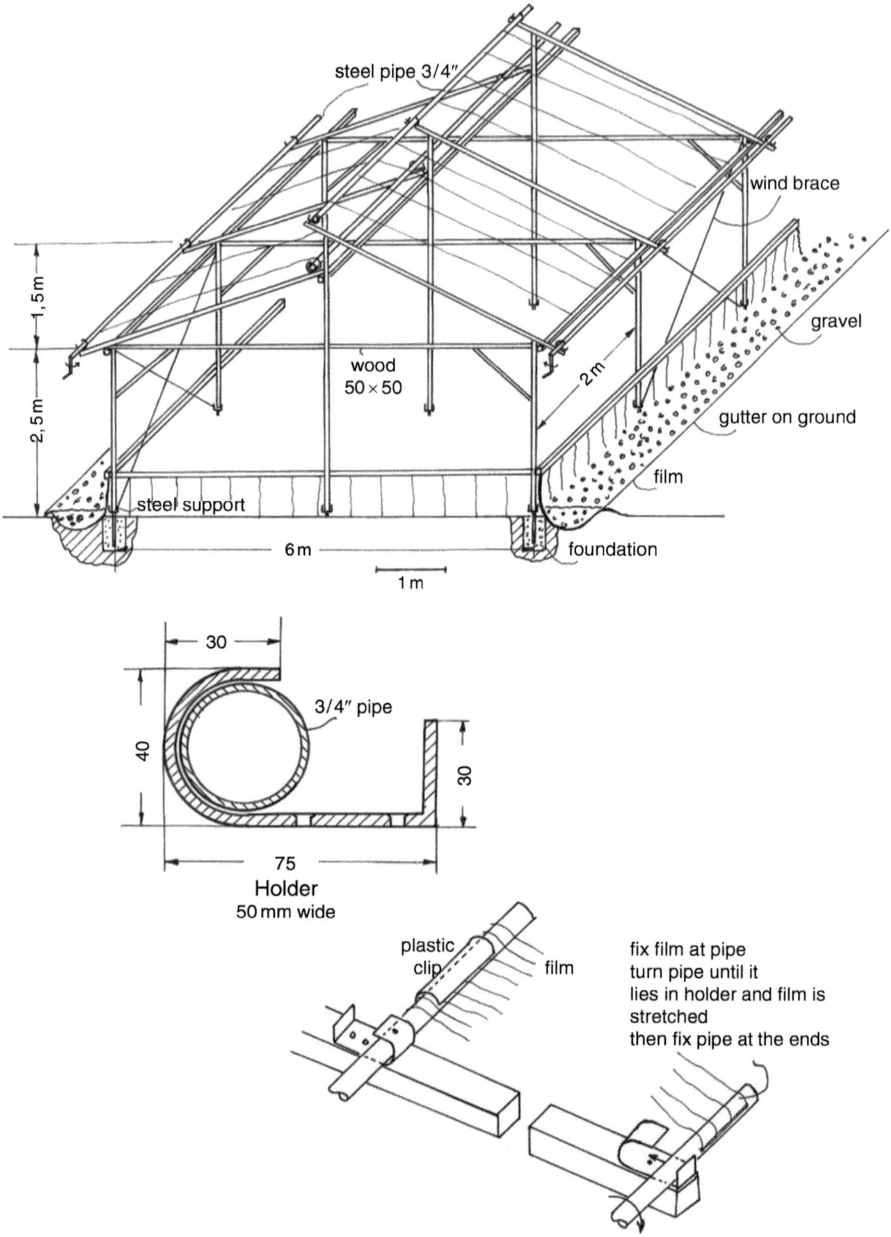

Fig. 5.33 An improved saddle-roof structure, where the plastic film is fixed and stretched on the roof by rolling it up on steel tubes positioned in holders. The plastic film can be replaced more easily and tightened again in hot seasons

Table 5.1 Dimensions of the components for the construction in Fig. 5.32

Designation	Cross section (mm)	Length (m)
1. Pole	150 Ø	3.0
2. Pole	150 Ø	5.0
3. Timber	76 × 50	5.0
4 Timber	50 × 50	3.25
5. Timber	40 × 40	3.5
6. Timber	50 × 50	4.0
7. Timber	40 × 40	4.25
8. Timber	40 × 40	
9. Timber	76 × 50	
10. Timber	40 × 40	
11. Timber	40 × 40	5.0
12. Wire for wind brace		

The costs were calculated in Indian Rupees (year 2000) for galvanised and non-galvanized steel according to Indian standard IS 1239 OF (1990), as well as for non-standardised, non-galvanised steel.

The total price of 6,140 Rupees for the galvanised structure is 1.4 times more than the non-galvanised structure of standardised profiles, and 3.1 times more than the mild steel structure of non-standardised profiles. Non-galvanised (mild) steel profiles must be painted very carefully a minimum of three times to prevent rust. In this case, the duration of life can be similar to galvanised steel. Galvanised steel profiles should never be welded together.

Some important recommendations for tropical greenhouses are (Mutwiwa et al. 2007):

- Large ventilation openings with more than 60% opening related to floor area covered by insect screens to block insect entry physically and optically (see Chaps. 9 and 10).
- Spectrally modified cladding material, to reduce heat load by blocking NIR transmittance (Sect. 7.3).
- UV blocking material to reduce pest population.
- A forced extraction ventilation when temperatures exceed the permissible maximum.

5.4 Greenhouses for Tropical Highlands and Subtropics

The greenhouses for tropical highlands and for subtropics have to have lockable ventilators that should close very tight without leaks to prevent heat loss during cold nights. Very efficient ventilation for hot periods is a prerequisite for good climate control. Greenhouses for tropical highlands should have ridge vents for natural ventilation, although they are relatively expensive. Greenhouses for the subtropics do not have ridge or roof ventilation very often. In this case, the width of the

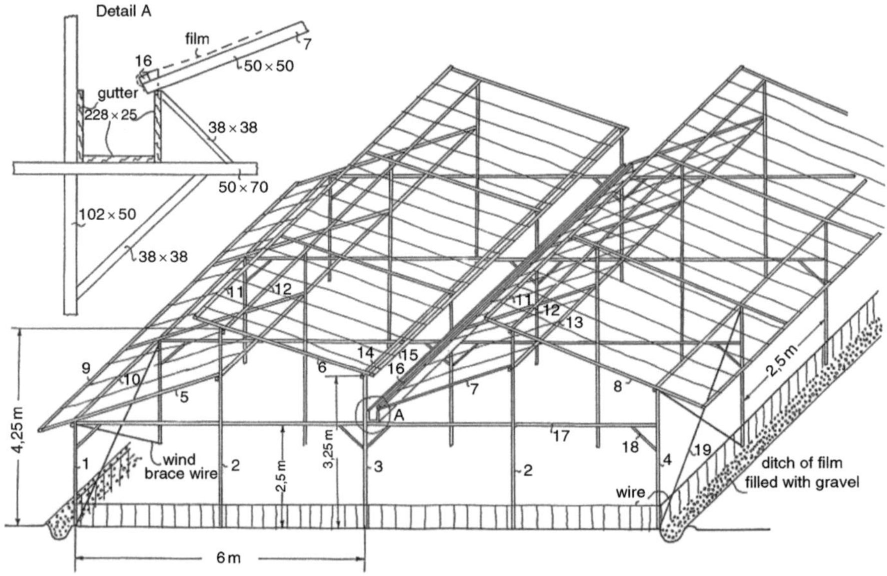

Fig. 5.34 A design for a double- or multi-span timber structure for the Seychelles, where imported timber profiles from South Africa were the cheapest available materials in the country, much cheaper than inland timber and steel tubes. Even the gutter is made of timber. The inner side of the gutter is lined with plastic film dimensions are given in Tables 5.2

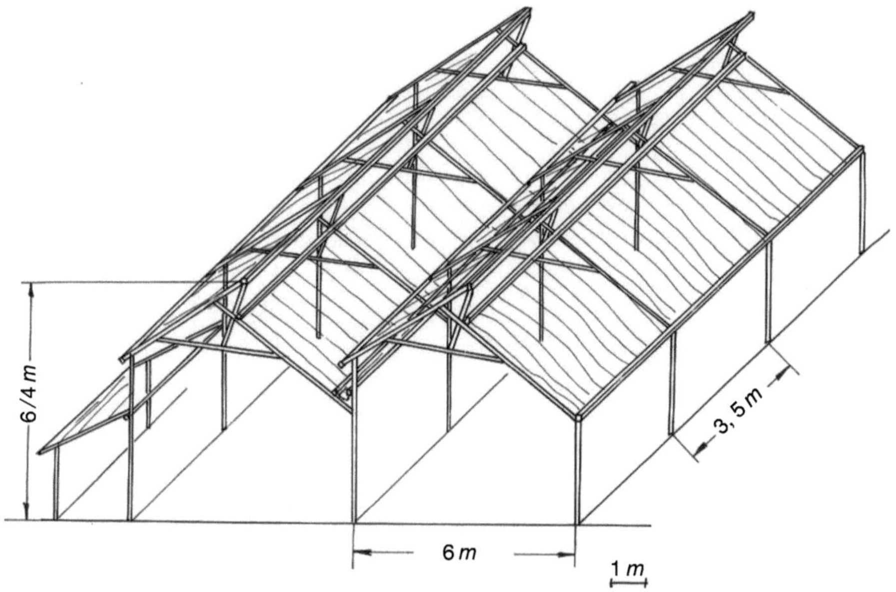

Fig. 5.35 Wooden structure from Kenya: 6.4 m ridge height, high volume and a canopy for rain protection on the weather side. The plastic film gutter is not advisable because of getting water pockets. The canopy at the side wall against sloping rain fall is highly recommendable

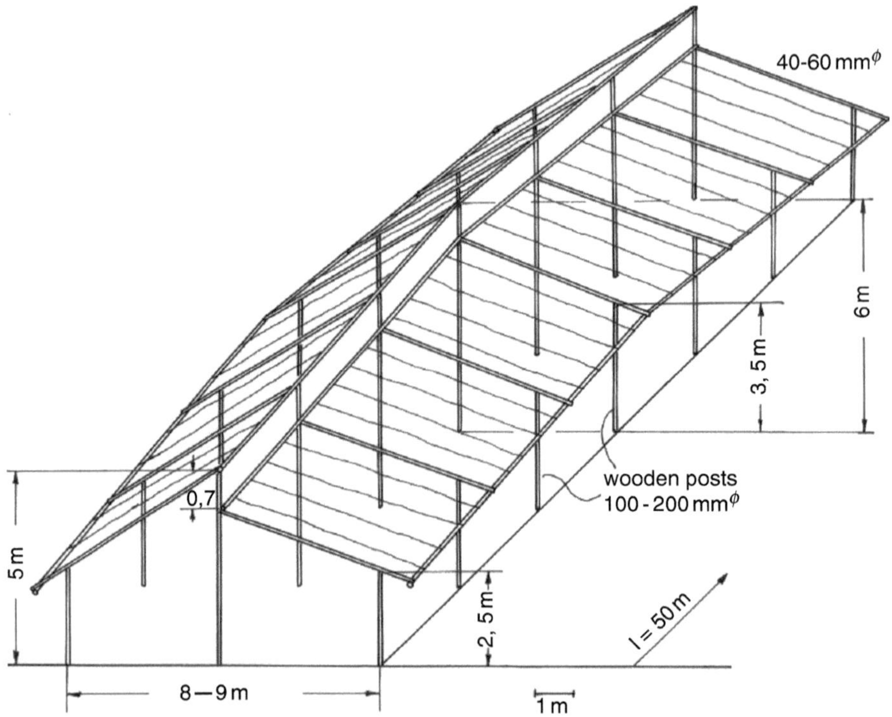

Fig. 5.36 Wooden structure in Colombia with a roof slope in longitudinal direction to lead off rainwater

Fig. 5.37 Greenhouse for roses in Malawi with open space only in the roof and gable area, to protect the flowers from wind and sloping rainfall

multi-span greenhouse units should be limited to guarantee sufficient ventilation by side-wall and gable vents only (see Chap. 9). Greenhouses in the USA have forced ventilation by fans because of cheap electricity.

Fig. 5.38 Steel tube structure in Thailand with 5 m gutter height, ridge ventilation and nets against birds

Gutters for rainwater collection are necessary for rainy seasons. The cladding material should be opaque to long-wave radiation because of heat loss at night (see Chap. 7). The plastic film should have No-Drop properties to avoid drop condensation and dripping (see Sect. 7.4). Ventilation openings should be designed with regard to the installation of insect screens (see Chap. 10) and energy-saving measures considered if heating is necessary in the subtropics. The design criteria are given in Chap. 4.

Various greenhouse types are available, from simple wooden structures to highly developed steel tube and aluminium structures with all equipment for climate control and energy-saving methods. One can observe that the height of greenhouses

Fig. 5.39 Steel tube greenhouse on farm in Bangalore, India, covered with insect screens. The ridge vents at the gable ends have alternate opening directions for better ventilation efficiency

is increasing with increasing state of the art and knowledge about efficient climate control, but the investment costs are also increasing (Figs. 5.43–5.56, Tables 5.6–5.10).

5.5 Greenhouse Constructions for Arid Regions

Greenhouses for arid regions have to protect crops from excessively high irradiation, too low temperatures in winter, wind, sandstorms, and too low humidity. Usually, they have forced ventilation by fans in combination with evaporative cooling (see Chap. 11). A heating system is necessary for cold nights with frost (see Chap. 12). The cladding material is shaded by outside nets for flower production in many cases. Permanent outside shading is less suitable for vegetable cropping in the main winter season. Moveable shading systems inside or outside have advantages for light control, in the early morning for example. Moveable inside shading can be used as thermal screens to reduce heat losses in winter. Shading systems are also necessary to reduce incoming radiation and to improve the cooling efficiency (Figs. 5.57–5.63).

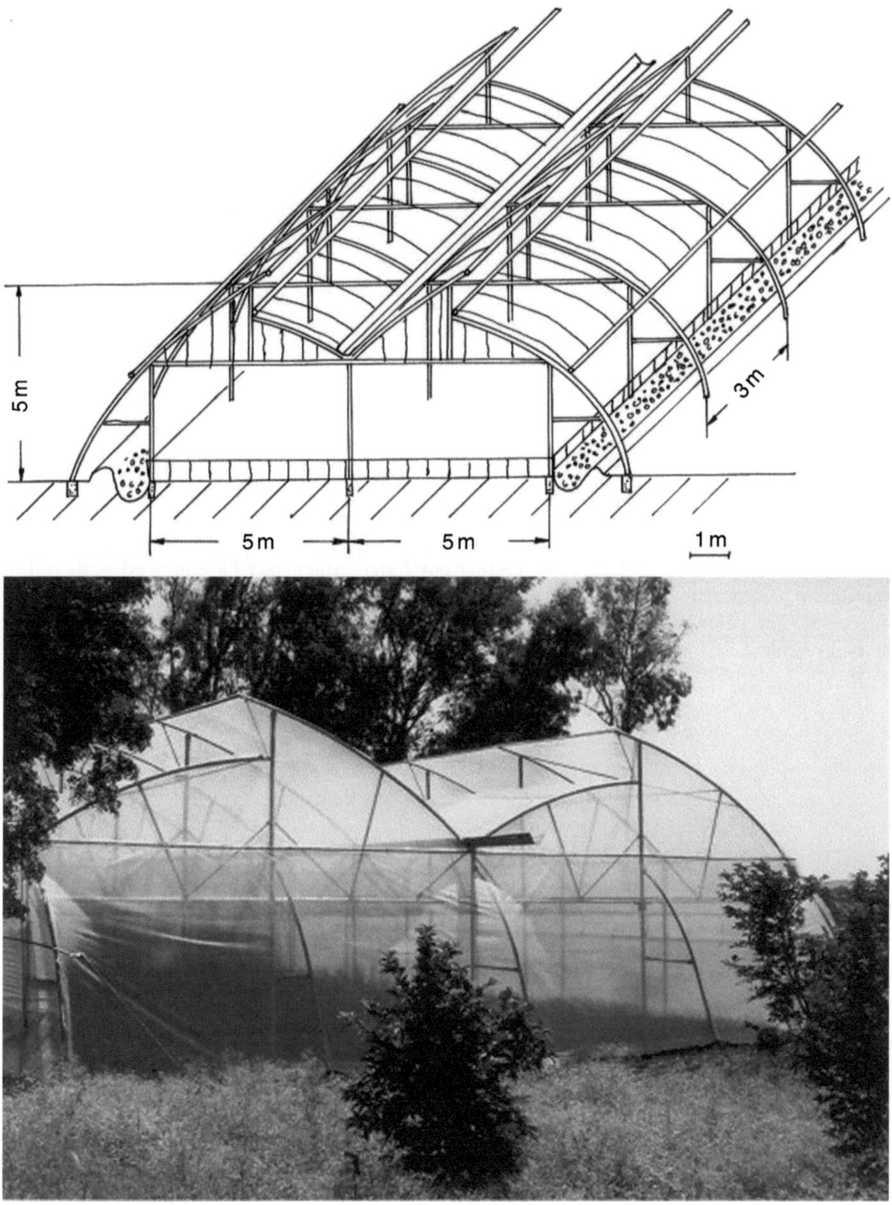

Fig. 5.40 Very stable steel tube structure from India, with bent roof and side wall tubes

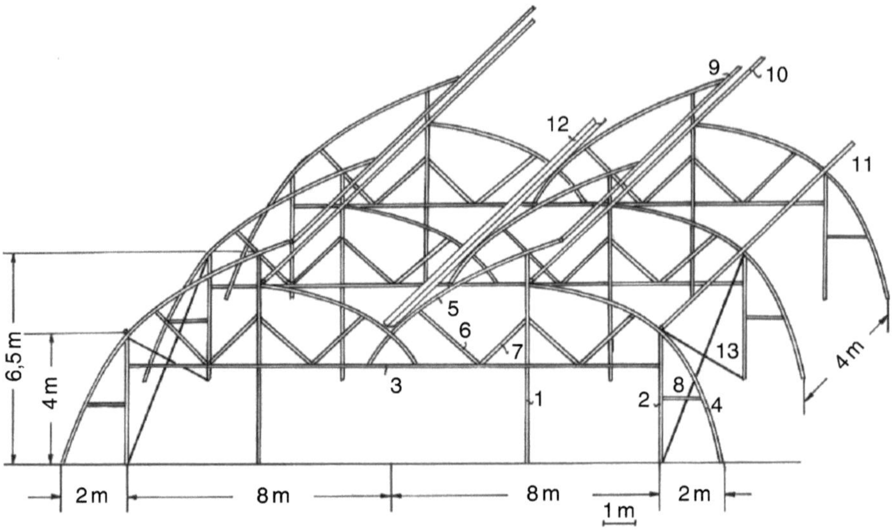

Fig. 5.41 Double-span steel tube structure from India, which can be equipped with roll-up ventilation at side wall and ridge for tropical highlands also. Remarkable is the position of the main stanchion below the ridge and not below the gutter. The dimensions of the components are given in Table 5.3

Table 5.2 Shows the dimensions of the components in Fig. 5.34

Designation	Cross section (mm)	Length (m)
1	102 × 50	2.25
2	102 × 50	4.25
3	102 × 50	3.25
4	102 × 50	3.0
5	50 × 50	3.9
6	50 × 50	3.8
7	50 × 50	3.0
8	50 × 50	4.8
9	50 × 50	
10	50 × 50	
11	50 × 50	
12	50 × 50	
13	50 × 50	
14	50 × 50	
15	50 × 50	
16	38 × 38	
17	50 × 70	12.0
18	38 × 38	0.75

Table 5.3 Dimensions of the components in Fig. 5.41

Designation	Diameter (inches)	Length (m)
1	2½″	6.5
2	2½″	4
3	2″	16
4	1½″	9
5	1½″	7.5
6	1½″	2.5
7	1″	2
8	1″	1.5
9	1¼″	
10	1¼″	
11	1¼″	
12	1 mm metal sheet	
13	Wind brace	

5.6 Screen and Shade Houses

Shelter structures in subtropical and tropical areas are covered sometimes only by plastic screens, nets or simple natural branches Figs 5.64–5.67. Teitel (2006) reported that insect-proof screenhouses completely covered by screens became popular in recent years in Israel and other Mediterranean countries because of the lower investment compared to other greenhouses.

Screenhouses can be used, for example, for all-year-round production at two different sites, whose harvesting periods are complementary. Vegetable production can be transferred to screenhouses in highlands when the temperatures are too high in coastal areas during summer in subtropical areas (Castilla and Hernandez 2007; Montero 2009) (Fig. 5.64).

The cladding materials of these shelters

- Reduce the impact of heavy rain fall
- Reduce too high global radiation and influence temperature and humidity
- Reduce the influence of heavy wind and hail on the crop
- Reduce the penetration of pest insects and birds

Photoselective nets and screens have been designed to get specific physiological responses, to transform direct light into scattered light, which improves the penetration of light into the plant canopy, and to influence pest control (Shahak et al. 2009) (see Sect. 7.3).

The *disadvantages* of those net or shade houses are:

- The plants are permanently wetted by rain and infested by diseases.
- There is no positive effect on spraying efficiency.
- There is no positive effect of fertilization and water-use efficiency. Fertilizers will be washed out.

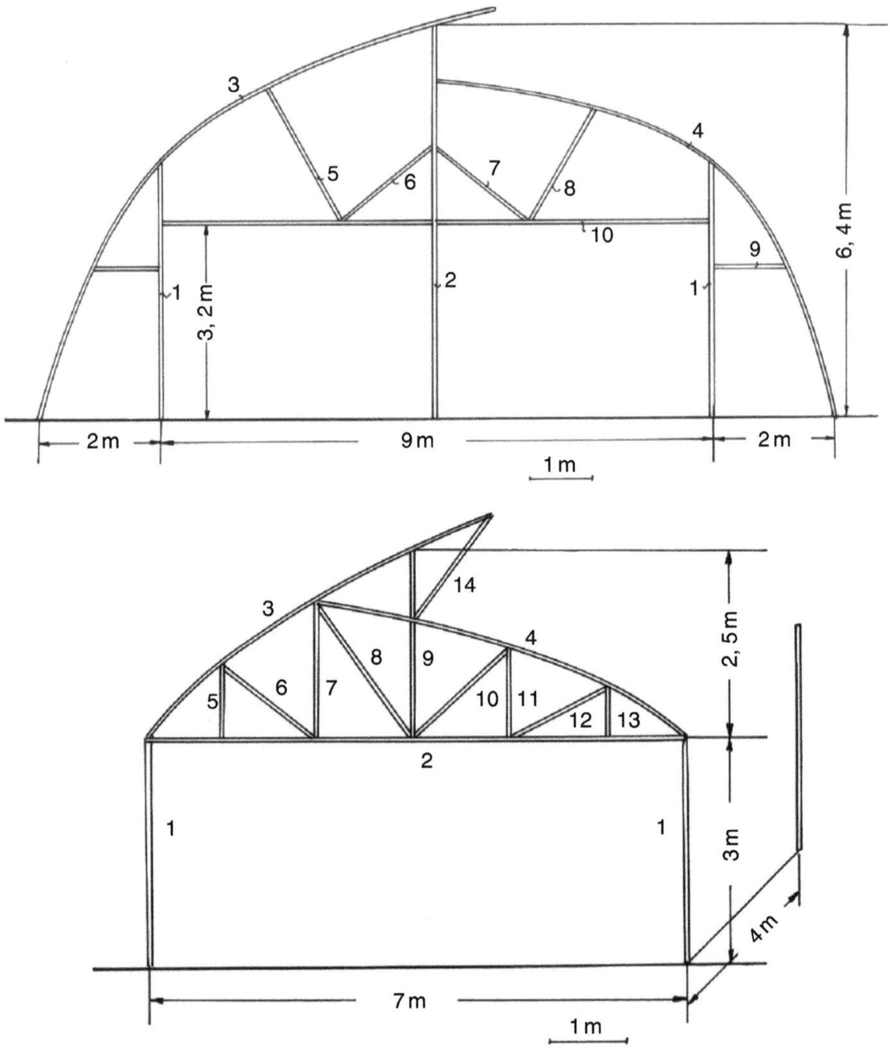

Fig. 5.42 A comparison of two different trusses made and used in India

Another disadvantage is the decreasing temperature at night in winter. To overcome this problem, many net houses are covered by 50 μm-thick plastic film in winter, with the consequence of reduced light transmittance by two cladding materials.

Screenhouses may be limited to regions with very low rainfall during the cropping season (Figs. 5.65, 5.66).

Table 5.4 Dimensions of the bent truss tubes in Fig. 5.42

Designation	Diameter (inches)	Length (m)
1	2½″	4.2
2	2½″	6.4
3	1½″	10.74
4	1½″	9.5
5	1¼″	2.5
6	1″	2.0
7	1″	2.0
8	1½″	2.0
9	2″	1.25
10	2″	9.0

Table 5.5 Dimensions of the welded truss in Fig. 5.42

Designation	Cross section (mm)	Length (m)
1 Tube	2″	3
2 Angle iron	35 × 35 × 4	7
3 Angle	30 × 30 × 3	5.5
4 Angle	30 × 30 × 3	5.25
5 Angle	30 × 30 × 3	1
6 Angle	30 × 30 × 3	1.5
7 Angle	30 × 30 × 3	1.75
8 Angle	30 × 30 × 3	2
9 Angle	35 × 35 × 4	2.5
10 Angle	30 × 30 × 3	1.75
11 Angle	30 × 30 × 3	1
12 Angle	30 × 30 × 3	1.5
13	30 × 30 × 3	0.5
14	30 × 30 × 3	1.75

Summarising, one can distinguish three main groups of screens and nets (see Sect. 7.3):

1. Shading materials, black and clear, that reduce solar radiation and wind influence.
2. Coloured screens that filter out colours of the light spectrum.
3. UV-blocking screens that reduce the penetration of UV-radiation.

Romacho et al. (2006) investigated screenhouses covered with clean and alternately green nets, both 6 × 6 threads/cm^2 (15 mesh), in the highlands of southern Spain. The transmittance of the screenhouses was about 60% and the temperatures were similar under both nets; slightly higher in the morning and slightly lower in the afternoon compared to outside temperature. The total tomato yield was between 5 and 7.2 kg/m^2 in two growing seasons.

Medany et al. (2009) compared sweet pepper crop growth in net-covered tunnel screenhouses and a PE-film-covered tunnel greenhouse in Egypt. The sweet pepper

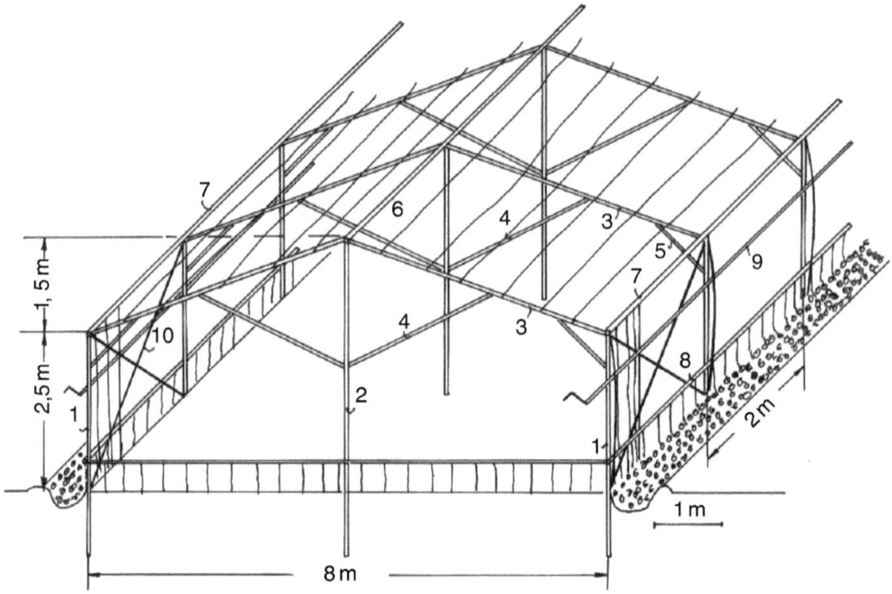

Fig. 5.43 An example for a simple wooden structure with 8 m span width and roll-up ventilation. The dimensions of the components are listed in Table 5.6

Fig. 5.44 A well-designed wooden structure in Albania

crop was transplanted in August ,and the cladding materials of the 9 × 40 m tunnels were:

- Black insect-proof screen, 40% shading, and additional 50 μm PE film in winter
- White insect-proof screen, 40% shading; and additional 50 μm PE film in winter
- PE film, 200 μm thick

The temperature difference inside to outside was 1–3°C in the screenhouses and 3–5°C in the plastic-film greenhouse. Early yield was highest in the black

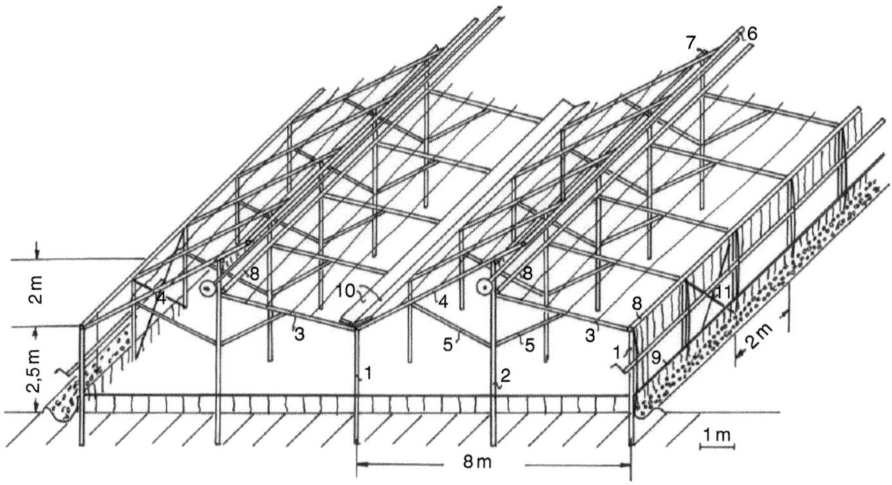

Fig. 5.45 Wooden structure with roll-up side wall ventilation and vertical ridge ventilation. Dimensions of the structure n are listed in Table 5.7

screenhouse, but total yield was the lowest. Total yield was highest in the white screen tunnel. The white screen cover plus 50 μm plastic film in winter was the best choice for sweet pepper under the climate conditions in Egypt.

Experiments were carried out in a commercial flat-top screenhouse, 3.2 m high, 60 by 110 m floor area, located in Israel, 32° 16′N. The cladding material was a 50-mesh screen, 50 threads per inch and 0.24 mm thread diameter (Tanny et al. 2003; Möller et al. 2003). The air exchange in the middle of the screenhouses could be expressed by the regression curve

$$N = 3.21 \times v_w + 12.75(1/h)$$

and the theoretical air exchange for a pepper crop in the open field with wind speeds between 1.5 and 3.5 m/s

$$N = 23.82 \times v_w(1/h)$$

where N = air exchange number and v_w = wind speed (m/s).

The reduction of ventilation rate was 51–71%. The temperature inside near the screen was similar to the outside temperature, and the temperature gradient from the upper level to the lower crop level was up to 5°C in the middle of the day (Tanny et al. 2003).Other measurements showed a relatively constant temperature gradient over the height (Möller et al. 2003).The mean temperature difference inside to outside was between −2°C and +2°C, and did not exceed +2.5°C (Möller et al.

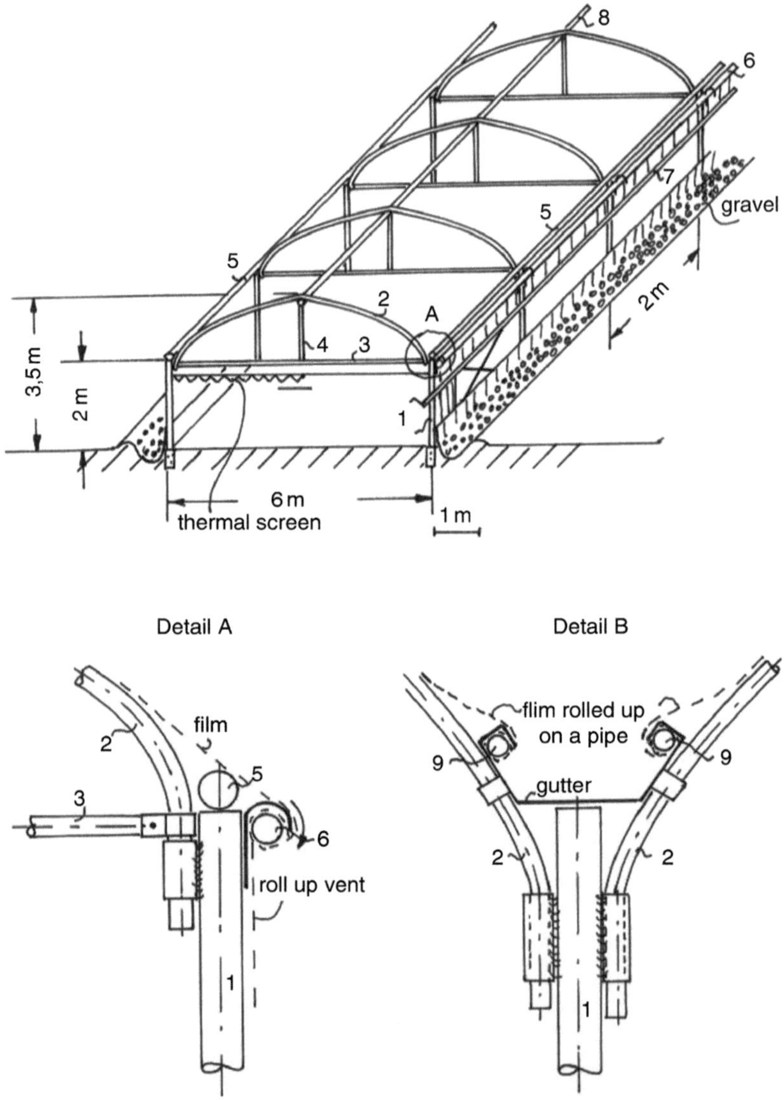

Fig. 5.46 Pointed-arched steel tube structure with a moveable thermal screen and gutter at side wall on the ground level. The plastic film of roof covering and roll-up ventilation at side wall are fixed and stretched by one steel tube (detail *A*). The plastic film is rolled up on a tube in a profile of the gutter (detail *B*). The dimensions of the components are listed in Table 5.8

2003). The outside wind speed was reduced by 75–95% above the pepper crop (Fig. 5.67).

The height of the screenhouse has an influence on the microclimate. Two screenhouses of 2 m and 4 m height respectively were investigated, covered with a screen of 60% shading and 3 × 2 mm hole size (Tanny et al. 2003):

Fig. 5.47 Multi-span greenhouse in Morocco. The ventilation opening may be not sufficient

Fig. 5.48 Multi-span structure with roll-up ventilation in Cyprus

- The net radiation was almost identical.
- The air temperature near the plants and the leaf temperature were higher in the lower screenhouse. The average daily air temperature difference between the two screenhouses was 1.5°C, and maximum difference at noon was 2.7°C.
- The vertical temperature gradient was about three times larger in the lower screenhouse.
- The absolute humidity was closer to the outside humidity in the higher screenhouse.
- The climate for workers is more comfortable in higher screenhouses.

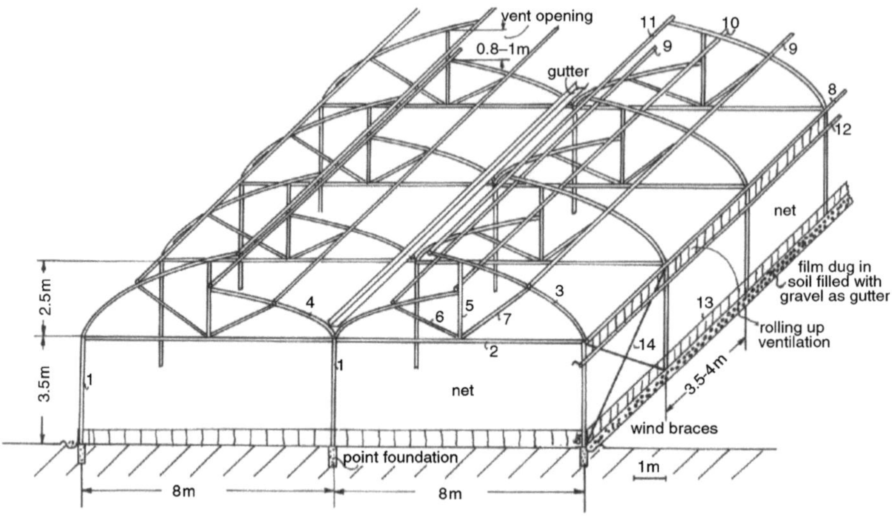

Fig. 5.49 Steel tube structure with side wall and ridge vents that are open in opposing directions in both spans for better ventilation efficiency. The roll-up ventilation at the ridge is not drawn in for better overall view. The components are connected by clamps. The dimensions are listed in Table 5.9

Higher screenhouses up to 5 m are more advantageous than lower heights, as 3.5 m, with regard to temperatures and humidity (Montero 2009).

Most of the protected cultivation on the Canary Islands is under screenhouses, but they do not protect the crops against rain and very low-humidity conditions. Farmers are therefore thinking about changing to plastic film greenhouses (Santos et al. 2006). A Parral-type screenhouse, covered by a screen with 10 $\times$ 14 threads/cm^2, 35 mesh, 36% porosity, has a mean transmittance of 70% for global radiation when the screen is clean. The transmittance decreases over time to lower than 60% by dust accumulation. Rainfall cleans the screen again to higher transmittance.

5.7 Specific Greenhouses

5.7.1 Parral Type, Almeria

One of the most widespread greenhouse types in the area of Almeria, South Spain, is the Parral type with flat roof or with low roof slope built by local workmen (Fig. 5.68). Wooden or steel poles, vertical in the middle and sloped at the side wall, are connected crosswise by wires at the top, and two wire grids are tightened over the construction. The plastic film is placed in between the grids. About 33% of the area is still covered by these flat-roof types (Castilla and Hernandez 2005).

Fig. 5.50 Shed roof with roll-up ventilation at vertical ridge in Israel and Tunisia

The Parral-type greenhouse has some disadvantages (Perez-Parra et al. 2004):

- The drop condensation at the flat roof reduces light transmittance and thus yield and quality (Fig. 5.69).
- Drops falling off the roof wet the crop and encourage diseases. Some growers stretch extra plastic film below the roof to prevent drop fall (Fig. 5.70).
- Ventilation efficiency by side walls only is not sufficient.
- The Almeria area has relatively low rainfall. Therefore, flat roofs may be sufficient, but for regions with higher amounts of rainfall such as Antalya, South Turkey, the flat roof is not suitable because water pockets form on the flat plastic film and may destroy the film, Fig. 5.71.

Fig. 5.51 Greenhouse structure with open ridge and roll-up vent at side wall in India. For tropical highlands, roll-up vents at ridge are preferable

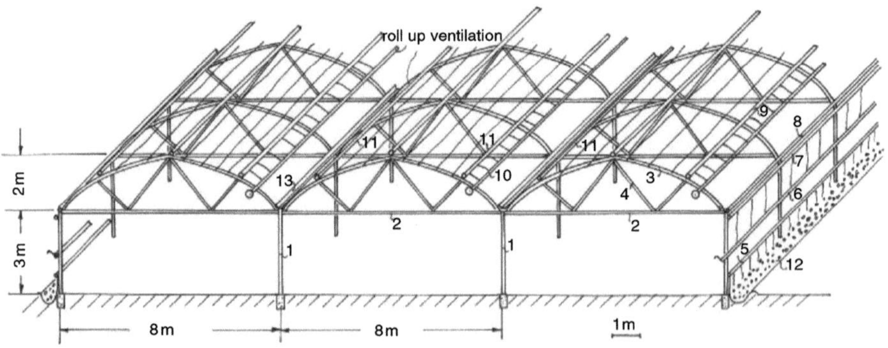

Fig. 5.52 Multi-span structure with roll-up ventilation at roof and side wall (see Sect. 5.2). Dimensions are listed in Table 5.10

Asymmetrical or symmetrical saddle-roof constructions with roof vents were improvements to the Parral type, done in particular for ventilation efficiency (see Chap. 9).

The yield of cucumbers was 25–50% higher in an asymmetrical east–west-oriented and unheated greenhouse with a roof slope of 45° to the South and 27° to the north, in comparison to a type with 11°/24° respectively (Castilla et al. 2001).

Fig. 5.53 Greenhouse structures with different ventilation openings at roof and ridge in Israel. The gutter height is 4 m minimum

Fig. 5.54 Greenhouse with flap ventilation and high side wall in Turkey. The side wall is double-inflated

The average greenhouse structure costs including plastic film for 1 ha in the South of Spain were (Castilla et al. 2005) (Figs. 5.72, 5.73)

Low roof slope structure	7.8 €/m^2
High roof slope structure multi-span	10.2 €/m^2
Arched shape multi-span	15.6 €/m^2

Fig. 5.55 Greenhouse structure with alternate roof openings. The roof opening is linked to the gutter, and opens at the ridge. The height of the gutter is more than 4 m

5.7.2 Greenhouses for Banana

Some Mediterranean countries grow banana in greenhouses, for example Tunisia and Morocco. These banana greenhouses have an extra height of up to 8 m. The greenhouses protect the bananas from low temperatures, and provide them a suitable climate. The constructions have to be secured against wind forces by several extra braces (Fig. 5.74).

5.7.3 Greenhouses for Terraces

Vegetables and fruits are cultivated on small terraces in some tropical countries, for example the Seychelles (Fig. 5.75). The terraces have a width of 2–4 m. Growers used to build very simple terrace shelters to protect the crop from too heavy rainfall (Fig. 5.76), but many of them were soon destroyed by heavy storms. Therefore, shelter structures for terraces have to be adapted to the individual form of the terraces and designed to be wind-resistant. Thus, more wind-resistant terrace greenhouses have been designed and introduced into practice (Fig. 5.77–5.80) (Von Zabeltitz 1996a, b). At first, the plastic film was nailed onto the structure, which is not to be recommended. New structures were designed, built using timber profiles (Fig. 5.78), where the plastic film can be fastened and stretched by rolling it up on steel tubes. In this way, the plastic film can be stretched at any time when it is stretched by temperature, and when it starts to flutter due to wind. The structure has gutters to collect and drain off rainwater (Figs. 5.79).

Fig. 5.56 Greenhouse in China. A round-arched structure is leaned against a north wall, and will be covered by straw mats at night

Table 5.6 Dimensions (mm) of wooden structure in Fig. 5.43

Component No.	Dimension (mm)
1	100 Ø or 76 × 127
2	100 Ø or 76 × 127
3	50 × 100
4	50 × 100
5	25 × 50
6	50 × 50
7	50 × 50
8	25 × 50 or tube 25 Ø
9	Steel tube 25 Ø
10	Wire for windbrace

Table 5.7 Dimensions of the wooden structure in Fig. 5.45

Component No.	Dimension (mm)
1	100 Ø or 76 × 127
2	100 Ø or 76 × 127
3	50 × 100
4	50 × 100
5	50 × 100
6	50 × 50
7	50 × 50
8	50 × 50
9	25 × 50
10	Gutter 25 × 250
11	Tube 25 Ø

Table 5.8 Dimensions of the steel tube design in Fig. 5.46

Component No.	Dimension (mm)
1	50
2	40
3	40
4	25
5	30
6	25
7	25
8	30
9	25
10	Gutter

5.7.4 Tent-type Construction

A simple plastic-film greenhouse structure suitable for different climates has been designed in the Institute for Horticultural Engineering Hannover (Von Zabeltitz 1985, 1990a) (Fig. 5.81, 5.82). It consists of two parts: the load bearing base

Table 5.9 Dimensions of the steel tube structure in Fig. 5.49

Component No.	Dimension (mm)
1	60
2	50
3	40
4	40
5	50
6	30
7	30
8	40
9	40
10	40
11	40
12	20
13	20
14	Wire for windbrace

Table 5.10 Dimensions of the construction in Fig. 5.52

Component No.	Dimension (mm Ø)
1	60
2	50
3	50
4	30
5	25
6	25
7	30
8	50
9	40
10	25–30
11	40
12	Gutter
13	1 mm metal sheet

structure and the roof structure. The base structure is a guy-wire construction similar to a tent, with inclined wire ropes or steel rods (*a*) connecting the gutter or upper end of the stanchions respectively to the foundation. Within the spans, the stanchions are connected crosswise at the upper ends by steel rods or wires (*b*). Longitudinally, the stanchions are connected by gutters. The stanchions are loaded only by pressure forces, the other components by tension forces. Only small foundations are necessary below the inside stanchions, while deeper foundations are positioned below the inclined steel rods (*a*). Assembly is simplified, and material costs are reduced.

Advantages:

- Efficient stability with reduced construction components
- Improved light transmittance
- Smaller foundations inside

Fig. 5.57 A greenhouse structure for desert regions in Kuwait, covered with rigid plastic sheets and equipped with fan and pad cooling

The greenhouse construction stood for 15 years and was resistant against wind and snow forces all the time.

The roof structure in Fig. 5.81 consists of plastic jacketed spring steel rods with a diameter of 20 mm covered with double inflated plastic film. The structure is very resistant to wind, storm and a snow load of 35 kg/m^2. With higher snow loads, the spring steel rods bent, but the structure was not destroyed. After removing the snow, the spring steel jumped to the original form.

Force measurements were carried out on the structure. Only pressure forces were measured on the stanchions. Pressure forces of 90–110 dN (deka Newton) were

Fig. 5.58 A round-arched structure covered with plastic film and permanent shading net, ventilated and cooled by fan and pad system

Fig. 5.59 Round-arched tunnels in the desert of Tunisia, with hedges and plastic fence as windbreak

measured at wind speeds of 3–6 m/s at the stanchions under the gutter. Even at high wind speeds from various directions, no tensile forces could be measured on the stanchions. Tensile forces of 330–480 dN were measured on the inclined rods (*a*) at both sides, at wind speeds of 5–6 m/s. The tensile forces on the horizontal rods (*b*) were 220–250 dN. Even in a gale, there was no damage to the construction. Different roof structures can be designed Fig. 5.83.

Fig. 5.60 Round-arched tunnel with fan and pad cooling at gable ends in Kuwait

The structure has a cladding of double inflated plastic film for stabilizing and energy-saving purposes. Even the roll-up vents on the side wall can be inflated. In the closed position, the double film is inflated and therefore very tight, without leaks at the edges. When the film is rolled up for ventilation, the excess air must have the possibility to escape during rolling up. There is a very simple flap valve with a counterweight installed, which keeps the valve closed against the air pressure and opens when the pressure increases during the rolling up of the vent, Fig. 5.84. This self-made valve has worked safely for years (Figs. 5.85). The roll up ventilation on both sides can be operated by out driving motor, Fig. 5.86. A commercial structure has been designed, Fig 5.87.

Fig. 5.61 Round-arched tunnel with fan and pad cooling at side wall in Kuwait

5.8 Life Cycle Assessment, LCA

"Life Cycle Assessment is a process to evaluate the environmental burdens associated to a product, process, or activity by identifying and quantifying energy and materials used and wastes released to the environment to affect environmental improvements." (Society of Environmental Toxicology and Chemistry, SETAG 1993).

Russo et al. (2005) defined the LCA as an "instrument which provides a quantitative estimate of all flows of matter and energy related to the realization of a product, providing an evaluation of the environmental compatibility and end result of each productive choice".

The LCA considers the entire life cycle of a product from resource extraction to waste disposal (Anton et al. 2003). The LCA method can be applied to horticultural production and to greenhouse structures and equipment, as well as to comparison of crop production in different countries.

The aim of the LCA is to compile the entire knowledge about environmental impact of products and processes during the life cycle, and to give the possibility to compare products (greenhouse construction) or even crop production in different countries (Van der Velden 2004; Schüsseler and von Zabeltitz 2004).

The procedures of the LCA are described in different standards (ISO 14040, ISO 14042, DIN EN ISO 14040, DIN EN ISO 14043). A LCA study can be divided into four steps (Schüsseler and von Zabeltitz 2004, Anton and Montero 2003):

1. The goal and scope definition describes the system under investigation (greenhouse construction), its function and boundaries.

Fig. 5.62 Round-arched tunnel greenhouses with fan and pad cooling and hydroponics in the desert of Quatar

2. The inventory analysis compiles the resource consumption associated with the system. The potential environmental impact of the various emissions and resource consumption is not considered in this phase.
3. The life cycle impact assessment evaluates the inventory data with regard to their potential to harm the natural ecosystem, human health and resources. Special impact categories and indicators are defined, for example energy consumption, climate change by CO_2 emission, toxicity, water consumption, soil pollution, and others.

Fig. 5.63 Shed roof structure covered with corrugated polyester sheets. Those sheets are not recommendable anymore, because the light transmittance decreases within a short time and the sheets are relatively expensive.

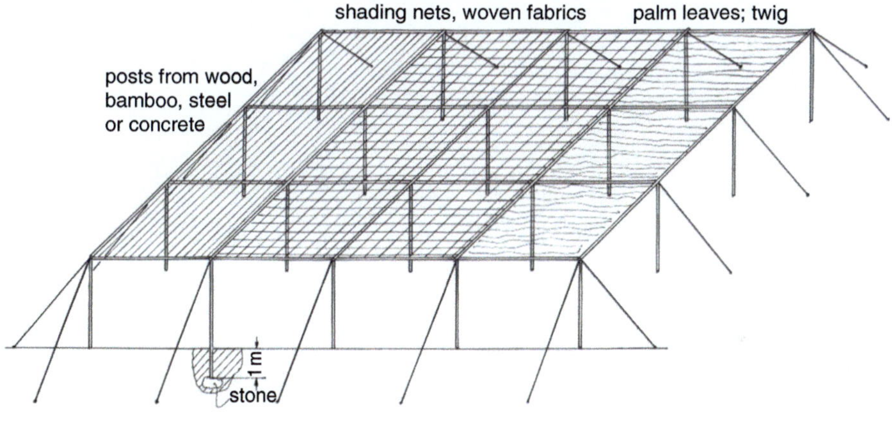

Fig. 5.64 Design of net and shade house

Fig. 5.65 Shade house in Southeast Asia

Fig. 5.66 Tunnel structures covered with nets to protect the crop from birds and insects

Fig. 5.67 Net houses in Thailand and India

Fig. 5.68 Parral-type greenhouse in Almeria with flat roof

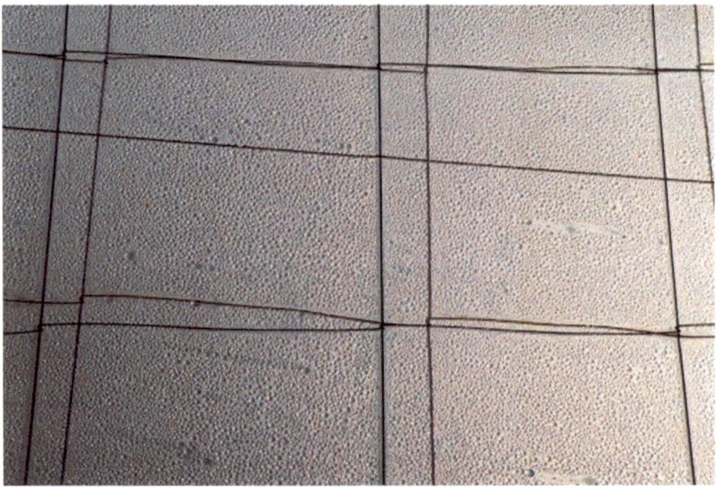

Fig. 5.69 Drop condensation on flat roofs of Parral type

4. In the interpretation phase, the results of the inventory and impact assessment are discussed. Conclusions will be drawn to define options for the improvement of environmental performance of the system under investigation.

One important factor for crop production in greenhouses should be "sustainable development that meets the needs of the current generation without undermining the ability for future generations to meet their own needs" (De Pascale and Maggio 2005).

Sustainable production can be judged by the following indicators (De Pascale and Maggio 2005):

- The use of non-renewable resources
- The use of renewable resources
- The level of pollution

Horticultural production produces performance and causes environmental burdens (Fig. 5.88) (von Zabeltitz 1997). Sustainable production means valence of performance > valence of environmental burdens.

The question is: how can we assess the environmental burden?

Greenhouse crop production has a smaller environmental impact than open-field production in many cases.

Munoz et al. (2008) compared the environmental impacts of tomato production in a greenhouse and open field in a Mediterranean region (41°, 31′N) near Barcelona, Spain. The greenhouse was a 5,000 m^2, six-span steel frame structure, 4 m gutter height, 5.5 m ridge height, with six roof and two side vents. No heating was in operation.

Fig. 5.70 Extra plastic film stretched below the roof to prevent wetting of plants by drop fall. This method takes a lot of necessary light transmittance

The phases of greenhouse production, transportation of materials, installation and waste management were considered as well as irrigation, fertigation and water pumping. All field operations by tractors and agricultural machinery, the energy for raw materials, repair and maintenance, as well as fuel consumption and production of fertilizers, were considered for open-field production.

The tomato yields were:

8.6 kg/m^2 in the open field
16.5 kg/m^2 in the unheated greenhouse

Fig. 5.71 Water pockets in the plastic film of flat roof

Most of the impact categories related to 1 kg of tomato production were lower in greenhouse production. A comparison brought the following results for tomatoes (Montero 2009; Montero et al. 2008a; Munoz et al. 2008):

Impact category	Greenhouse	Open field
Global warming (kg CO_2/kg)	7.44 E−02	5.01 E−02
Energy consumption (MJ/kg)	0.94	1.19
Water consumption (L/kg)	24.4	42.84

The production of the greenhouse structure (steel and concrete) had the greatest influence in the global warming category, measured in kg CO_2 equivalent per kg tomato. Forced ventilation and heating increase the environmental impact considerably.

In addition to higher productivity in comparison to open-field production, and other advantages such as water-use efficiency, there are noticeable negative environmental impacts through greenhouse crop production, for example:

- Negative landscape impact in regions with high density of greenhouse crop production like Almeria, Spain, Ragusa, Italy, and Antalya, Turkey (De Pascale and Maggio 2005).
- Waste of plastics
- Energy consumption

The plastic consumption for protected crops in Italy is about 80,000 tons per year, and the estimated consumption for plastic covering is 2.2-3 tons per ha and year in the Campania region (De Pascale and Maggio 2005). The consumption for plastic film for recovering in the Almeria area is about 1.1 tons per ha and year. In Germany, the plastic film consumption for recovering greenhouses is about

Fig. 5.72 The Parral-type constructions were improved to asymmetrical saddle-roof types with different roof slopes

1.5–1.8 tons per ha and year, and altogether 600–720 tons per year for 400 ha of plastic film greenhouses.

Possibilities for reducing the waste of plastic are recycling and burning, and the use of long-life plastic film. About 30% of the plastic used in greenhouse industry is recycled (La Malfa and Leonardi 2001). The possible use of low-grade recycled plastic is limited. In Germany, most of the plastic will be burned for energy consumption. Burning for energy consumption is recommendable as second-hand use for the raw material oil.

Fig. 5.73 Industrial manufactured high-arched greenhouses are used in particular for flowers

The consumption of chemicals for pest management is normally higher in warmer climates than in temperate climates. An estimated amount of 47 kg/ha of active ingredients is used for the most intensive Italian greenhouses versus 31 kg/ha in the Netherlands (Stanghellini et al. 2003). A remarkable reduction of pesticides can be achieved by IPP (Integrated Production and Protection) management.

A comparison of the environmental input of greenhouse crop production in Netherlands and Spain has been evaluated (Van der Velden and Janse 2004). The energy consumption and crop protection for various vegetable crops was examined for the Spanish regions Almeria and Murcia versus the Netherlands. The physical production rate is significantly lower in Spain than in the Netherlands:

	Physical production (kg/m^2)		
	Tomato	Sweet pepper	Cucumber
Almeria	9	6	9
Murcia	8	8	
Netherlands	50	26	70

This is caused by a shorter cultivation period, limited light availability in winter, low temperatures in unheated greenhouses, unsuitable air humidity and limited use of greenhouse technology.

The primary fuel consumption per kg of vegetable in the Netherlands is estimated to be higher by 13 times for tomatoes, 14–17 times for sweet pepper and nine times for cucumber than in Spain.

Not only energy and consumption of chemicals, but even the transport has to be considered. Figure 5.89 shows the total consumption of energy (MJ/kg tomato) for production and transport of tomatoes from different countries to Frankfurt in Germany, as well as the consumption of active ingredients (g/kg) (Verhaegh 1996).

Fig. 5.74 Greenhouses for bananas

Fig. 5.75 Vegetables and fruits cultivated on terraces in the Seychelles

Fig. 5.76 Simple but not wind-resistant terrace shelters in the Seychelles

Fig. 5.77 Wooden structure for terrace shelter with nailed-on plastic film in the Seychelles

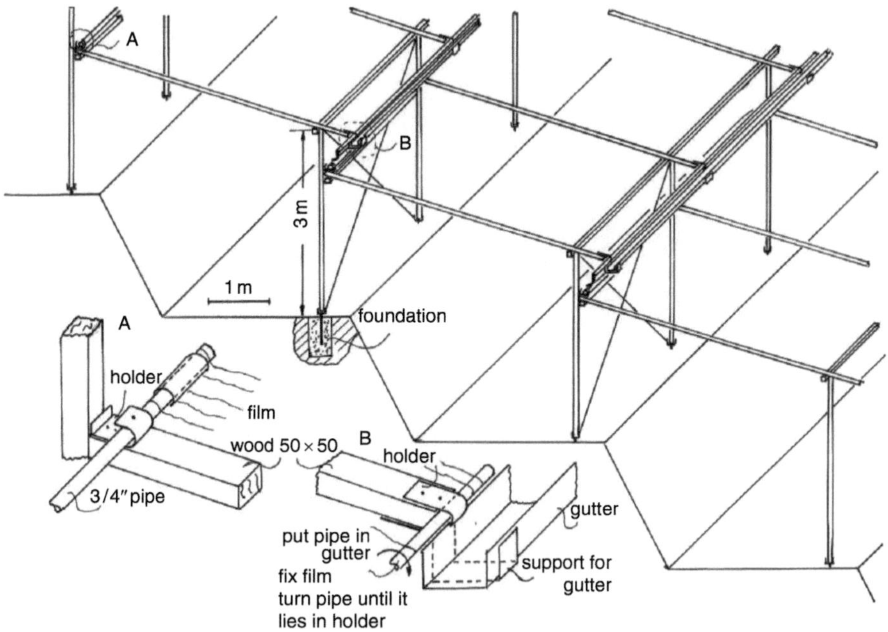

Fig. 5.78 Wooden terrace greenhouse structure with gutter and devices for fastening and stretching the plastic film

The energy consumption in the Netherlands is highest because of the need for heating in winter; energy for transport is only remarkable for flight transport from Israel. The total energy consumption for tomatoes from southern countries with transport by truck is relatively low. The consumption of active ingredients for pest control is much higher in southern countries than in temperate climates.

The total energy consumption per stem of roses from the Netherlands is comparable to that from southern countries with transport by air freight. The consumption of active ingredients is about 11 times higher in the Morocco than in the Netherlands (Fig. 5.90).

One can make some suggestions how to improve crop production in southern countries, for example to use a little bit more heating to reduce the chemical consumption for pest control.

The environmental compatibility of plastic-film greenhouse structures is higher than for glass houses. This can be judged by the LCA method.

Different glass and plastic-film greenhouses have been investigated by a mass and energy balance (simplified LCA) (von Zabeltitz et al. 1992, Schüsseler and von Zabeltitz 2004).The greenhouses evaluated are described in Fig. 5.91.

The mass and energy balance was calculated uniformly for 2,000 m^2.

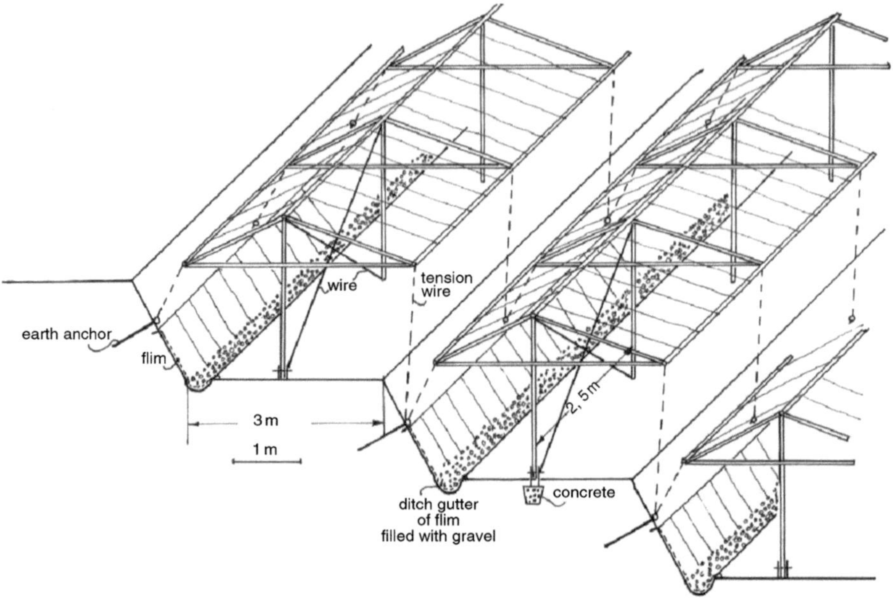

Fig. 5.79 Umbrella-type for terrace shelter structures, fastened by tension wires and earth anchor. The rainwater can be collected by a plastic film gutter dug into the soil. The plastic film is stretched over the structure and fastened at the edges by self-made clips from PVC water tubes cut open in longitudinal direction

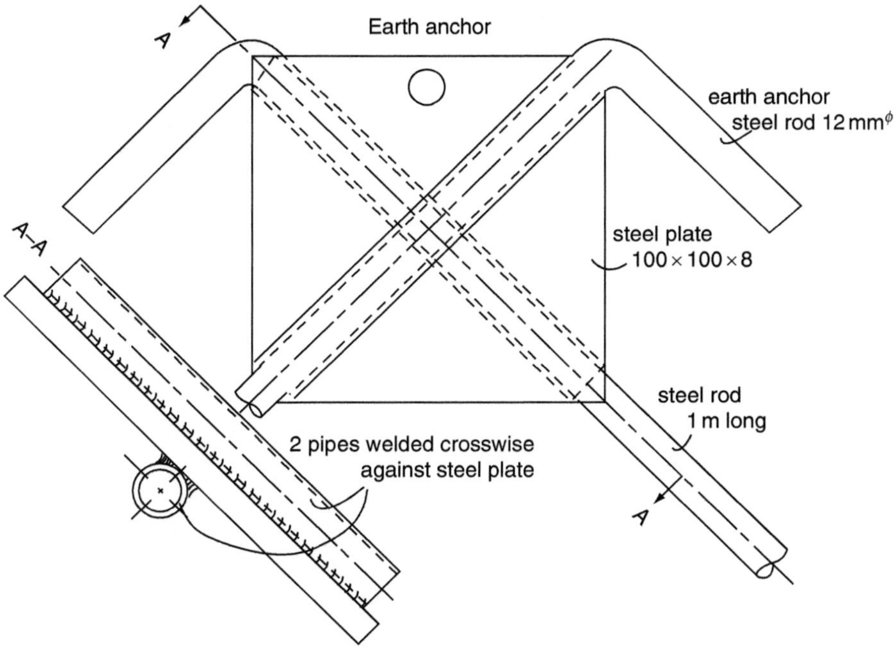

Fig. 5.80 Some self-made earth anchors for fastening the tension wires. Commercial earth anchors may be expensive or not available. Those earth anchors can withstand relatively high wind forces

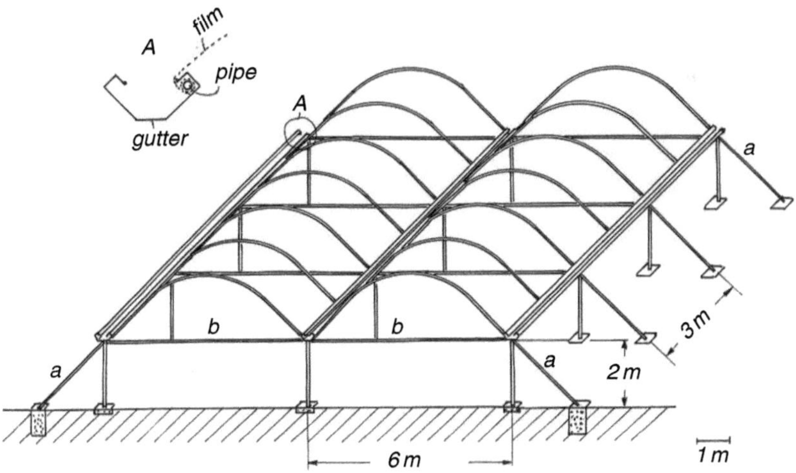

Fig. 5.81 Tent-type structure, Hannover

Fig. 5.82 The plastic-film greenhouse, Hannover

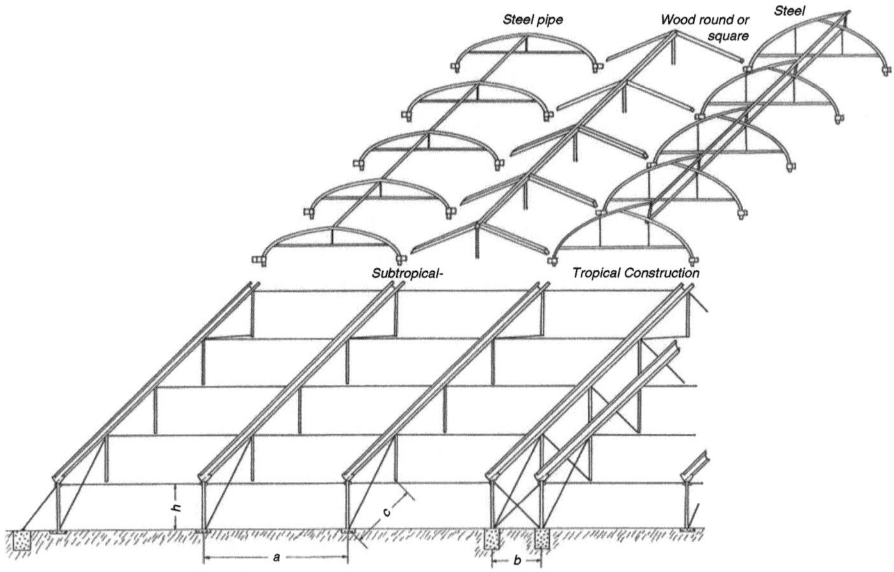

Fig. 5.83 The independent roof structure can be designed in various forms with different profiles and can be fixed to be wind-resistant to the gutter or the stanchions respectively, for example for subtropical or tropical regions

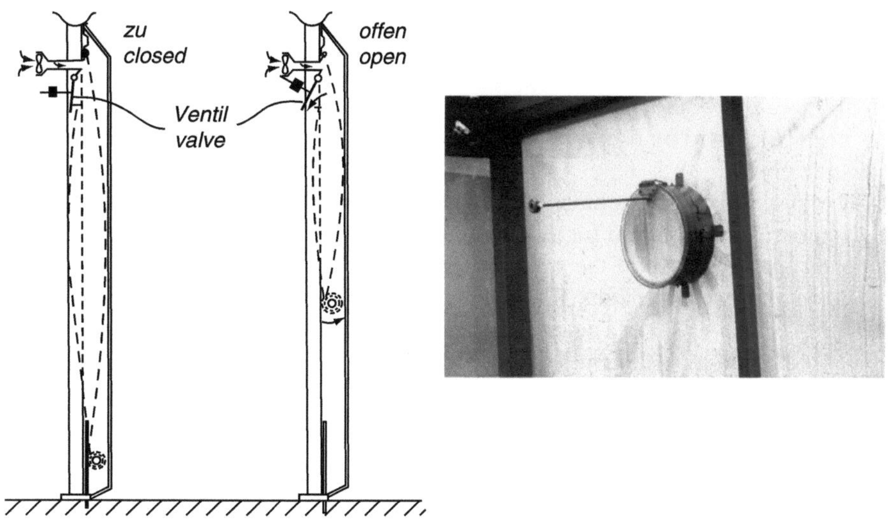

Fig. 5.84 Roll-up ventilation with double-inflated plastic film

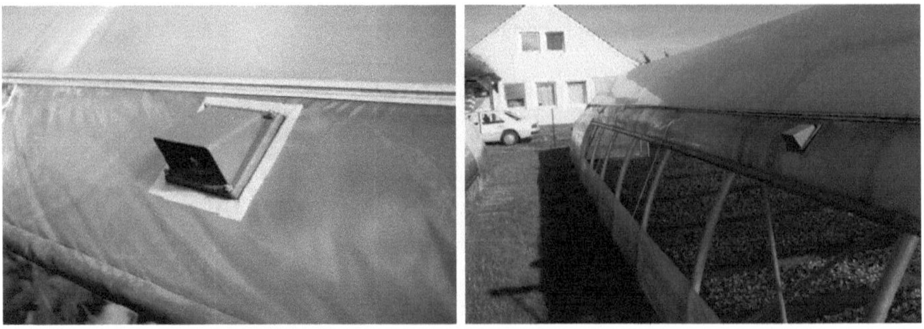

Fig. 5.85 Meanwhile, commercial valves for the double-inflated film are available

A: Wide span glass greenhouse, steel-aluminium construction.
Width: Five spans of 9.67 m each; Length: 42.9 m; Floor area: 2,075 m^2.
B: Venlo type glass greenhouse, steel-aluminium construction.
Width: Five spans of 9.6 m each; Length: 40 m; Floor area: 1,920 m^2.
C: Plastic film greenhouse I, single film, steel aluminium construction.
Width: Four spans of 8.7 m each; Length: 56 m; Floor area: 1,949 m^2.
D: Plastic film greenhouse II, air cup film, steel construction.
Width: Two spans of 10 m each; Length: 100 m: Floor area: 2,000 m^2.
E: Plastic-film greenhouse III, air inflated double film and single film, steel construction.
Width: Five spans of 10 m each; Length: 40 m; Floor area: 2,000 m^2.

5.8.1 Results

Figure 5.92 shows the material requirements of steel, aluminium, and zinc, which has been calculated by use of design drawings, parts lists and information from the manufacturer.

The glass greenhouses have the highest requirements of steel, and a relatively high amount of aluminium for purlins and bars. The amount of zinc for galvanizing depends on the surface of the steel components.

The energy requirements for the energy balance have been taken from the BUWAL study (1991). The energy and mass fluxes are calculated from raw material extraction to the end of the product, including transport and disposal. An energy amount of 7.6 kWh/kg is necessary for steel production, and an amount of 32.7 kWh/kg for the aluminium production, taking into account a recycling rate of 45% for steel and 35% for aluminium (Fig. 5.93).

The energy requirement for glass greenhouses is higher than for plastic-film greenhouses.

The energy equivalents for glass and low-density polyethylene LDPE are (Buwal1991):

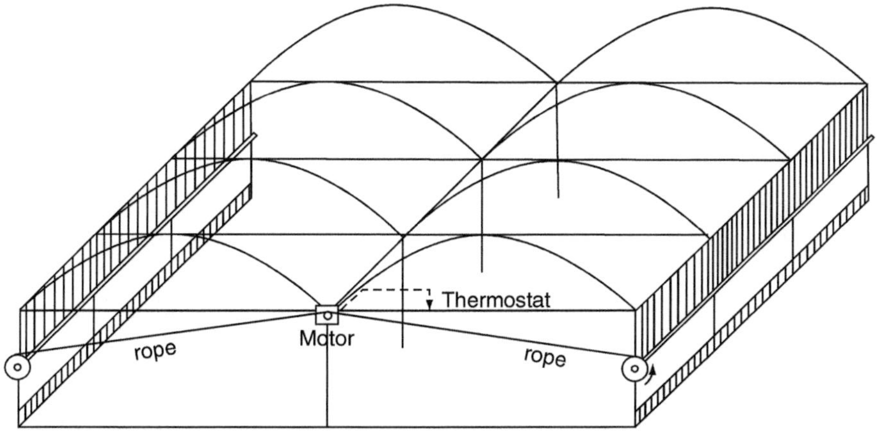

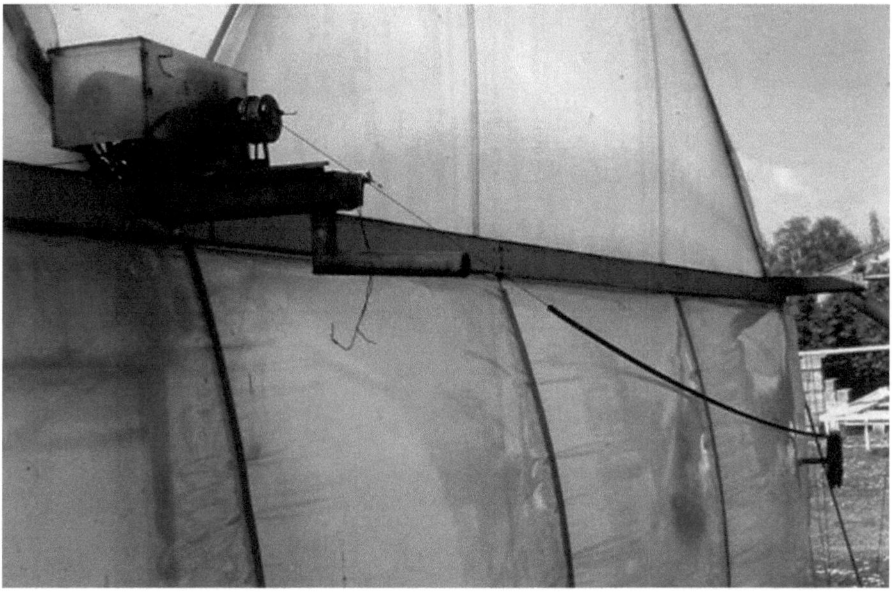

Fig. 5.86 The roll-up ventilations on both side walls can be operated by one driving motor and controlled by a thermostat

Glass (45% recycling)	2.2 kWh/kg
LDPE (without energy recovery by burning)	19.1 kWh/kg
LDPE (with energy recovery of 80% of waste)	13.2 kWh/kg

The energy requirement per kg of material is much higher for plastic material PE than for glass, but calculating the energy for 1 m^2 of greenhouse floor area reverses

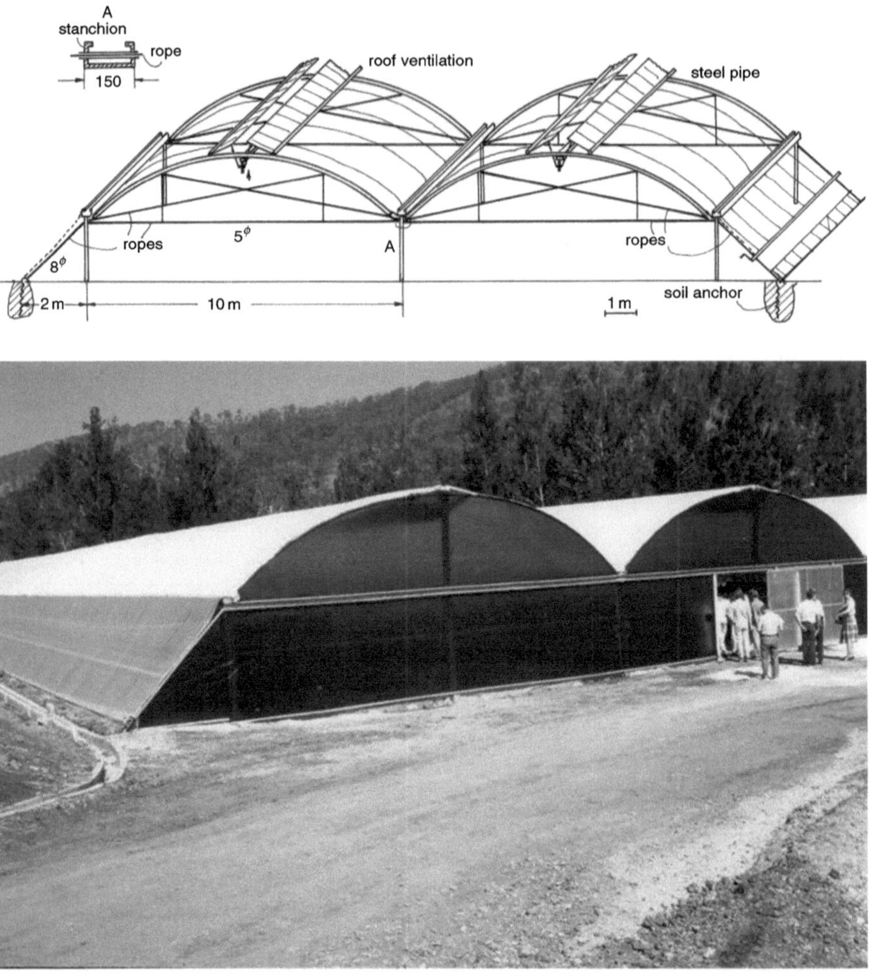

Fig. 5.87 Australian design with inclined steel ropes at the side walls for stabilizing the structure. The plastic film will be rolled up at the inclined ropes for ventilation. The gutters and steel tubes inside the greenhouse are connected and guyed by ropes

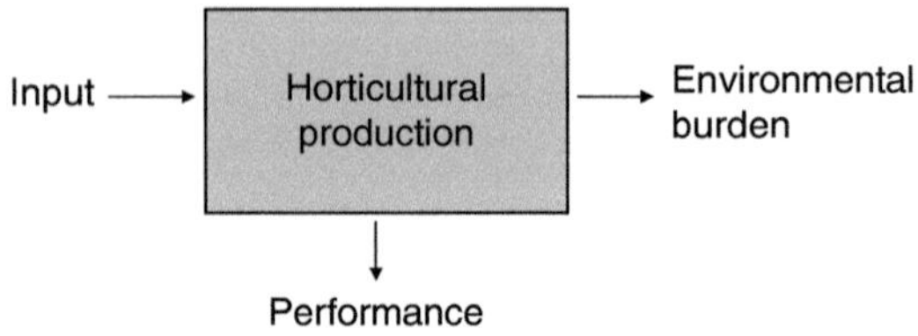

Fig. 5.88 Valences of horticultural production

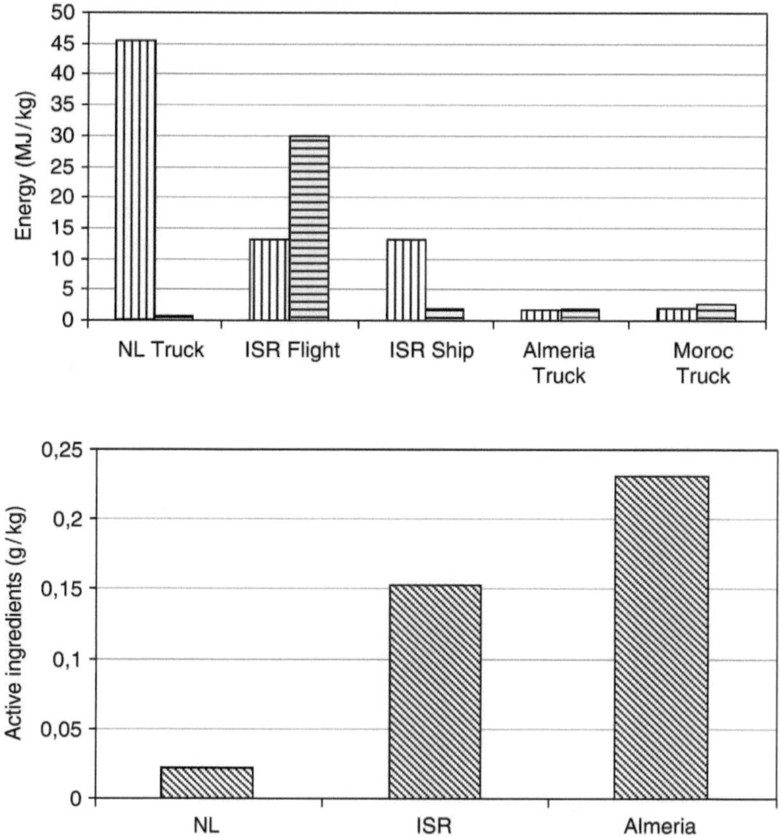

Fig. 5.89 Consumption of energy and active ingredients for production and transport of tomatoes from different countries to Frankfurt in Germany (Verhaegh 1996). *NL* Netherlands, *ISR* Israel, *Moroc* Morocco

the conditions, because the material requirement for plastic film is lower than for glass (Table 5.11). The energy requirement for glass is about 22 kWh/m^2, and for single plastic film 5–6 kWh/m^2.

The energy balance for plastic-film greenhouses is more favourable than for glass greenhouses, in spite of the repeated changes of plastic film (Fig. 5.94).

Assumptions: 12-years life span of the greenhouses.

Glass: one cladding for 12 years
PE film: four changes in 12 years
Air cup film: three changes in 12 years

The energy consumption for heating the greenhouses is much higher in 12 years than for manufacturing the structure. The percentage for manufacturing the

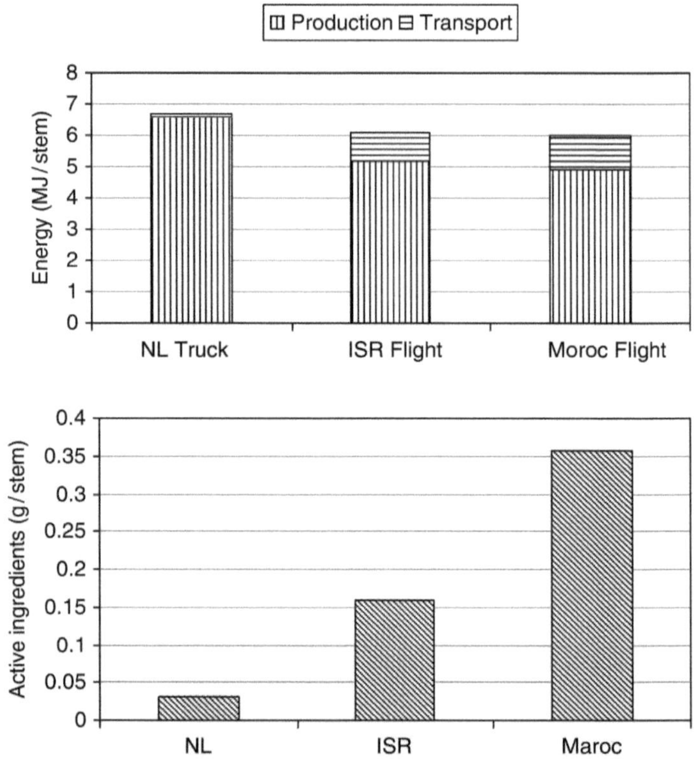

Fig. 5.90 Consumption of energy and active ingredients for production and transport of roses from different countries to Frankfurt in Germany (Verhaegh 1996). *NL* Netherlands, *ISR* Israel, *Moroc* Morocco

structure is less than 3% for German conditions in comparison to the energy heating over 12 years.

The conclusion is: a higher amount of material for energy-saving measures, such as double-inflated plastic film and air cup film for plastic-film greenhouses results in much higher energy saving for heating (30–40%).

The water consumption for the production of steel and aluminium is 0.2–0.4 m^3/m^2 floor area. That is less than the irrigation water for 1 year.

Russo and Scarascia Mŭgnoza (2005) used the LCA method to compare greenhouse constructions:

A: Zinc coated steel and aluminium structure covered by 5-mm glass, life span 10 years. This greenhouse had the highest environmental impact.

B: Round-arched tunnel, zinc-coated steel covered by 200 μm PE plastic film, life span 2 years, five times recovered in 10 years. The environmental impact was remarkably lower than for the glass greenhouse.

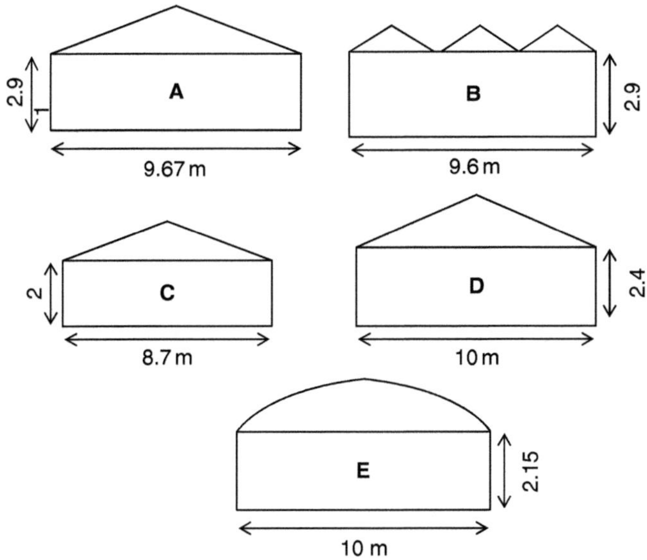

Fig. 5.91 Greenhouse structure investigated by a mass and energy balance. Gutter heights in meters

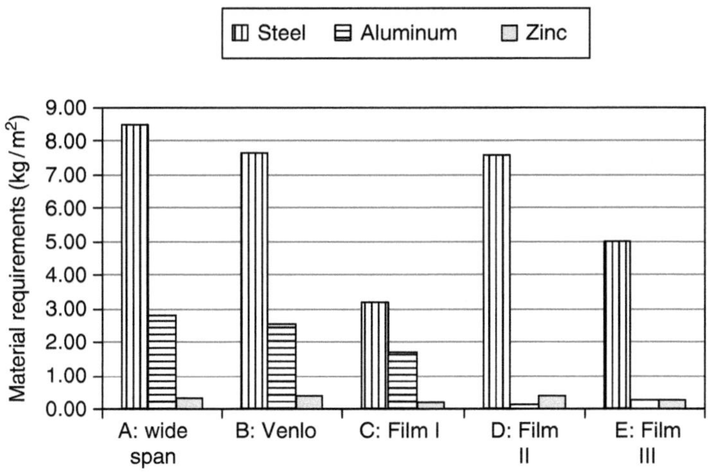

Fig. 5.92 Material requirements of steel, aluminium and zinc per m^2 floor area of the different greenhouses

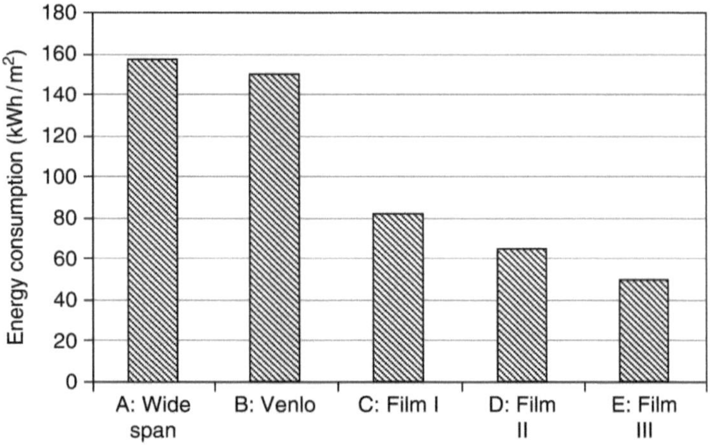

Fig. 5.93 Energy requirement per m^2 floor area for the construction materials steel, aluminium and zinc of the different greenhouses

Table 5.11 Material and energy requirements for greenhouse cladding materials

	A: Wide span, glass	B: Venlo, glass	C:Film I, Single PE film	D:Film II, air cup film	E: Film III, single PE film	E: Film III, double PE film
Material (kg/m^2)	10.2	10.0	0.32	0.66	0.24	0.48
Energy for production (kWh/m^2)	22.5	22.0	6.1	12.6	4.6	9.2
Energy for production, minus 80% recovery (kWh/m^2)			4.3	8.8	3.2	6.4

C: Chestnut wood structure, 5 years life span, covered by 200 μm PE plastic film. This greenhouse had the lowest environmental impact.

Three methods of cooling have been evaluated by using the LCA method (Anton et al. 2006) in three Parral-type greenhouses, 630 m^2 floor area each, with a sweet pepper crop:

- Forced ventilation by three fans,
- Fog system with high pressure nozzles,
- Whitewash shading.

The total production costs for sweet pepper of the forced ventilation system and the fog system were 51% and 13% respectively higher compared to whitewash, set to 100%.

The environmental impact due to CO_2 and SO_2 releases was about 14 times higher for the forced ventilation than for the fog system, due to the higher electricity consumption for forced ventilation.

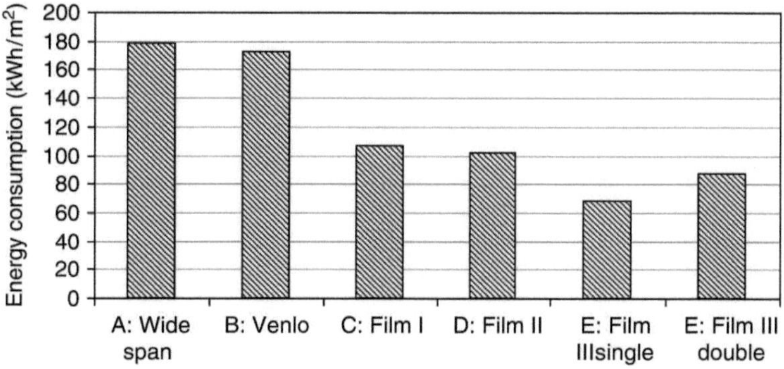

Fig. 5.94 Total energy balance for the greenhouses in Fig. 5.91

The fog system could be justified from an environmental and economical point of view if the increase in production were 9 and 6% respectively higher in comparison to the whitewash. The forced ventilation system could not be justified in the Parral-type greenhouse.

Chapter 6
Light Transmittance of Greenhouses

Light is one of the most important climatic factors for the photosynthetic activities of the plants. Crop growth is directly related to the available solar radiation. Optimizing the transmittance of solar radiation is a very important factor for crop growth in subtropical areas (Soriano et al. 2009); the light transmittance of greenhouses depends on:

- Type, thickness and spectral transmittance of cladding material by absorption and reflection (Chap. 7)
- Number of layers, single or double cladding
- Condensation on the cladding material (Chap. 7)
- Shading effect by greenhouse structural components
- Type of roof shape and roof inclination
- Orientation of the greenhouse
- Sun elevation, latitude, season of the year, time of day
- Share of direct and diffuse global radiation

It is relatively difficult to obtain accurate data about the greenhouse light transmittance.

Values of greenhouse transmittance for global radiation under practical conditions higher than 70% are the exception. The transmittance for single-covered greenhouses ranges between 55 and 70%, with the highest values in summer and the lowest in winter. The transmittance for double-covered greenhouses is 50–60% (Baille 1999b). A survey about the results on light transmittance measurements is given by von Elsner et al. (2000a, b) and von Zabeltitz (1999). Different roof shapes and roof inclinations have been investigated by Nisen (1969) (Figs. 6.1 and 6.2), and Kirsten (1973) (Fig. 6.3). Some new results of modelling the transmittance of plastic film greenhouses for different latitudes in Mediterranean areas have been published by Soriano et al. (2009).

The total transmittance T of a greenhouse is defined as the ratio of transmitted global radiation q_{rg} to the outside incoming radiation q_{ro}

C. von Zabeltitz, *Integrated Greenhouse Systems for Mild Climates*,
DOI 10.1007/978-3-642-14582-7_6, © Springer-Verlag Berlin Heidelberg 2011

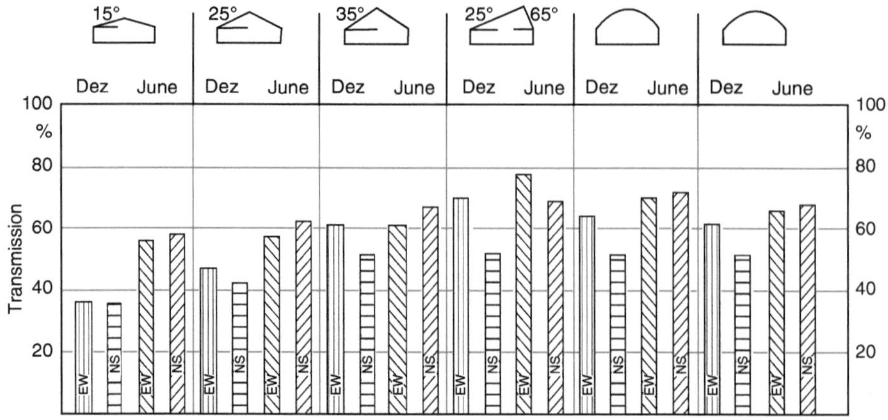

Fig 6.1 Mean transmittance of various greenhouse types with East–West (EW) and North–South (NS) orientation in December and June (Nisen 1969, von Zabeltitz 1999)

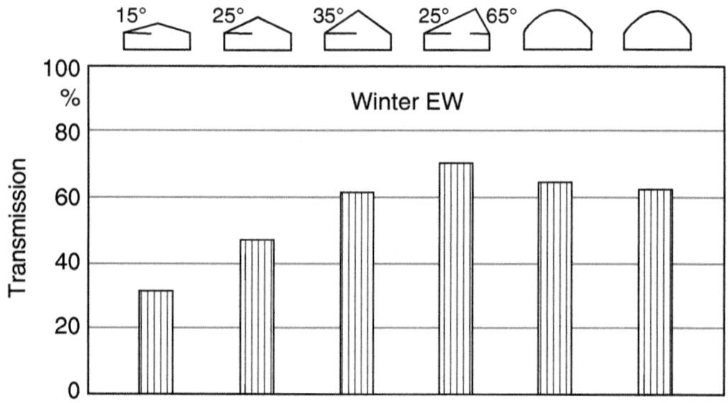

Fig. 6.2 Comprehensive description of the transmittance for EW orientation in winter (Nisen 1969)

$$T = q_{rg}/q_{ro}$$

The following conclusions can be drawn from Figs. 6.1 and 6.2:

- Light transmittance increases with roof inclination in saddle-roof structures.
- Light transmittance in EW-oriented greenhouses is higher in winter and lower in summer than in NS-oriented greenhouses.
- A saw-tooth or shed roof with the steeper and shorter roof side to the south in the Northern Hemisphere has a better transmittance than a saddle roof, but the roof surface area is larger.
- Greenhouses with curved roofs have better transmittance than greenhouses with saddle roof and 25° roof inclination.

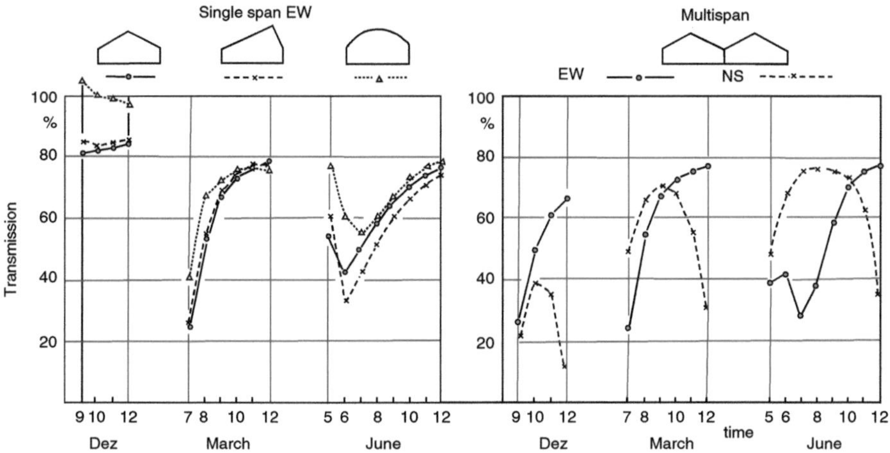

Fig. 6.3 Calculations for the transmittance for different greenhouse shapes at different times of the day (Kirsten 1973). A reduction of 10% by dirt and structural components is taken into consideration

Table 6.1 Light transmittance (%) of a Venlo-type greenhouse (Bot 1983)

Date	Orientation	
	East–West	North–South
21 December	45	35
21 February	58	53
22 March	59	62
21 April	60	67
21 June	65	70

The following conclusions can be drawn from Fig. 6.3:

- The best transmittance in winter is given by the curved roof of single-span greenhouses with an EW orientation, followed by the saw-tooth and saddle roof.
- The transmittance is a little bit better under a saddle roof in summer than under a saw-tooth roof.
- The transmittance through multi-span saddle-roof structures is higher in winter with EW orientation than NS.
- There is more light in the greenhouse with NS orientation in the morning in spring and summer.

Average daily figures for the transmittance of a Venlo-type greenhouse were calculated by Bot (1983) (Tables 6.1 and 6.2).

Weimann (1985, 1986) investigated EW-oriented plastic-film greenhouses covered with various single and double plastic films (Table 6.3).

Bredenbeck (1985) measured the light transmittance of three nearly identical greenhouses oriented in a NS direction. The greenhouses were covered with single glass, double glass and double acrylic sheets (Plexiglas Stegdoppelplatte, 16 mm).

Table 6.2 Mean measured values for the light transmittance in plastic film and glass greenhouses are given by von Zabeltitz (1986a)

Cladding material	Orientation	Summer	Winter
Single glass	NS	55–60%	48–55%
	EW	60–70%	55–65%
Single PE film	EW	65–70%	50–65%
Double PE film	EW	50–60%	45–55%

Table 6.3 Light transmittance of different plastic-film greenhouses (Weimann 1985, 1986)

System	Mean transmittance (%)	
	Summer	Winter
Inflated double PE (0.2 mm) 1 Year old	53–55	52–53
Inflated double PE No-Drop (0.18 mm) 1 month old	55–62	58–60
Inflated double PE-EVA		55–58
Inflated single PE-EVA (0.18 mm) 1 year old		65
Single PE standard (0.2 mm) 1 year old		62

Table 6.4 Transmittance of different greenhouses (Bredenbeck 1985)

Greenhouse	Winter (%)	Summer (%)
Single glass greenhouse	55	60
Double glass greenhouse	42	49
Double acrylic, treated	60–64	60–64

Table 6.5 Influence of greenhouse orientation on daily light sum (Wh/m^2 day) in the Netherlands (Waaijenberg 2006)

Week number	East–West	North–South
2 January	379	293
4 January	426	322
6 February	578	530
10 March	1,243	1,226
14 April	1,955	2,104
20 June	2,720	2,969

The glass panes were 0.6 m wide, while the Acrylic sheets were 1.2 m wide, needing fewer construction components in the roof area. The acrylic sheets were treated inside and outside by a No-Drop wetting agent (5% Sun Clear solution). Measurements were taken continuously every 10 min by solarimeters. Mean values for the transmittance are given in Table 6.4. The transmittance of the treated double acrylic sheet was comparable to single glass, but had a energy saving of 40% in winter. After those results had been published, the manufacturer of the double acrylic sheets developed the well-known No-Drop double acrylic sheets for greenhouse cladding.

Waaijenberg (2006) reported the influence of greenhouse orientation on the daily sum of radiation (Wh/m^2 day) in the Netherlands (Table 6.5). The transmittance is higher in winter with east–west (+23%) than with north–south orientation. The north–south-oriented greenhouses get more light inside in summer.

One trend to improve greenhouse design in subtropical regions is the increasing of roof slopes for better light transmittance and runoff of condensed water

(Sect. 7.4), as well as higher structures for better inside climate conditions (Baille 2001; Castilla et al. 2004).

Castilla and Lopez-Galvez (1994) investigated so-called low-cost Parral-type greenhouses in Spain:

G1: Asymmetrical east–west oriented single-span greenhouse, 21 m wide, 18° roof slope to the north, 8° roof slope to the south, 2.43 m height to the eave and 4.5 m ridge height, ventilators at side wall and ridge.

G2: Conventional flat-roof Parral-type greenhouse, 2.33 m height, ventilators at side wall.

All ventilators were covered with insect screens, and the cladding material was 0.2 mm PE film.

The transmittances of the greenhouses are given in Table 6.6.

The solar radiation transmittance was higher in winter and spring during the main cropping season in the greenhouse with slight roof slopes than in the one with the flat roof. Only in midsummer is the transmittance under the flat roof higher.

The temperature difference inside to outside was always higher in the flat-roof greenhouse. The asymmetrical greenhouse with roof inclination was more profitable than the flat-roof greenhouse.

Castilla et al. (1999) got similar results with asymmetrical greenhouses. Steeper roof slopes improve the light transmittance in winter, resulting in better yields of vegetables (Montero and Anton 2003).

Increasing light transmittance has a positive effect, even in subtropical countries. Different plastic-film cladding materials on two identical round-arched greenhouses, three spans of 6 m each, 19.2 × 12 m total floor area, 2.5 m gutter height, 4 m ridge height, have been used to investigate the influence on flowers and tomatoes (Anton et al. 2005; Montero et al. 2005; Montero and Anton 2003). The plastic cladding materials were:

GH1: Coextruded three-layer plastic film. The light transmittance of the greenhouse was $T = 64\%$ from June to October.

GH2: Tetrafluor–ethylene copolymer, 60 μm thick. Light transmittance of the greenhouse $T = 81$–83% from June to October. The transmittance of GH2 remained high in spite of dust accumulation.

The measured temperatures and humidity were similar in both greenhouses.

The high radiation transmittance in GH2 resulted very positively in early blooming time, quality and quantity of the geranium crop.

The increase of light led to higher yields of the tomato crop in summer and winter. The final yields of tomatoes were 15 and 27% higher in GH2 for winter and summer crop respectively.

Table 6.6 Solar radiation transmittance (%) of two greenhouses G1 and G2 in the south of Spain

Month	Greenhouse G1	G2
December	70.6	62.5
March	70.7	66.3
June	71.7	74.2
September	64.3	65.5

These results are in contradiction to the often expressed opinion that light in subtropics is above crop requirements even in winter.

Therefore greenhouse manufacturers, plastic film manufacturers and growers should pay more attention to the light transmittance of greenhouses to improve economic results.

Soriano et al. (2004) measured and calculated the light transmittance of greenhouse scale models with different saddle roof slopes for the Mediterranean region all year round (37° North altitude). The scale of the models was 1:15. Each model had three spans, 110 cm long, 40 cm wide, and was covered with glass panes. East–west orientation is better than north–south orientation for maximum light transmittance in autumn and winter. Table 6.7 shows the seasonal light transmittance for different roof slopes.

The symmetrical saddle roof with 27° roof slope had the highest transmittance in winter, and the most uniform transmittance throughout the year. The seasonal variation in transmittance was highest in the asymmetrical saddle roof 18°/8°.

Soriano et al. (2009) calculated the direct solar radiation transmittance of multi-span plastic-film greenhouses (Parral-type) with different roof slopes and orientations in Mediterranean areas for latitudes of 30°N, 37°N, and 45°N. Most of the global radiation in Mediterranean regions is direct radiation. The cladding material was a multilayer plastic film (PE, EVA, PE No-Drop) of 200 μm thickness.

The following regions can be found around the investigated latitudes

30°N	37°N	45°N
South Morocco	North Tunisia	North Italy
North Egypt	South Spain	North Japan
South Israel	South Sicily	North USA
Kuwait	South Turkey	
South Iran	North India	
North Mexico	North Japan	
South USA		
South Japan		

The maximum transmittance at latitude of 30°N in winter has been calculated as 76.4% for roof angles of 35°S/25°N and 76.3% for roof angles 35°S/30°N. The maximum transmittance at latitude 37° in winter has been calculated as 74.2% for roof angles 35°S/30°N. Taking into consideration the problem of drop condensation (Sect. 7.4), symmetrical roof angles of about 30° may be optimal for greenhouses in subtropical regions (Table 6.8).

Table 6.7 Mean seasonal transmittance throughout the year for four roof slopes (Soriano et al. 2004)

Roof slopes (°)		Seasonal transmittance (%)		
South-facing	North-facing	Summer solstice	Equinox	Winter solstice
18	8	74.9	69.8	59
36	55	69.7	66.3	56.7
45	27	71.3	67.7	66.6
27	27	71	68.5	70.1

Table 6.8 Calculated solar radiation transmittance (%) on December 21 at different latitudes and different roof slopes (Soriano et al. 2009)

Roof slope	Latitude and orientation					
South/North (°)	30°		37°		45°	
	EW	NS	EW	NS	EW	NS
	Transmittance (%)					
10/10	65–70	60–65	60–65	55–60	60–65	50–55
20/20	70–75	60–65	70–75	55–60	65–70	50–55
30/30	75–80	60–65	70–75	55–60	65–70	50–55
35/35	75–80	60–65	70–75	55–60	65–70	50–55
25/30	70–75	60–65	70–75	55–60	65–70	50–55
35/30	75–80	60–65	70–75	55–60	65–70	50–55
30/35	70–75	60–65	70–75	55–60	65–70	50–55

Ageing and dust deposition on cladding materials and screens have a remarkable influence on the light transmittance. The reduction of light transmittance by dirt accumulation, depending on the material, can be as high as 30% (see Sect. 7.2).

Chapter 7
Cladding Material

Global radiation is a minimum factor in mild winter climates, and therefore high transmittance of greenhouses is a prerequisite for optimized production in winter (Baille 1999). Convenient roof slopes are necessary for the improvement of light transmittance, avoidance of drop condensation and dripping (von Zabeltitz 1988c). Commercial cladding materials have different characteristics for spectral transmittance and energetic demands, and are available for various properties. Those characteristics have to be taken into account in the choice of cladding materials in different climatic regions. The characteristics of the cladding material determine the quality of light transmitted into the greenhouse, and radiation transmittance can be improved qualitatively and quantitatively (Castilla and Hernandez 2007). Various characteristics have been discussed for different climatic regions (Table 7.1).

7.1 Materials

Cladding materials for greenhouses should have:

- High transmittance for visible light, photosynthetic active radiation PAR, with wavelengths 400–700 nm
- Low transmittance for long-wave radiation, FIR, in the range of wave lengths from 3,000–20,000 nm
- Low reduction of light transmittance by global radiation (long duration of life)
- Low ageing by UV-radiation
- No drop-wise condensation on the inside of the roof, but condensation as a liquid film (No-Drop properties)
- Low accumulation of dirt and dust
- Strength against wind

Some scientists recommend and prefer lightscattering material, so-called diffusivity or haze of the material (Hemming 2005; Waaijenberg 2006). Diffuse material

C. von Zabeltitz, *Integrated Greenhouse Systems for Mild Climates*,
DOI 10.1007/978-3-642-14582-7_7, © Springer-Verlag Berlin Heidelberg 2011

Table 7.1 Importance of the criteria of plastic cladding materials (Baille 1999b), supplemented for subtropics, PAR (photosynthetic active radiation), FIR (far infrared) and NIR (near infrared) radiation

Property	Arid climate	Subtropical climate	Tropical climate
Anti-dust	High	High	Low
Scattering of direct radiation	High	Medium	Low
No-Drop	Low	High	Medium
High PAR transmittance	Medium	High	High
FIR blocking	High	High	Low
NIR blocking	High	High in summer low in winter	High

reduces the danger for plants getting burned by direct sun radiation, and spreads the light better inside the greenhouse and into the plant canopy. But the total transmittance of PAR is reduced in most of cases, and light is a minimum factor in winter and under overcast weather conditions. Therefore the transmittance for PAR should be as high as possible, in particular for vegetable crops.

A further problem is drop condensation on the inner side of the cladding material (see Sect. 7.3.5).

The main cladding materials for greenhouses are:

- Glass
- Rigid plastic sheets
  - Acrylic double-wall sheets (PMMA)
  - Polycarbonate double-wall sheets (PC)
- Plastic film
  - Polyethylene, PE, with UV stabilisation
  - PE-IR with absorber for FIR radiation
  - PE-EVA (ethylenevinylacetate), coextruded film with low transmittance for FIR radiation
  - PVC (polyvinylchloride), mostly used in Japan, with low transmittance for FIR radiation
  - ETFE (ethylenetetrafluorethylene). A film with a high light transmittance of 93–95%, high transmittance for UV radiation, low transmittance for FIR radiation, and a duration of life of more than 10 years, but a limited width and a relatively high price of more than ten times compared to PE film.
  - Screens or nets for screenhouses

Table 7.2 shows the transmittance for direct and diffuse light of greenhouse cladding materials.

Plastic film is the most applied cladding material in countries with tropical, subtropical and arid climates. Only Turkey has a remarkable but decreasing amount of glass greenhouses. Rigid plastic sheets and glass will be used in some arid regions.

PE plastic film with different additives for different properties is the cheapest cladding material for greenhouses. But the film has to be changed relatively often

Table 7.2 Transmittance for direct and diffuse light of greenhouse cladding materials (Waaijenberg 2006; von Zabeltitz 1986a, 1999)

Material	Thickness (mm)	Direct light (%)	Diffuse light (%)
Glass	4	89–91	82
PE film UV	0.1–0.2	89–91	81
EVA film	0.18	90–91	82
PVC film	0.1–0.2	87–91	
ETFE film	0.1	93–95	85
PC double sheet	12	80	61
PMMA double sheet	16	89	76

because light transmittance and stability is reduced by dust, heat and global radiation. The duration of life is limited.

PE film for greenhouses should have basically UV-stabilisation, reduced infrared transmittance and No-Drop properties.

Drop condensation can reduce the transmittance remarkably. This light loss results in considerable yield losses. Water drops fall down on the crop and increases the danger of disease infestation. Condensation itself is desirable to avoid too high humidity, but condensation should appear as film condensation (see Sect. 7.3.5).

The spectral transmittance of the cladding material and the condensation properties have an influence on plant growth, on quality and on infestation by diseases. The cladding materials can be treated with regard to spectral transmittance.

The spectrum of the global radiation is divided into the following ranges of wavelengths (nanometre nm):

7.1.1 Ultra Violet Radiation (UV)

Radiation below 300 nm does not reach the surface of the earth.

UV-B radiation	380–315 nm
UV-A radiation	315–380 nm

7.1.2 Photosynthetic Active Radiation PAR or Visible Light

Visible light	380–780 nm
Violet	380–400 nm
Blue	400–500 nm
Green	500–600 nm
Red	600–700 nm
Far red	700–780 nm
Near infrared NIR	780–3,000 nm

7.1.3 LongWave Radiation, Far Infrared FIR

The transmittance of FIR radiation with wavelengths of 3,000–20,000 nm increases heat transfer by radiation from the crop through the cover, with the consequences of lower air temperature, lower plant temperature and temperature inversion (lower inside than outside temperature). If plant and air temperature sink below the dew point, temperature condensation and fog may occur and the danger of diseases increases.

Glass is opaque to FIR radiation. Thermic film with IR absorber, or co-extruded PE and EVA film, should be used for greenhouses in subtropical and arid climates, to keep the temperature as high as possible in unheated and heated greenhouses during cold nights. Table 7.3 shows the transmittance for FIR.

Fog can form in unheated greenhouses covered with No-Drop film when temperature inversion occurs. Temperature inversion means inside temperature drops below outside temperature. If radiation is going from soil and plants through the cladding material to the clear sky at night, the inside temperature sinks below the outside temperature. The plastic film is warmed up by higher outside temperature, and the condensate at the cladding film evaporates. The dew point of the inside air is lower than directly at the film, and the vapour condenses as fog. Fog formation under No-Drop film occurs more often, because the water film condensate has a bigger surface for evaporation than the droplets on standard PE film. Fog and temperature inversion can be avoided by early ventilation.

7.2 Influence of Dust and Dirt

Dust and dirt on the greenhouse cladding material reduce light transmittance remarkably. The light loss depends on the cladding material and on the region (industrial or rural). It is difficult to exactly quantify the light loss by dust. It can vary in the range of 5–40%.

The influence of dust on light transmittance can be higher with No-Drop film than with standard PE film. A light loss of the films on a greenhouse after 18 months exposure of 5.8% for standard PE film and 6.7% for PE No-Drop film has been measured by Gbiorczyk (2003a, b).

Table 7.3 Transmittance for long-wave radiation FIR (Waaijenberg 2006; von Zabeltitz 1986a, 1999)

Material	Thickness (mm)	Transmittance (%)
Glass	4	0
PE UV film	0.2	40–60
PE IR film	0.2	25–35
EVA film	0.18	30–40
ETFE film	0.1	15–20
PC double sheet	12	0
PMMA double sheet	16	0

The light transmittance of some used samples of plastic film (thickness 0.2 mm) coming from desert regions in North Africa were measured in Hannover:

Light transmittance (wavelength 400–760 nm)

PE film, new	86.2%
PE film, 1 year old, dirty	79.8%
PE Film, 1 year old, cleaned	85.2%
PE film, 3 years old, dirty	56.0%
PE film, 3 years old, cleaned	85.2%

Light losses of 29% due to dirt reduce the crop growth in winter remarkably.

The possible dust accumulation on plastic film is higher than on glass because the electrical charging of plastic film is higher. Rain can remove dust from glass easier than from plastic film. But the shading effect from structure components is higher in glass greenhouses due to the numerous bars for the glass panes. The calculated shading effect for direct radiation in winter is 20% for plastic film, and 40–50% for glass (Jaffrin and Morison 1994). The shading effect for overcast sky is <5% for plastic film and <10% for glass.

Light loss due to dust is about 5–30%, together with drop condensation up to 40% (Jaffrin et al. 1994).

Light loss under plastic film due to dust accumulation and drop condensation can reverse the advantage of the lower shading effect, so that plastic-film greenhouses may have lower light transmittance than glass greenhouses under unfavourable conditions.

Regular cleaning of the cladding material outside every year or every few months is very important in dusty and dry areas, together with avoidance of drop condensation through the use of No-Drop film and adequate inclination of the roof (see Sect. 7.3.5) for the advantage of plastic-film greenhouses (Fig. 7.1, Tables 7.4–7.5).

Dust accumulation increases the shading effect of nets and screens during dry seasons (Table 7.6) (Shahak et al. 2004). The dust ratio factor is the shading (%) at the end of the season divided by the shading (%) at the beginning of the season. Dust also reduces the effect of the nets to modify the spectrum.

Montero (2009) reported a reduction of transmittance in a 35-mesh screenhouse from 73 to 56%.

7.3 Influence of Selective Spectral Transmittance

Cladding materials with special spectral transmittance can have various effects on greenhouse climate, plant behaviour and pest control. One has to distinguish the influence of the different spectral ranges and different cladding materials. The main influencing factors of the spectral ranges (see Sect. 7.1) are:

- UV radiation
- Different colours of visible PAR (photosynthetic active radiation)

Fig 7.1. Plastic film greenhouse with heavy dust accumulation in an industrial region

Table 7.4 The influence of dust for standard PE film and film with anti-dust additive. The anti-dust additive causes lower transmittance than the standard PE film (Manor et al. 2005)

Transmittance after	Standard PE	PE with anti-dust additive
1 week	85%	81%
5 weeks	81.6%	77%

Table 7.5 Influence of dirt on diffuse light transmittance (Gratraud 1990; von Elsner et al. 2000a, b)

Material	Light loss by dirt
Coextruded PE-EVA film after 12 months	13.2%
Glass after 12 months	4.3%

Table 7.6 Dust effect on shading factor of different nets (Shahak and Gussakovsky 2004). Maximum measured data in Israel 2003

Net	Shading end of April (%)	Shading September (%)	Dust ratio
Grey knitted	34	39	1.14
Red knitted	29	35	1.21
Pearl knitted	33	52	1.58
White knitted Anti-hail	16	27	1.69
Insect-proof 25-mesh woven	27	46	1.70
Clear/white anti-hail woven	12	30	2.50

- Non visible NIR radiation
- Long-wave IR radiation

The main differences in cladding materials are:

- Rigid materials glass and plastic sheets
- Plastic film
- Nets and screens

The transmittance for PAR radiation should be as high as possible in winter at low sun altitude.

Figure 7.2 shows the ideal spectral transmittance of covering materials in the various spectral ranges.

The various spectral ranges have the following influences (Schultz 1996a, b; Schultz 1997; von Zabeltitz 2004; Hoffmann 1999a–e; Bessler 2000; Ludolph and Bessler 2002).

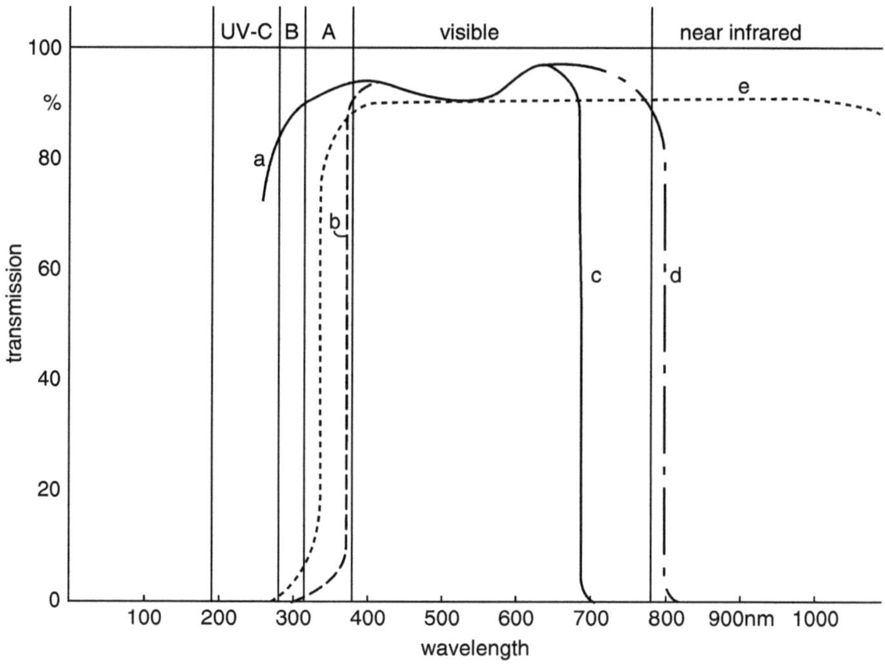

Fig. 7.2 Spectral transmittance for different desired, ideal materials. Curve *a*: total transmittance for UV radiation. Curve *b*: opaque to UV radiation. Curve *c*: no or low transmittance for near infrared radiation NIR to improve the energy balance inside the greenhouse. Curve *d*: visible light. Curve *e*: conventional PE film

7.3.1 Transmittance and Blocking of UV Radiation

7.3.1.1 Transmittance for UV-B Radiation

- Influence on colouring of leaves and blossoms
- Reduction of leaf area and stem elongation of flowers
- Hardening of seedlings before transplanting in the open air
- Little delay of blossom with low radiation and low temperature is possible
- Danger of leaf burning by radiation

Materials for transmitting UV-B radiation:

- Special Glass (Planilux, Diamond)
- Special PE film
- ETFE film (F-Clean, see 7.1)

7.3.1.2 UV-Blocking of Cladding Material

UV radiation is important for the orientation of insects. Without UV radiation, they are disoriented and can not find the plants. There is lower infestation by virus diseases, in particular by thrips, aphids and white fly under UV-blocking materials. UV-blocking materials protect the crops from infestation by insects and the spread of viruses as follows (Antignus and Ben-Yakir 2004):

- Fewer insects penetrate greenhouses covered with UV-blocking material
- The flight activity inside the greenhouse will be reduced

Antignus et al. (1996) reported on reduction in the infestation of vegetable crops grown in greenhouses under UV-absorbing films by *B. tabacci*, western flow thrips and aphids. They tested UV-absorbing and UV-transmitting plastic films in walk-in tunnels 6 × 6 m, 2.7 m height. Figure 7.3 shows the number of white flies (*B. tabacci*) captured on yellow sticky traps under the different plastic films.

Figure 7.4 shows the infection of tomato plants by TYLCV (tomato yellow leaf curl virus disease) in walk-in tunnels under UV-absorbing and UV-transmitting plastic film. Seventy days after planting, all plants grown under UV-transmitting film were infected by TYLCV, whereas only 50% were infected under UV-absorbing film.

Experiments in commercial walk-in tunnels (5 × 50 m) used for growing herbs resulted in far fewer white flies, thrips, and leaf miner flies under UV-blocking PE film than under standard PE film. The quality and yield were much better, and the number of pesticide applications could be reduced by 50–80% in greenhouses covered with UV-blocking film (Antignus and Ben-Yakir 2004). The infestation of tomatoes grown under UV-blocking film by insect-born viral diseases, tomato yellow leaf curl virus (TYLCV) was only 1% compared to 80% under standard film.

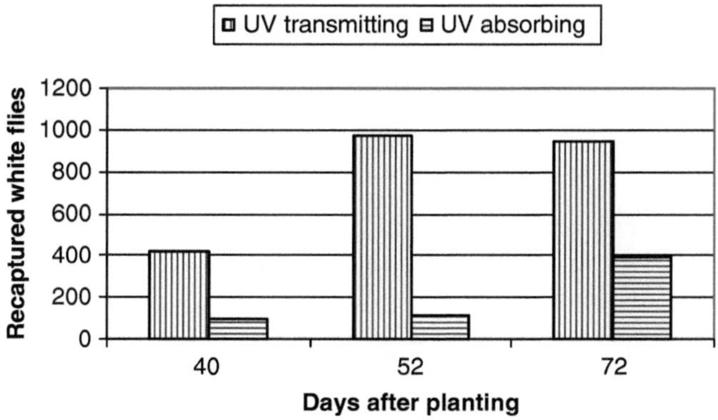

Fig. 7.3 White flies recaptured under UV-absorbing and UV-transmitting film (Antignus et al. 1996)

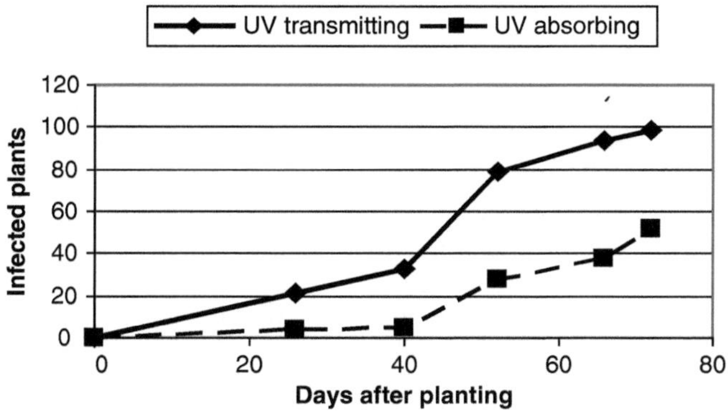

Fig. 7.4 Disease infection by TYCLV of tomato plants grown under UV- absorbing and UV-transmitting plastic film (Antignus et al. 1996)

PVC film, which has low transmittance for UV radiation, has higher protection against insect pests than standard PE film.

The entry of unfiltered light through ventilation openings reduces the efficiency of UV-blocking roof cladding. Therefore, the vent openings should be covered by insect-proof screens (Costa et al. 2002).

Special experiments were carried out in the Institute for Horticultural Engineering Hannover (Mutwiwa 2004; Mutwiwa et al. 2005), where white flies (Trialeurodes Vaporariorum) had the choice to fly in plastic tunnels alternately covered with UV-transmissive film and with UV-blocking film. Four tunnels were placed crosswise opposite each other, covered alternately with the films and connected to a black compartment in the middle, where the white flies were released. Two "tunnel crosses" with four tunnels each were built up. The arrangement of the tunnels provided one tunnel covered with each type of plastic film in each of the four cardinal points.

The transmittance (%) of the plastic films (Hyplast, Belgium) was:

	UV-transmitting	UV-absorbing
Light (250–2,500 nm)	87	85
UV	78	10
PAR	86	89
NIR	88	89

UV intensity inside the tunnels:

Sunny day	3.4 ± 0.3 W/m^2	0.13 ± W/m^2
Cloudy day	1.91 ± 0.14 W/m^2	0.03 ± W/m^2

The global radiation intensity did not differ significantly in the tunnels, and the recorded temperature differences were very small.

Figure 7.5 shows the number of white flies recaptured in the tunnels after releasing in the middle compartment. Both the type of plastic film as well as the position of the tunnels (flight direction of the insects) affected the number of recaptured white flies. More than 90% of the insects were recaptured under the UV-transmitting plastic film. The highest number was in the eastern UV-transmitting tunnel. That means UV-transmitting film can reduce the population of white fly (*T. vaporariorum*), and UV-transmitting film supports infestation by pest insects.

The question is still the effect of UV-absorbing film on beneficial insects.

Another choice experiment was carried out in two tunnels, 0.5 by 0.5 m cross-section and 1.8 m long with a black box in the middle, where the white flies were released and where they had the choice of flying in either tunnel (Costa and Robb 1999). Yellow sticky traps were placed in both tunnels. One of the tunnels was covered with standard PE film, the other one with UV-blocking PE film.

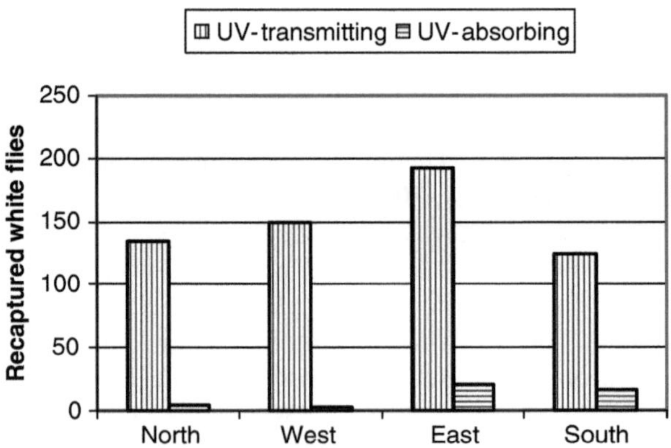

Fig. 7.5 Recaptured white flies under UV-absorbing and UV-transmitting plastic films in the choice experiment (Mutwiwa 2004, Mutwiwa et al. 2005)

The standard PE film transmits a part of the UV radiation below 380 nm and blocks UV radiation below 360 nm, whereas the UV-blocking film is nearly opaque to UV radiation. Under the standard PE film, 85–94% of the white flies were captured, but only a few under UV blocking film. Under standard film, 90–98% of thrips were caught.

Field studies in round-arched plastic tunnels, 4 m wide, 3 m high, and 8 m long, covered with UV-blocking and UV-transmitting film were carried out (Costa et al. 2002).The side walls of the tunnels were open to a height of 1.5 m for ventilation. The tunnels were covered with PE standard film, blocking UV radiation <360 nm, and UV-blocking film, blocking UV radiation <380 nm. The standard film (<360 nm) transmitted significantly more UV radiation than the UV blocking film (<380 nm). The number of aphids and thrips captured on traps and caught on *Chrysanthemum* plants were significantly lower under the UV-blocking film (<380 nm) than under the standard film (<360 nm). But the number of white flies did not significantly differ under the two plastic films, in contrast to the first experiment in the small closed tunnels. The reason for that is the open vent, where unfiltered light can penetrate the greenhouse.

The conclusion is:

If UV-blocking cladding material ought to be used for fighting pest insects, then no natural light should be able to enter the greenhouse, through open vents for example. UV radiation can enter the greenhouse when the vents are open, so that pest insects can orientate again and infest the plants. Special insect screens with reduced UV transmittance (Bionet with 50% reduction of UV transmittance) can hinder the orientation of the insects even under open ventilation conditions (Fig. 7.6).

Antignus et al. (1998) carried out further experiments with new developed insect screens. Those screens with reduced UV transmittance (Bionet) combine the

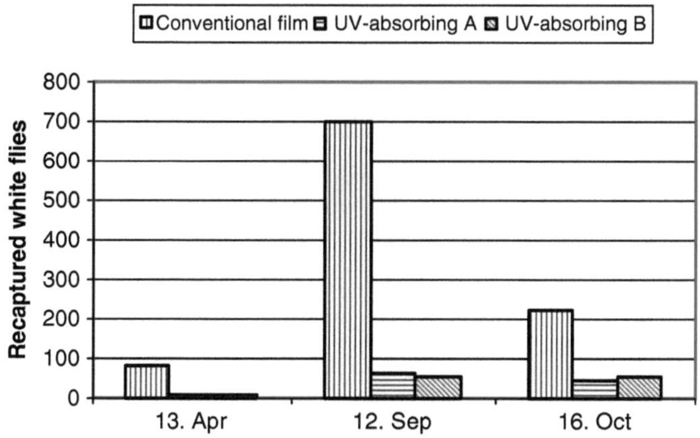

Fig. 7.6 Recaptured white flies (Bemisia tabacci) in a large-scale greenhouse on 160 leaves (Rapisarda et al. 2006). Three 1,250 m^2 greenhouses (24 × 52 m) covered with two UV-absorbing films, one conventional film and 40-mesh screens in front of the vents

mechanical and optical properties of insect screens. The Bionet screens filter out a part of the UV radiation. 50-mesh anti-virus screen blocks about 50% of the UV radiation compared to conventional screens, 30-mesh anti-insect screen blocks 18% and 16-mesh screen 15%. Walk-in tunnels, 6 × 6 m, were covered with conventional screens and with the UV blocking Bionet screens. The 50-mesh screen had a high effect on the population of white flies (Fig. 7.7) and on the infestation by TYCLV (Fig. 7.8). The 16-mesh screens had no effect on preventing the penetration of white flies. No difference could be found under 30-mesh Bionet and conventional screens. All mesh types failed to prevent the penetration of western flower thrips to the walk-in tunnels.

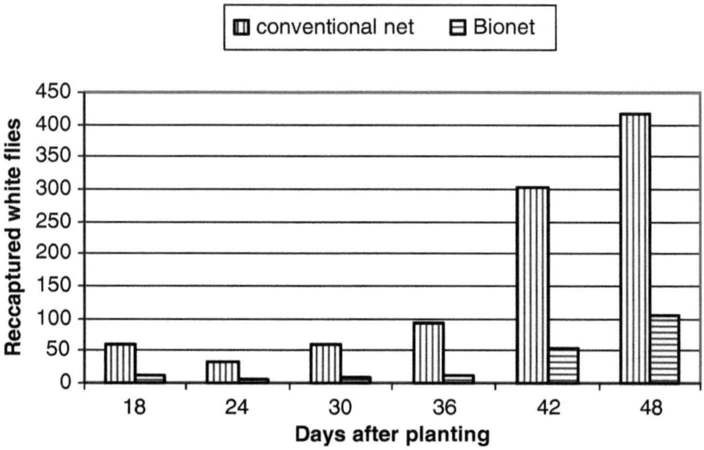

Fig. 7.7 White fly population under 50-mesh UV-absorbing net (Bionet) and UV-transmitting net (Antignus et al. 1998)

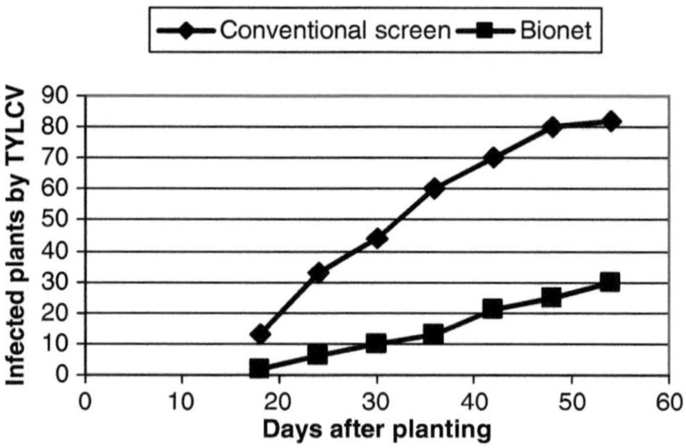

Fig. 7.8 Plants infected by TYLCV under Bionet screen (Antignus et al. 1998)

7.3.2 Photoselective Plastic Films and Screens

The different ranges of the PAR spectrum have the following effects:

7.3.2.1 Blue part of PAR (400–500 nm)

A low ratio of UV to blue radiation hinders the sporulation of some fungi.

7.3.2.2 Red to Far Red Ratio

A high ratio of red (600–690 nm) to far red (>700 nm) has the following influences:

- Reduction of stem elongation and increasing compactness
- Shorter main shoots of plants and more side shoots
- Increasing yield of some vegetable varieties

Materials with an absorption of far red radiation and high transmittance of PAR radiation are favourable to the compactness of flowers. The reduction of internodes length of 30–40% is higher, as with the transmittance of UV-B radiation with 10–15% (Hoffmann 1999a–e). On the other hand, the formation of blossoms can be hindered without red radiation.

Photoselective plastic films and screens will be used as cladding materials for green- and screenhouses as well as insect screens in greenhouses. The objectives are (Shahak et al. 2008):

1. Special physiological responses
2. Increasing radiation use efficiency. The transformation of direct light into scattered light improves the light penetration into the inner plant canopy
3. Special effects on plant pests, beneficial insects or diseases
4. Saving water due to reduced radiation and wind speed

Black nets are usually used for shading purposes. They reduce the light intensity by 40–80%, but they do not influence the light spectrum and amount of scattered light.

White, clear or transparent nets with shading of 8–25% are used to protect the crops from environmental hazards, birds and pests.

So-called coloured photoselective nets, containing photoselective pigments or additives, modify the light quality for special physiological effects, and scatter the incoming light (Shahak et al. 2009).

Coloured nets have the following characteristics (Shahak et al. 2004):

Blue: a wide peak of transmittance in the blue/green range, 400–540 nm.

Yellow: transmittance of light from 500 nm up.

Red: transmittance of light from 590 nm up.

Table 7.7 shows the effects of different nets in the various spectral ranges of solar radiation. The blue/red ratio is largest under blue nets and smallest under red and yellow nets. The differences are much higher in scattered light, which penetrates better into dense canopies.

The following responses of horticultural crops to photoselective nets and screens are reported from Israel (Shahak et al. 2008):

Pepper (*Capsicum annuum*) is commercially grown in screenhouses covered by black nets (30–40% shading) to avoid sun burns, to save irrigation water and to produce high-quality fruits. The fruit yield of pepper grown under photoselective nets was 1.15–1.35 times higher than under black net covering. The yield under red nets was highest, followed by pearl and yellow nets.

Foliage crops and cut flowers gave positive effects in screenhouses, covered with photo selective nets.

Additional shading by pearl and red nets on plastic-film greenhouses instead of using black nets in Israel increased the weight of lettuce heads by 20–50% depending on the variety. The shading of plastic-film greenhouses by red nets instead of whitewash increased the yield of pepper and tomato crops in south Spain.

Table 7.8 shows the penetration of pests into screenhouses covered by photoselective nets (Shahak et al. 2008; Shahak et al. 2009).

Table 7.7 Measured shading (%), scattering (%) and ratios red/far red, blue/red PAR/UV for various coloured nets for total and scattered light (Shahak et al. 2004)

Net	Shading (%)	Scattering (%)	Red/far red		Blue/red		PAR/UV	
			Total light	Scattered light	Total light	Scattered light	Total light	Scattered light
No net	0	13.3	0.73	0.82	0.92	1.63	24.6	10.1
Red	55.4	35.6	0.71	0.71	0.63	0.42	33.1	27.0
Yellow	48.6	44.1	0.70	0.70	0.53	0.31	40.2	41.0
Blue	51.9	26.0	0.73	0.66	1.26	3.06	35.1	27.2
Pearl	54.2	62.0	0.68	0.66	0.80	0.81	33.4	31.3
Grey	46.4	20.1	0.72	0.73	0.87	1.26	29.3	15.7
Black	48.4	11.4	0.74	0.74	0.86	1.17	29.2	10.7

Table 7.8 Pest penetration into screenhouses covered with photo selective nets

Cladding type	Colour	Reference cladding	Reduction of pest penetration		
			White flies	Thrips	Aphids
50-mesh screen white	White UV-blocking	Transparent	2–5 fold	3–10 fold	2–5 fold
Shade nets (18–35% shading)	Pearl	Black	None	None	3–4 fold
	Yellow	Black	2–3 fold	None	2–3 fold
	Blue	Black	None	2–3 fold	None
	Red	Black	None	None	None

The UV-blocking white insect screen had the following transmittance compared to transparent standard screen

	White insect screen UV-blocking	Transparent insect screen
UV transmittance (280–380 nm)	40–50%	80–90%
Visible light transmittance (380–800 nm)	60–70%	85–95%
Reflection of solar radiation (400–750 nm)	2.5 times higher	

The use of photoselective screens as covering material on screenhouses in southern Spain decreases the population of white flies (Bemisia tabacci) and the incidence by TYLCV (tomato yellow leaf curl virus). The yield can be improved. But photoselective screens could not decrease the incidence of thrips (Gazquez et al. 2009). Screens which reduce the transmittance of global radiation below 60% have a very negative effect on the yield.

Shahak et al. (2008b) carried out experiments with basil and lettuce in tunnel greenhouses covered by plastic film and additional photoselective nets on top of the film. The shading effect of the film and the net together was about 65%. Table 7.9 shows the head weight of lettuce grown under plastic film and photoselective nets compared to open field and under aluminised screen, normally used as thermal screen in greenhouses. Production was significantly increased under pearl and red screens.

Studies with pepper were carried out in screenhouses, 2.5 m high (Shahak et al. 2004; Shahak et al. 2009). The results are shown in Table 7.10.

The influence of dust can be found in Sect. 7.2.

7.3.3 Near Infrared (NIR) Blocking and Reflecting Materials

The greenhouse temperature often becomes too high in summer, and hinders optimal plant growth. Growers apply whitewash on the cladding material or use

Table 7.9 Lettuce head weight (different varieties grown under plastic film and photo selective screens in Israel (Shahak et al. 2008)

Nets:	No net	Aluminised screen	Blue	Pearl	Red
Head weight: (g/head)	250–258	236–311	242–282	295–418	320–393

Table 7.10 Fruit yield of pepper (different varieties) in photoselective screenhouses compared to black shade nets

Net:	Black (30–40% shading)	Pearl	Red
Fruit yield (ton/ha)	97–116	118–134	127–136

internal moving shading systems. But both of these reduce the PAR, and can cause reduction of quality and yield (Runkle et al. 2002).The NIR part of the global radiation (800–2,500 nm) is inefficient for the photosynthesis of plants, but has an influence on the heat balance of crop and greenhouse. It warms up the greenhouse and crop, and contributes to transpiration. NIR radiation is not desirable with high irradiation in summer, but useful and desirable with low irradiation in winter, when the temperature is below the biological optimum and when heating is necessary. The necessary heat energy can be increased by about 10% under NIR-blocking material in temperate climates (Hemming et al. 2006). NIR-reducing cladding material can have advantages under high light conditions because it reduces undesired NIR radiation, and thereby inside greenhouse temperature, and does not reduce the important PAR.

The NIR radiation can be reduced by different methods:

- Pigments for absorption or reflection
- Newly-developed plastic films
- Moveable screens
- Special shading material
- Conventional whitening with special pigments

The NIR radiation can be absorbed or reflected by the materials (Hemming and Waaijenberg 2001; Von Elsner 2005a, b; Waaijenberg 2006; Lopez-Marin et al. 2008).

NIR-absorbing material can have influence on the inside temperature, because the cladding material is warmed up by absorption and emits itself one part of heat radiation to the outside and one part to inside. The latter one contributes to increasing inside temperature. NIR-reflecting material is desirable (Hemming et al. 2006a; Strauch 1985a).

Permanent filtering material may be unsuitable for unheated greenhouses in Mediterranean climate, where temperatures decrease in winter and in heated greenhouses as well. A permanent NIR-reflecting material increases the energy requirement for heating, and lowers the mean temperature in unheated greenhouses in winter. Lower inside temperatures need less ventilation, and thereby affect the CO_2 concentration negatively. Moveable NIR-reflecting material may be more suitable for subtropical climates or the improvement of ventilation efficiency respectively (Kempkes et al. 2008; Kempkes et al. 2009).

Experiments with different methods of covering were carried out in a three-span greenhouse with separated spans of 240 m^2 each in south Spain (Lopez-Marin et al. 2008). The covering materials were

E1: Standard PE film, 200 μm thick
E2: Standard PE film, 200 μm thick, plus whitening
E3: NIR-blocking film, 200 μm thick

Table 7.11 Climate conditions and yield of sweet pepper in 240 m^2 greenhouse span under different cladding materials (Lopez-Marin et al. 2008)

Experiment	Solar radiation (W/m^2)	Temperature (13–16 h) (°C)	Total yield (kg/240 m^2)
E1	436.1	37.8	250.9
E2	337.6	34.4	271.2
E3	358.5	34.7	269.0

Table 7.11 shows the summarised results of the experiments. The NIR-blocking film had similar results to whitening, but was better than standard film.

The effects of NIR-reflecting material with different ratios of reflection for subtropical climates were simulated by Kempkes et al. (2009). The simulated NIR reflection was set to 0 (reference), 0.5 and 1.

The results for the winter period (December–March) were:

- The mean maximum temperature decreased by 3.6°C in comparison to no NIR reflection with 100% reflectivity and 1.4°C with 50% reflectivity.
- The mean minimum temperature also decreased by 1.1°C with 100% NIR reflectivity, caused by lower day temperature.
- The lower air temperature lowered the ventilation requirement, and thereby affected the CO_2 level. The CO_2 level decreased with increasing reflectivity.
- Transpiration was reduced.

The results for the summer period (April–July), where a whitewash that had 40% reflectivity of NIR and PAR was the reference:

- The overall transmittance decreased with whitewash more than with NIR reflecting material of 50 and 100%. The mean maximum air temperature under white wash was lowest. It decreased by 2.5°C, 5.5°C, and 6.6°C from reference of 0% reflectivity to 50%, 100% reflectivity and whitewash respectively.
- The CO_2 concentration at daytime decreased below outside concentration, caused by less ventilation requirement.
- The transpiration decreased with increasing NIR reflection.

Shading paints with entrapped NIR-reflecting materials have been investigated for use in tropical greenhouses (von Elsner 2005b; Mutwiwa et al. 2008). Two greenhousesof the same type shown in Fig. 10.7, 20 by 10 m, 3.8 m gutter height, 6.4 m ridge height, 228 m^2 ventilation area, were covered with 200-μm PE film, and one of them additionally with a shading paint containing NIR-reflecting pigments (Mutwiwa et al. 2008). The experiments were carried out near Bangkok (Thailand) with tomatoes during dry season (November–March) and wet season (May–September). Table 7.12 shows the climate conditions during the experiments. The use of those NIR-reflecting pigments reduces heat load, solar radiation and PAR. The total yield of tomatoes was reduced by 16–17%, but the non-marketable yield by 9.9 and 15.6% in dry and wet season respectively. It may be more advantageous to design very effective ventilation systems with the relatively low solar radiation in tropical areas than to use NIR-reflecting materials.

Table 7.12 Climate conditions in tropical greenhouses covered with and without NIR reflecting shading paints

	Dry season temperature (°C)	Dry season global radiation (W/m^2)	Wet season temperature (°C)	Wet season global radiation (W/m^2)
Outside	27.8	373.3	31.5	430.8
Without pigments	28.5	206.7	31.3	237.1
With pigments	26.8	188.6	30.7	222.1

7.3.4 Diffuse Light

Larger vegetable crops such as tomato or cucumber with a high leaf area index receive much more light at the upper leaves than the leaves inside the canopy. More uniform light distribution over all parts of the plants can improve growth and yield. This can be realised by:

- Reflective material as ground cover, like white plastic film.
- Transformation of direct light into diffuse light by photoselective film or screens for example (see Sect. 7.3.2). Diffuse or scattered light can penetrate deeper into a plant canopy than direct light.

Diffusing covering materials can improve the production, in particular in periods when direct light is high in summer, but the total amount of PAR should not be decreased too much.

Theoretical calculations showed the following results (Hemming et al. 2006b; Hemming 2005):

- If 100% of incoming light can be transformed into diffuse light, the production can be increased by 5–6% per year compared to normal glass in the Netherlands.
- A covering material that transforms 80% of incoming light to diffuse light and reduces the PAR to 85% has no advantage.

7.3.5 Drop Condensation

The air humidity in greenhouses is increased by the transpiration of plants and evaporation of soil. Condensation on the inner surface of the greenhouse cladding material is a sink for dehumidification, and thus an important and necessary factor for climate control. The condensation behavior is very different on various cladding materials from drop condensation to film condensation. Drop condensation as well as dripping from the roof is a problem. Drops at the inner roof surface considerably reduce ight transmittance by total reflection of the incoming light. Dripping off the roof on the plant canopy increases the danger of infestation by diseases and reduces the quality through water spots.

Condensation is necessary, but drop condensation and dripping should be avoided by the following measures:

1. So called No-Drop material with film condensation behaviour instead of drop condensation.
2. Minimum roof inclination so that condensate can run off the cladding material.

Figure 7.9 shows the drop condensation at the horizontal roof area of a round-arched greenhouse. If wind causes the film to flutter, the drops drip down. Figure 7.10 shows the transition from drop condensation with dripping to steeper

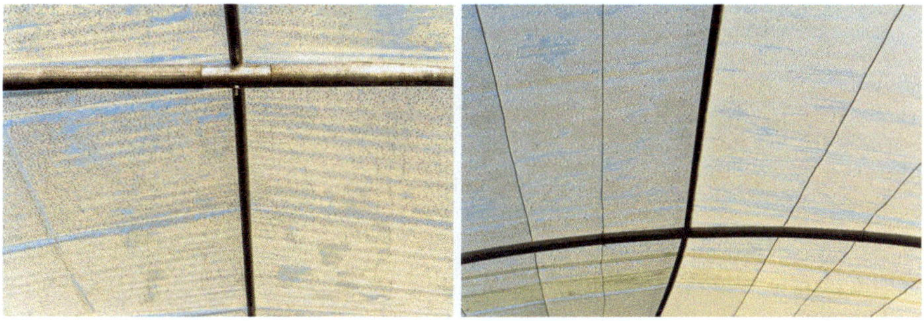

Fig. 7.9 Drop condensation in a round-arched tunnel greenhouse.

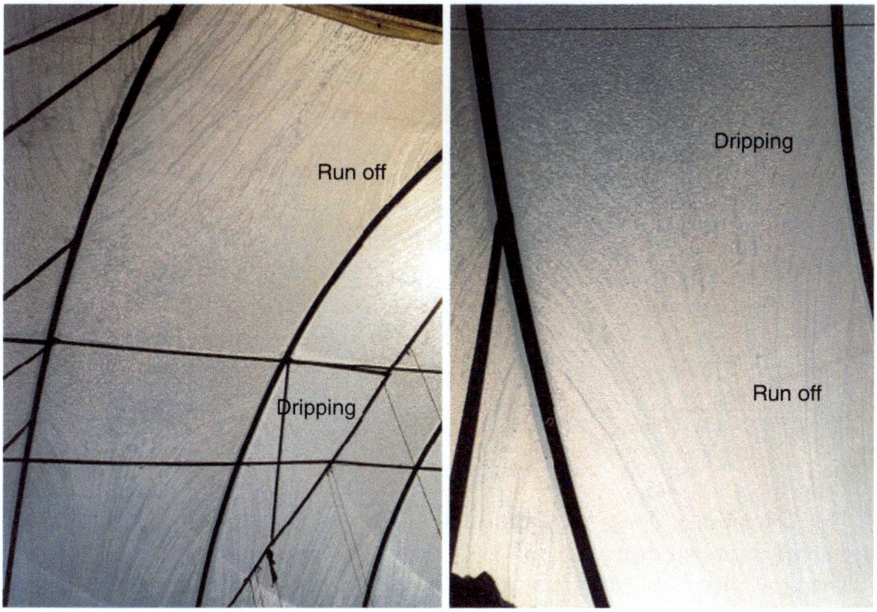

Fig. 7.10 Drop condensation and run-off condensation under a round-arched roof

Fig. 7.11 Dripping from wires stretched under the plastic film

inclinations with run-off of droplets, and Fig. 7.11 dripping from wires under the plastic film.

Dripping

The dripping of condensed water from the roof depends not only on the cladding material but strongly on the roof inclination for all materials. Figures 7.12 and 7.13 demonstrate the condensation behaviour of PE film and glass at various roof inclinations (Fähnrich et al. 1989). At 0° inclination, even glass and PE No-Drop film show drop condensation.

Gborczyk investigated the behaviour of condensation for different plastic films in a hot-box experiment (Gbiorczyk 2003a, b, Gbiorczyk and von Elsner 2004). The dripping and run-off condensate have been measured and described by a

Sliding factor = run-off condensate/evaporated water

Dripping factor = dripped condensate/evaporated water

At a roof inclination of 15°, the dripping factor for standard PE film was 16 and for PE No Drop film 3.

Critical roof angles were defined, where an increased dripping below and a run-off of condensed water above these angles is expected.

The critical inclination is:

Standard three-layer copolymer LDPE and EVA film	10°–14°
Commercial No-Drop film	7°–14°

The condensate does not slide and run off below those roof inclinations. It remains fixed and drops down.

The percentage of dripping at roof inclinations above 15° is much higher with standard PE film than with No-Drop film. The No-Drop film has good condensation behaviour without dripping at roof angles of 14–40°. To prevent dripping, the

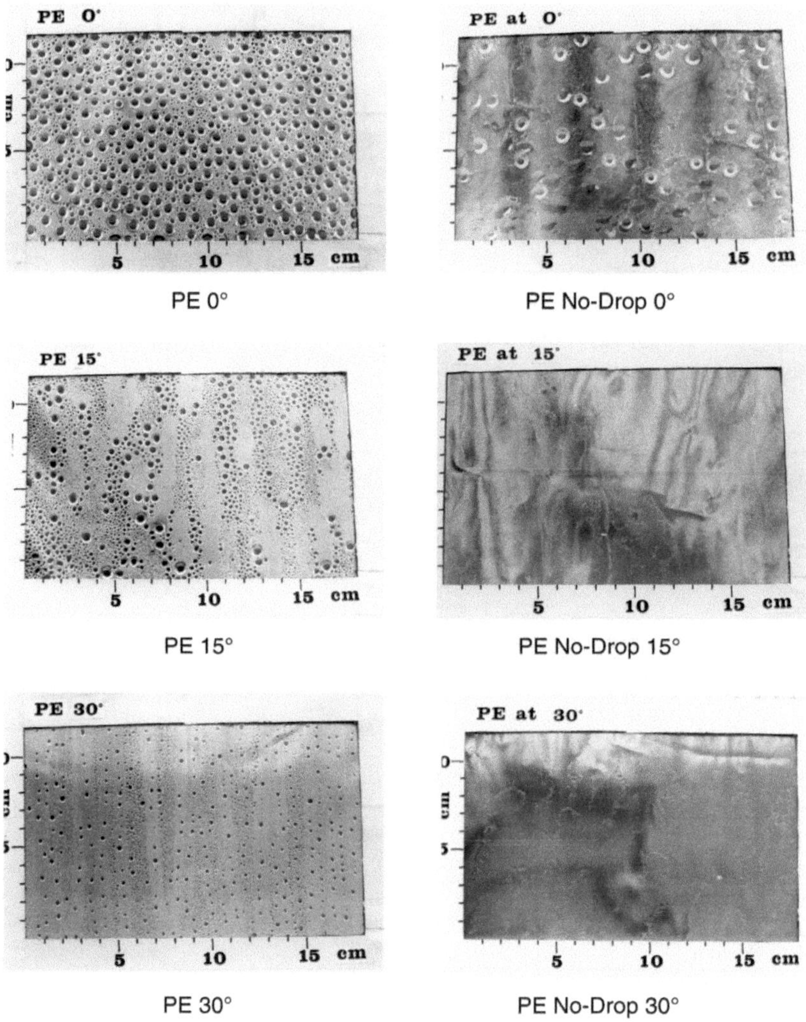

Fig. 7.12 Condensation behaviour of PE film at various roof inclinations (Fähnrich et al. 1989)

roof inclination must be higher than the critical inclinations, even for No-Drop films.

Light Transmittance

Table 7.13 shows the light transmittance of the materials with and without condensation (Fähnrich et al. 1989). The following results can be noticed:

- Heavy condensation reduces light transmittance, even on glass, due to flat long drops. But those drops do not fall down. They run off.
- Standard PE film shows a remarkable reduction of light transmittance due to micro- and large-drop condensation.

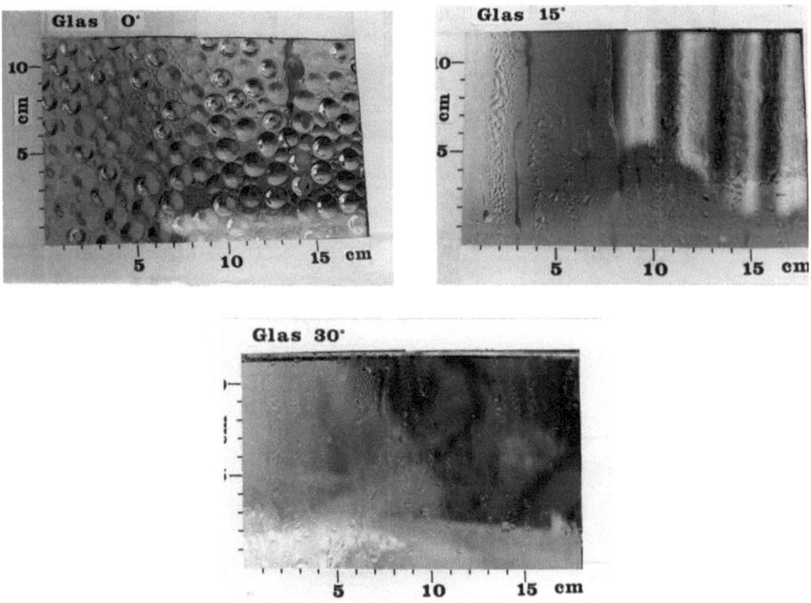

Fig. 7.13 Condensation behaviour of glass at various roof inclinations (Fähnrich et al. 1989). Even glass has drop condensation at 0° inclination

Table 7.13 Light transmittance of cladding materials measured in a hot box with and without condensation (Fähnrich et al. 1989)

Material	Roof inclination (°)	Light transmittance (%)	
		With condensation	Without condensation
Glass	15	90.8	82.7
	30	89.9	81.9
Standard PE film	15	90.8	82.7
	30	89.9	81.9
No-Drop PE film	15	87.1	88.1
	30	86.4	88.3

- Light transmittance is not reduced by No-Drop film.
- Light transmittance is much less reduced on No-Drop double acrylic sheets than on standard double acrylic sheets.

Further experimental and theoretical research work has been done concerning the influence of condensation on light transmittance of plastic films, reviewed by Gbiorczyk (2003a, b). She reported the results for direct irradiation, for diffuse irradiation, and for natural light in small tunnels and greenhouses.

Table 7.14 shows some of the results for the change of transmittance in greenhouses due to condensation.

Table 7.14 Reduction of light transmittance of natural light due to condensation (Gbiorczyk 2003a, b; Schultz 1997)

Author	Experiment (1) or simulation (2)	Measurement	PE standard (%)	PE No-Drop (%)
Lieftink and van Osten (1986)	(1)	Greenhouse	−1.7	Light increase
		Diffuse light		
Pearson et al. (1995)	(1)	Horizontal	−13.4	−3.1
Geoola and Peiper (1994)	(1)	Small tunnel	−8.0	−1.9
Pollet (2002)	(2)	Greenhouse overcast sky	−13 to 20	No change
Pollet (2002)	(2)	Greenhouse diffuse light	−8 to −21	No change

Jaffrin et al. (1994) measured the light transmittance with and without condensation in multi-span greenhouses covered with double-inflated plastic film

Film	Transmittance (%)
Standard PE double-inflated	
Without condensation	61
With condensation	52–53
No-Drop double-inflated	
Without condensation	57–60
With condensation	59–60

Weimann (1986) measured the light transmission in a two-span plastic-film greenhouse covered with double-inflated PE film and double No-Drop film

	Transmittance (%)	
	PE standard	PE No-Drop
November	54	59
December	53	59
January	52	56
February	50	51

Schultz (1997) measured the light transmittance in small plastic tunnels covered with improved plastic films. The mean transmittance over the measurement period in summer was

Film	Transmittance (%)	
	Dry	Condensation
Standard PE	71	62.4
PE No-Drop 1	75.5	79.8
PE No-Drop 2	69.8	76

Chapter 8
Greenhouse Components, Mounting, Installation and Maintenance

When the farmer has chosen a greenhouse structure, he has to observe the mounting and installation by the manufacturer very carefully. He should look around and check all components and connections of the structure, the fastening of the cladding material, the tightness of ventilators, doors and screens as well as the climate control systems. Even if the structure in principle fulfils the requirements and specifications, mistakes during the mounting can weaken the whole structure and can make the investment useless.

After the structure is ready for operation, the farmer has to take responsibility for continuous maintenance, so that the greenhouse can be operated as long as possible.

Examples of how to do and how not to do will be shown.

8.1 Foundation

The concrete foundation of a greenhouse should fulfil the following requirements:

1. It should safely sustain and transmit the loads of the greenhouse to the ground. These are pressure forces as well as uplift forces by wind suction.
2. The footing of the foundation should rest on undisturbed soil at a depth of about 500–600 mm below the ground surface. It is not necessary to build continuous foundations below the side walls and gables, but only stable concrete point foundations below the vertical stanchions.

The European standard for greenhouses EN 13031-1 (2001) gives some instructions for concrete point foundations and the possible tolerances (Fig. 8.1):

"The position of the prefabricated foundation block within the foundation hole shall be such that:

- Its centre lies within a circle with radius equal to D/5 or 100 mm, whichever is less, of the centre of the foundation hole.
- The distance between the face of the pile and the face of the foundation hole is at least 50 mm or D/8, whichever is larger".

C. von Zabeltitz, *Integrated Greenhouse Systems for Mild Climates*,
DOI 10.1007/978-3-642-14582-7_8, © Springer-Verlag Berlin Heidelberg 2011

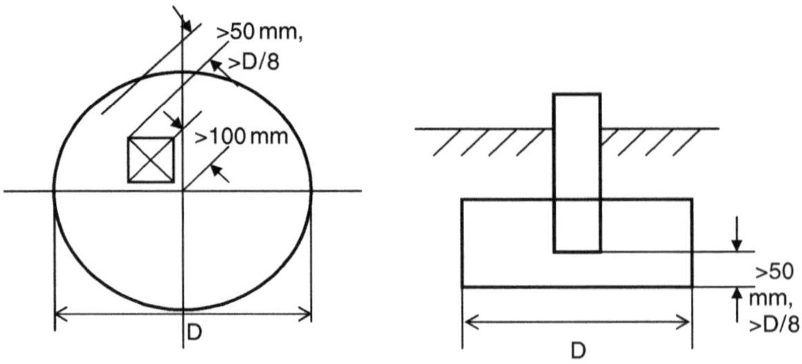

Fig. 8.1 Concrete foundation according to European standard EN 13031-1 (2001)

The concrete point foundation consists of the concrete base with the diameter D at a depth of 500–800 mm and the concrete pile as holder for the stanchions.

The Indian Standards IS 14462:1997 and ASAE EP460 DEC01 give average diameters of point foundations for greenhouses which can be taken also for sub-tropical and tropical conditions.

Greenhouse span (m)	Distance of stanchions (m)		
	2.4	3.0	3.7
	Diameter of foundation (mm)		
6 m	300	300	300
8.5 m	300	380	380
9.5 m	300	380	380

For windy areas, it is recommendable to install foundations with a diameter up to 500 mm.

Low-cost greenhouses often do not have concrete foundations. The steel pipes or wooden stanchions are very often inserted directly into the soil. This method is dangerous, because uplift wind forces can damage greenhouses very easily. This happens very often with round-arched tunnel greenhouses.

Figure 8.2 shows some well-designed concrete foundations (also see Figs. 8.3–8.5).

8.2 Connections and Clamps

All steel components of the greenhouse structure should be connected by screws or clamps. Welding is not recommended after galvanising the components.

The stable connection of the steel components by clamps is very important for the wind resistance of the greenhouse structure.

The clamps must not slide on the tubes, but have to be tightened firmly. After the mounting has been finished, one has to check whether all clamps, screws and bolts are screwed and fixed tightly.

Some example of how to do and how not to do are shown in the following figures:

Fig. 8.2 Some well-designed concrete foundations. The vertical columns should be inserted on or in a pipe which is part of the concrete foundation block. The columns should be fixed by screws or bolts and not welded

8.2.1 How to Do and How Not to Do the Clamp Connections

(See Figs. 8.6–8.10)

Fig. 8.3 Vertical posts inserted directly in the foundation are not recommendable

Fig. 8.4 Even timber stanchions should be inserted in concrete foundations to stabilise the structure

8.2.1.1 Welding

All galvanised steel components of the structure should be connected by clamps, screws or bolts as far as possible and not welded. Welding destroys the protective galvanised surface, and the steel component starts to get very rusty quickly. If

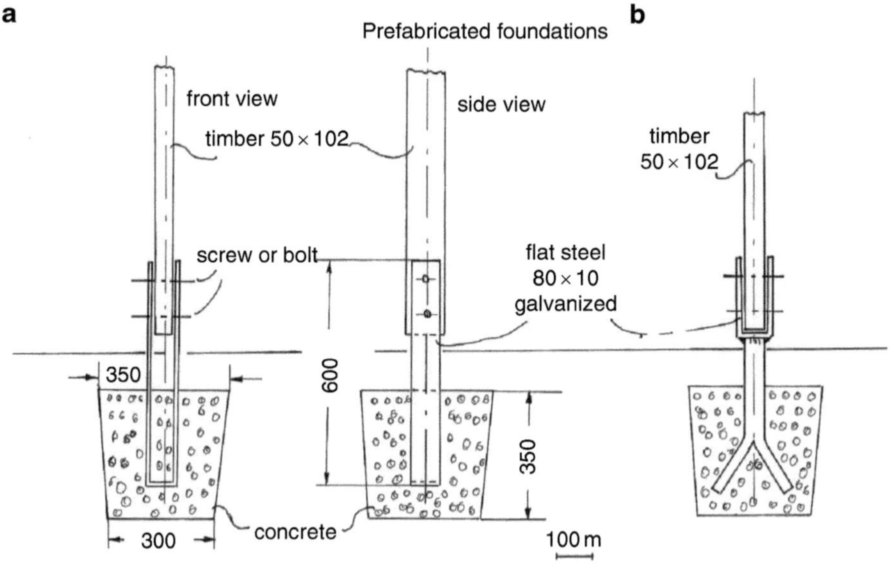

Fig. 8.5 Prefabricated foundations can be used for timber structures, for example. The timber stanchion should not be inserted directly into the soil or into the concrete footing, but should be fastened in a steel or wooden holder, which is inserted in the concrete footing. In that way, rotting of the structure can be avoided

Fig. 8.6 Tight clamps for steel components

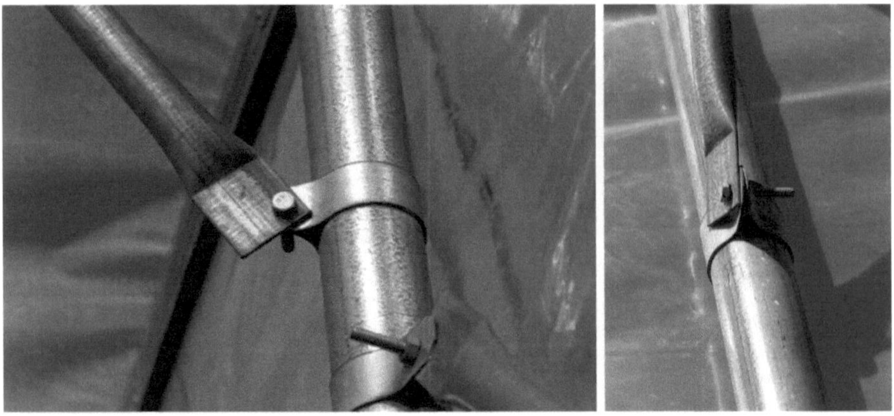

Fig. 8.7 The clamps are screwed tightly together, but the borehole in the brace is near the edge, so that the brace cannot be fastened firmly to the clamp. It is better to bevel the brace and to bore the hole in the middle

welding is unavoidable the welding, must be painted carefully. Otherwise, rust occurs immediately (Fig. 8.11).

8.2.1.2 Gutters

See Figs. 8.12 and 8.13.

8.3 Fastening and Stretching of the Cladding Material

Plastic film as cladding material on greenhouses has to be fastened to the structure and stretched tightly. The film must not be able to flutter due to wind forces. If the film starts to flutter, it will be destroyed quickly.

The plastic film has to be changed every 2–4 years, depending on the quality. This work is costly, and should be done quickly. That means the film should be fastened by detachable devices and not by nailing. Nailing is still used in several countries, in particular with wooden structures. If there is no other possibility, the plastic film should be fixed by screwing two laths together with the plastic film in between (Fig. 8.14).

Plastic film can be fastened by fastening devices made of plastic, aluminium, or steel (Fig. 8.15) or by rolling it up on a tube (Fig. 8.19). The film has to be fastened in a longitudinal direction at gutters and side wall as well as at the roof and gable ends (Figs. 8.16–8.23).

Fig. 8.8 The connectors are absolutely not acceptable. This is not only a matter of the design, but is caused by the mounting at the farm. The workmen worked badly

After having been fastened to the structure, the films have to be stretched tightly to prevent fluttering by wind forces. This can be done by the following methods:

- Fastening devices
- Rolling up the film on a steel pipe in the gutter and at the side wall
- Stretching additional plastic ropes over the structure (Fig. 5.20)
- Inflated double film

Double-inflated film is a very favourable method to stretch the film on the structure, with good resistance to wind and snow. Two plastic films will be fastened airtight on all sides and inflated by small fans with a pressure of 40–50 Pascal (4–5 mm water column). Double-inflated film can save 30–40% of heat in case of heating. The small fan can inflate side wall and roof elements at the same time. It

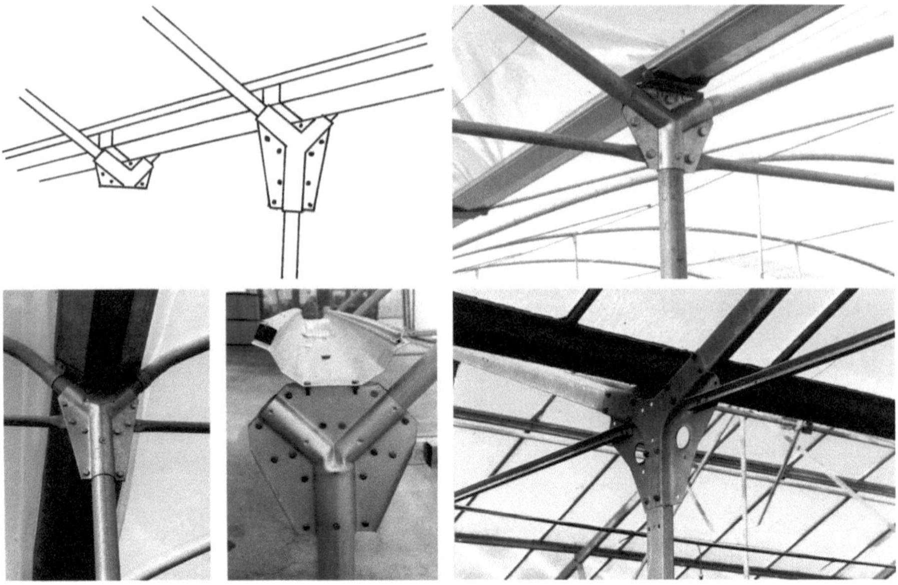

Fig. 8.9 Connectors for stanchions and roof pipes. The connectors are pressed or punched and screwed together

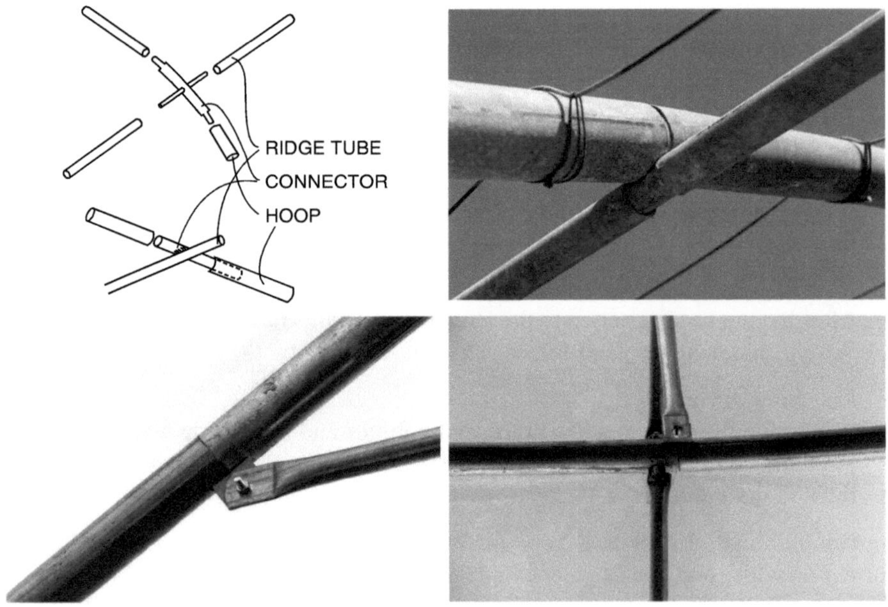

Fig. 8.10 The crossing of ridge tubes and hoops can be connected by special connectors inserted into the tubes, and screwed tightly

Fig. 8.11 Galvanised steel components welded together (see Sect. 8.2.1.1)

Fig. 8.12 All parts of the gutters have to be connected tightly by rubber tapes or by silicon to avoid dripping of rainwater

Fig. 8.13 Leaky gutter connection, bare holes or rusted-through holes cause water penetration into the greenhouse, flooding, and destruction of the crop

Fig. 8.14 Nailing the plastic film onto the structure is time-consuming and not recommendable

is very important to draw the air from outside. If humid air is drawn from inside, drop condensation occurs on the colder outside film. Light transmittance will be reduced remarkably, and algae will grow (Figs. 8.24–8.25). For protection of film see Figs. 8.26–8.27.

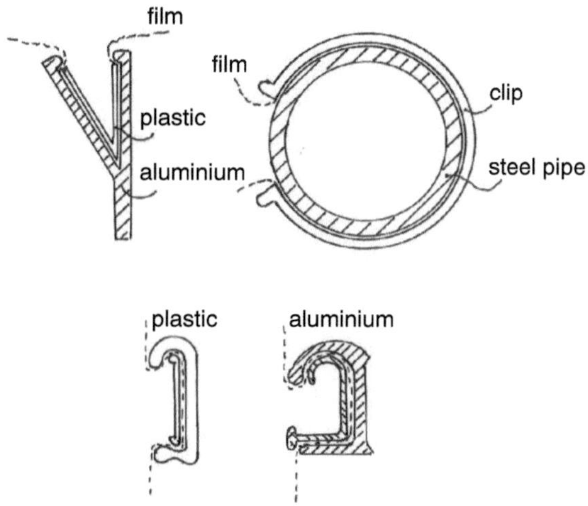

Fig. 8.15 Examples of fastening devices for plastic film

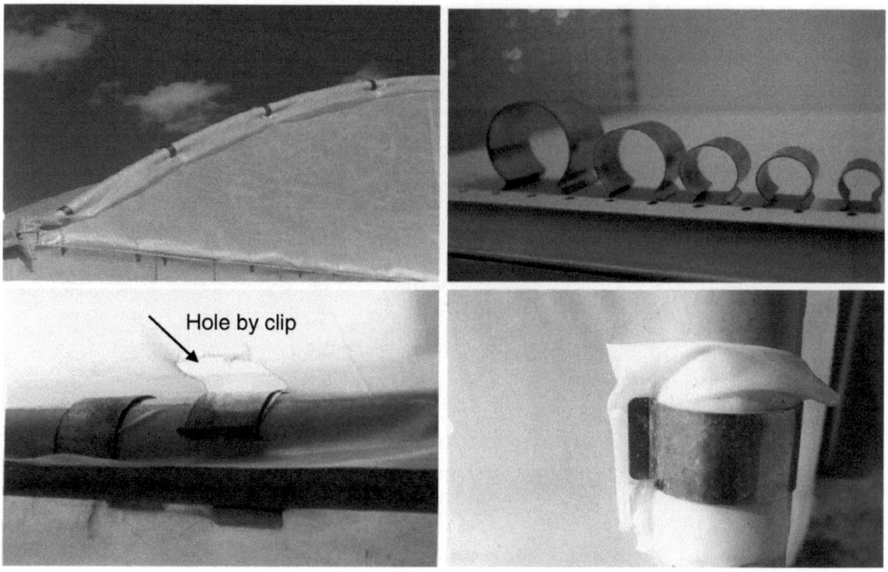

Fig. 8.16 A simple solution to fastening the film by commercial clips of spring steel. There is a danger that the plastic film can be destroyed by sharp edges of the clip. A tape of plastic should be put between the clip and the covering plastic film to protect the cladding film from destruction

8.4 Leaks

Leaks in the greenhouse structure must be avoided, wherever they occur at doors, ventilation openings, plastic-film fastenings, etc., for the following reasons:

Fig. 8.17 Prefabricated plastic clips

Fig. 8.18 Self-made clips for fastening of film or nets. One can use pieces of plastic water tubes which are cut open in a longitudinal direction and put over the construction component with the film or net in between, if no commercial fastening clips are available. This can be done even with timber profiles

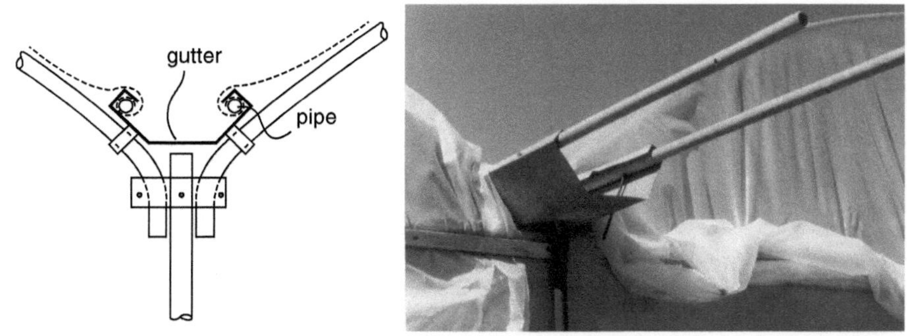

Fig. 8.19 Fastening and stretching of plastic film by rolling it up on a tube in the gutter. Either the gutter is formed especially for this reason, or a special holder can be fixed in the gutter. The sharp edges of the gutter should be protected by small plastic profiles to prevent cutting of the film. The advantage of this method is the fastening and stretching of the film at the same time

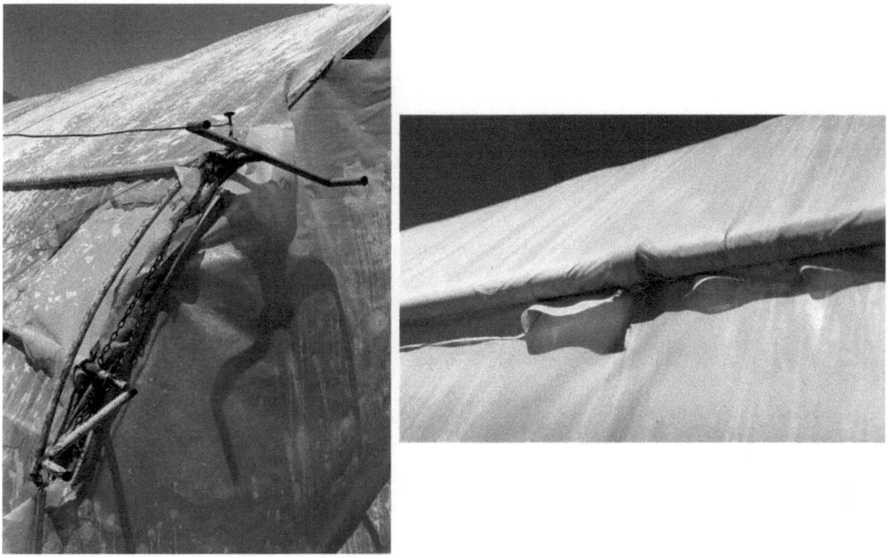

Fig. 8.20 One can fasten and stretch the plastic film from the roof side and side wall by rolling them up on one tube only. This method is a very elegant one

1. Solar energy will be stored in the daytime and will keep the air temperature in unheated greenhouses some degrees above outside temperature at night. If the warm air can escape through leaks, the temperature drops very quickly in the evening.
2. If there are holes in the plastic film caused by installation of the fastening clips, those holes are the starting point for damage to the plastic film by wind forces.

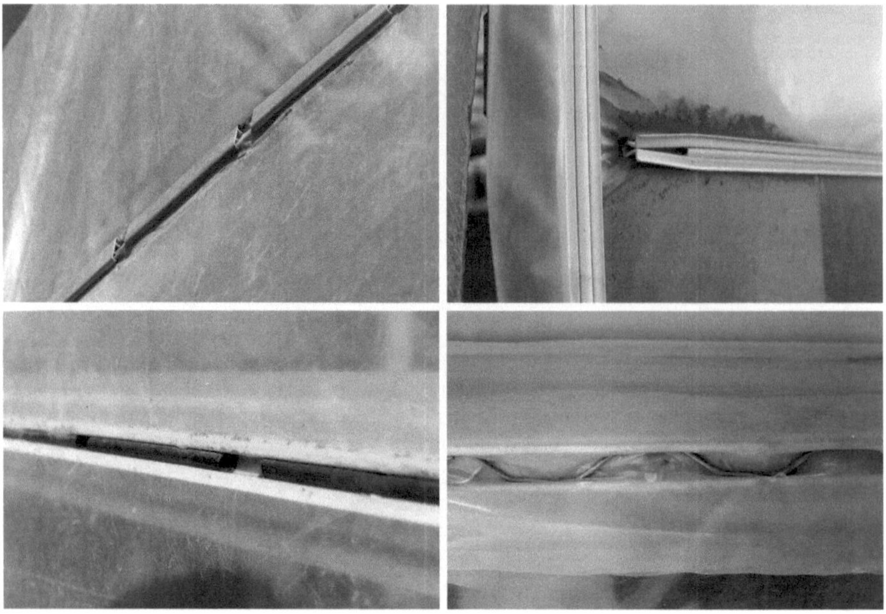

Fig. 8.21 A special method for fastening the film is to push the film into steel profiles using plastic profiles, spring steel or even rubber or plastic tubes

Fig. 8.22 Sometimes the plastic film is stretched across the structure with overlapping at the film ends. The possible fluttering of the film edges by wind is a danger. The film has to be fixed very tightly, for example by ropes

3. Leaks in the structure, vents and insect screens are not permissible when integrated production and protection (IPP) is used, with useful insects inside the greenhouse and when pest insects must be kept out.
4. Leaks in gutters and cladding material cause rainwater penetration, crop flooding and disease infestation (Fig. 8.28).

Fig. 8.23 Wires are fastened in a longitudinal direction at the roof components, very often to stabilise the plastic film, but the plastic film rubs on the wire and those points are the beginning of early destruction. Therefore, wires should be avoided if possible, even to prevent dripping from the roof (see Sect. 7.3.5)

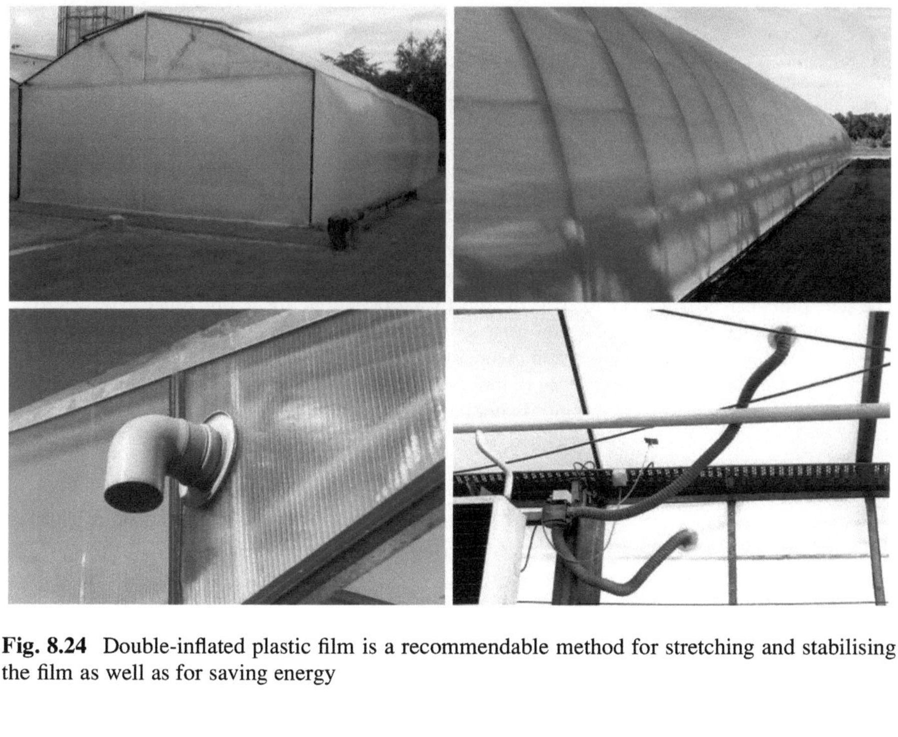

Fig. 8.24 Double-inflated plastic film is a recommendable method for stretching and stabilising the film as well as for saving energy

Fig. 8.25 Double-inflated film can also be used to repair and to use old structures, for example old glass houses. The old cladding material will be removed and double-inflated film installed

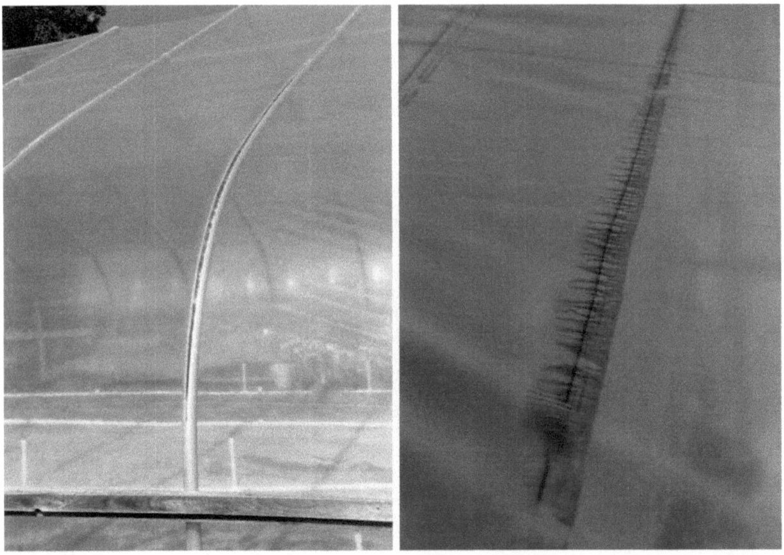

Fig. 8.26 Direct contact of the plastic film with steel tubes is a problem, in particular on hot days. The steel pipes can be heated up to 70°C by direct solar radiation. The plastic film can be destroyed earlier. In addition, the plastic film rubs on the steel tubes and the wires, and becomes black. This causes destruction of the film. It is recommendable to put some plastic tapes between film and steel pipes (Fig. 4.4). Another possibility is to paint the plastic white at the contact point to the steel pipes to prevent them from becoming too hot

If nails or screws are forgotten in the gutter during mounting of the structure, then the galvanised layer of the gutter will be destroyed and holes occur within a very short time so that rainwater can flow through. A corroded hole may be closed by a wooden stopper in one of the gutters (Fig. 8.29).

For leaks in insect screens, see Chap. 10.

Fig. 8.27 Protection of the plastic film above steel tubes and wires by plastic tapes

Fig. 8.28 Leaky doors where insects can penetrate and heat energy will be lost at night

8.5 Windbreaks

Wind action by high wind speed can cause damage to structure and plants, erode uncovered soil, and transport sand and soil through leakage into the greenhouse. The heat loss of a greenhouse also depends on the wind speed. Natural windbreaks by hedges and shelterbelts, as well as artificial wind screens, reduce the wind speed and thereby the impact of wind. Various rows of trees and hedges as well as fences

Fig. 8.29 Forgotten nails and screws cause rusting-through of the gutter. One corroded hole is closed by a wooden stopper

of plastic material can be used for protection from too high wind speeds. Windbreaks should not be completely airtight, but must have a certain permeability (Fig. 8.30) (Dierickx et al. 2001a; von Zabeltitz and Baudoin 1999).

The major factor influencing the wind speed reduction is the open area of the windbreak or screen. The open area is:

$A = \Sigma A_i / A_{tot}$ (%)

ΣA_i (m^2) = sum of single hole sizes.

A_{tot}(m^2) = Total area of the screen.

The air flow resistance R_c is:

$R_c = (v_w - v_r)/v_w$ (%)

v_w (m/s) = Freeflow wind speed without screen

v_r (m/s) = Reduced wind speed with screen.

Fig. 8.30 Hedges as windbreaks in desert regions

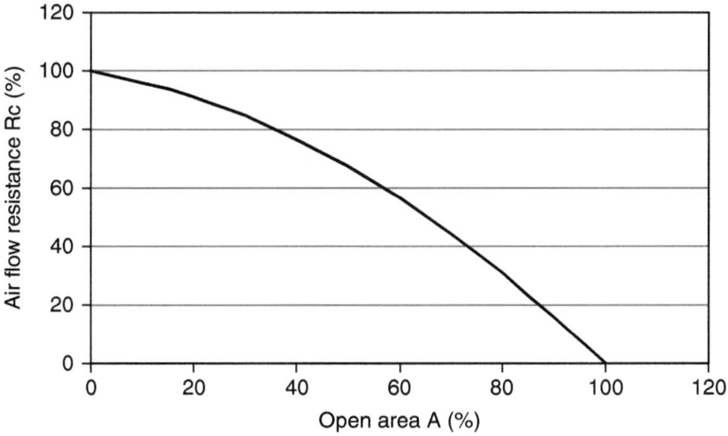

Fig. 8.31 Relationship between air flow resistance R_c and open area A (Dierickx 1998)

The relationship between R_c and A can be expressed by (Dierickx 1998), Fig. 8.31: $R_c = -0.0070 \times A^2 - 0.3 \times A + 100$.

Various geotextile materials were tested as windscreens with regard to their reduction of wind speed at different distances from the screen and at different heights above ground in wind-tunnel experiments (Dierickx et al. 2001a; Dierickx

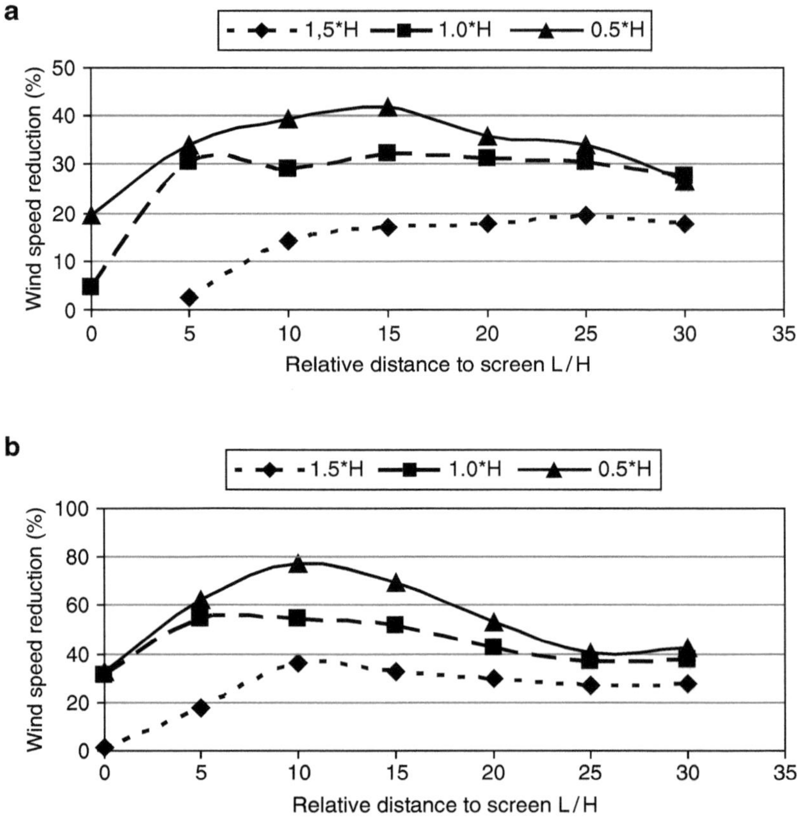

Fig. 8.32 Wind speed reduction (%) depending on relative distance to screen *L*/*H* for the heights 0.5*H*, 1.0*H*, and 1.5*H* above ground surface. (**a**) Open area $A = 57.5\%$ and air flow resistance $R_c = 59.6\%$. (**b**) $A = 49.1\%$, $R_c = 68.4\%$

et al. 2001b). Figure 8.32 shows the wind speed reduction (%) for two windscreen materials ($A = 57.5\%$ and $A = 49.1\%$) depending on the relative distance to the screen *L*/*H* and the height above ground, where *L* is the distance from the screen and *H* the height of the screen. Although the difference of the air resistance is very small, the windbreaking effect behind the screen B is much higher (Figs. 8.31–8.32).

Figure 8.33 shows the wind speed reduction for various tested materials between the curves for materials with an open area of $A = 24$ and 49.1% respectively and $R_c = 88.8$ and 68.4% respectively for the heights of 1.0*H* and 1.5*H*. All materials with an open area higher than 50% and a flow reduction >68% create wind speed reductions of more than 30% in relative distance to screen of 10*H*–20*H*, even at a height of 1.5 times the screen height. The open area of the screen should not be lower than 20% (R_c about 90%), because turbulent flow stream will be produced and thereby insufficient local wind speed effects.

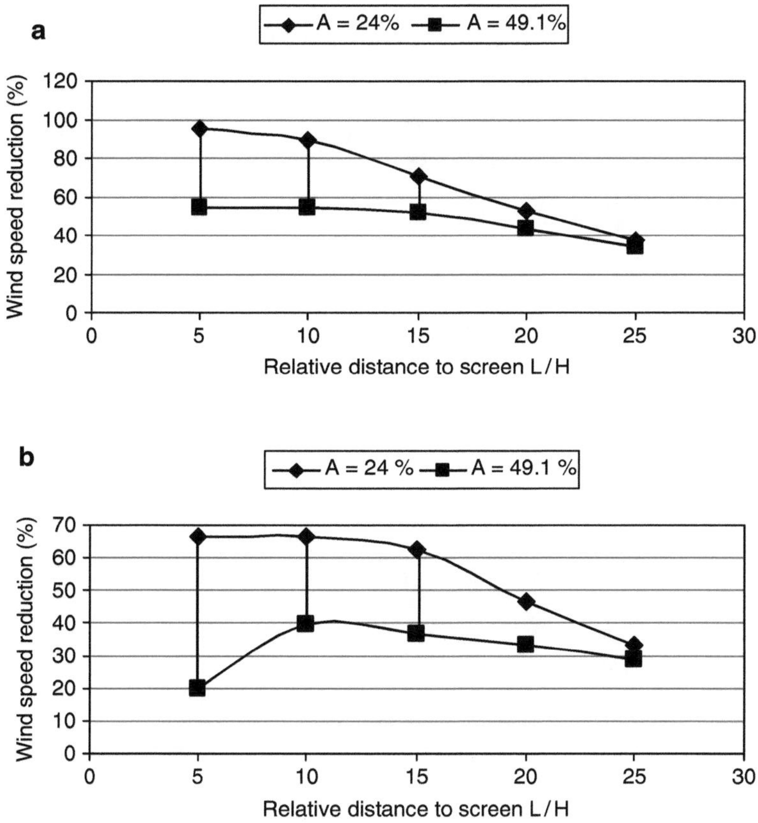

Fig. 8.33 Wind speed reduction for different material between the two curves with $A = 24\%$, $R_c = 88.8\%$ and $A = 49.1\%$, $R_c = 68.4\%$. (**a**) Height above ground 1.0H. (**b**) Height above ground 1.5H

A screen of 2.5 m height can theoretically protect a greenhouse of about 4 m height at a distance of 25–50 m from the screen.

The wind direction is not always perpendicular to the windscreen. Therefore, it may be important to know the effects on wind speed reduction, if the windscreen has an oblique position to the wind direction. Wind-tunnel experiments were carried out with screens of different positions to wind direction and different open areas (Dierickx et al. 2002).

Figure 8.34 shows the wind speed reduction at different heights and distances from the screen for a woven plastic material that has an open area $A = 62\%$ and an air flow resistance $R_c = 53.8\%$. The wind-flow angles oblique to the screen are 90° (perpendicular), 60° and 30°.

Wind screens are most effective when their position is perpendicular to the wind direction (90°). The wind speed reduction decreases with increasing angle oblique to the wind direction. Even an acceleration of wind speed may occur in some

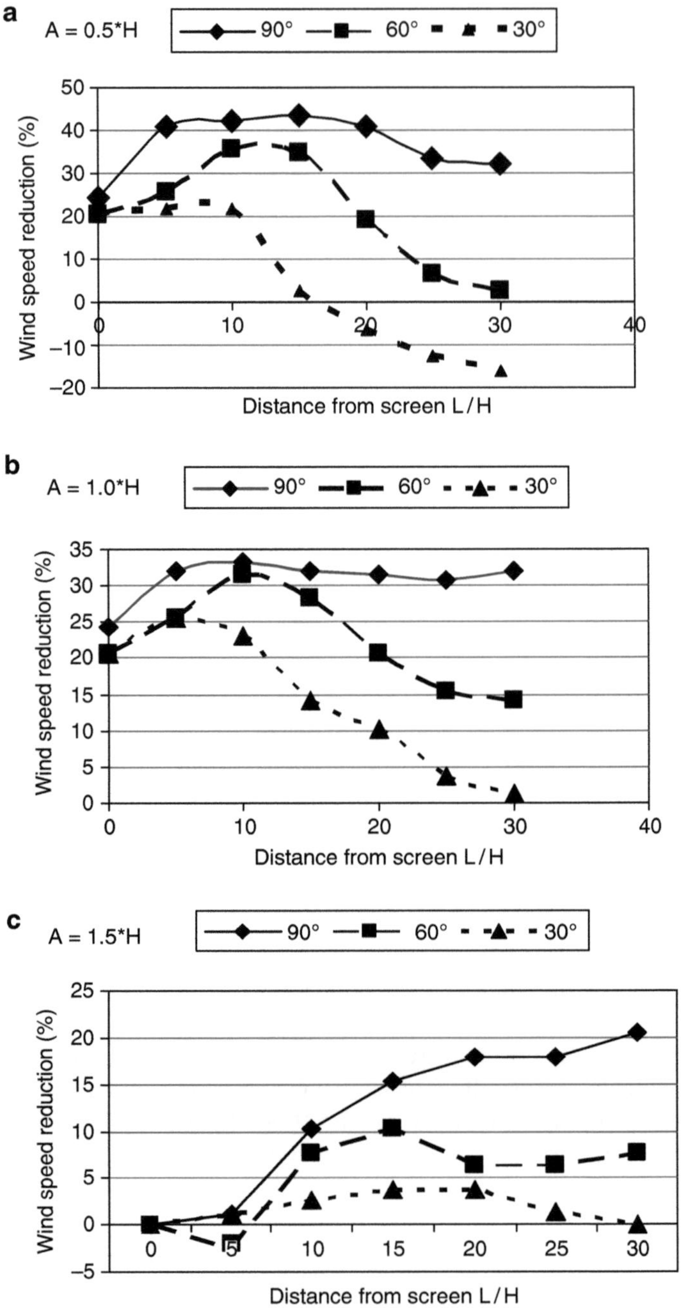

Fig. 8.34 Wind speed reduction for a plastic material, $A = 6.2\%$, $R_c = 53.8\%$, with different inclinations of wind direction, 90°, 60°, 30°, to the screen. (**a**) Height above ground 0.5*H*. (**b**) Height above ground 1.0*H*. (**c**) Height above ground 1.5*H*

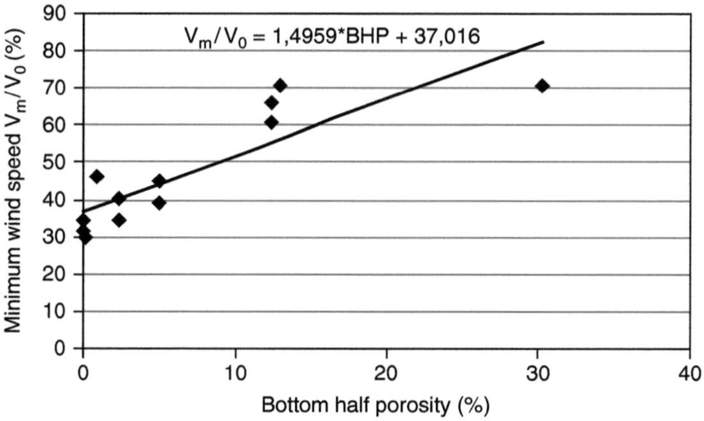

Fig. 8.35 Relationship between mean minimum wind speed v_m/v_0 and bottom-half porosity BHP (Loeffler et al. 1992)

positions of wind screen to wind direction (30°, $L/H > 15$). Wind screens with an open area >60% are more effective in oblique wind.

A rough surface on the windward side of the windbreak is less effective for wind reduction than a smooth surface, because of the formation of turbulent flows (Dierickx et al. 2003).

Inclined wind screens, particularly those with an inclination of 30° to the vertical position, are comparable to wedge-shaped natural windbreaks. If the width of the inclined screen is the same as for the vertical screen, the vertical screen is more effective, because of the smaller vertical height of the inclined screen. Inclined screens with the same vertical height as their vertical position will be more efficient than the vertical ones, but the screen width is larger, and as a result more costly. For natural windbreaks this phenomenon may be important (Dierickx et al. 2003).

If the open area of the wind screen is not evenly spread over the screen height, the wind speed reduction may vary. Wilson (1987) reported that the difference in wind speed reduction is nearly the same beyond a distance of $L = 7H$ for a windscreen of 50% open area and a windscreen which is more dense or more open near the ground surface.

Natural shelterbelts or hedges can be sufficient windbreaks, but the shading effect for greenhouses has to be considered. The effectiveness of various shelterbelts has been measured by Torita and Satov (2007) (Table 8.1).

Other natural shelterbelts have been evaluated by Loeffler et al. (1992) (Table 8.2). They measured optical porosity by evaluating black and white photographic silhouettes and using a digitizing technique (Kenney 1987). The full-height porosity from ground surface to the top of the trees, as well as the bottom-half porosity of the lower half of the tree height, were quantified. The measured values of v_m/v_0 and L_m are given in Table 8.2.

The wind speed reduction behind the pines was much less than behind the cedar and spruce windbreaks, due to large gaps in the pine wind break.

Table 8.1 Dimensions of some shelterbelts in the field site and wind speed reduction values

Shelterbelt	Height H (m)	Width (m)	Height to crown base (m)	Tree density (trees/ha)	v_m/v_0 (%)	L_m (m)	L_{70} (m)
Spruces	11	37	2	1,500	35.4	4.1H	9.2H
Broad-leaved wood with leaves	13	37	7	1,900	30.4	2.7H	8.3H
Broad-leaved wood without leaves	13	37	7	1,900	66.2	5.2H	2.6H
Spruces, pines, birches	10.7	30	3.7	1,500	26.7	4.2H	13.6H
Spruces, birches, pines	11	50	2	1,800	20.8	2.7H	10H
Spruces, birches	12	40	2	2,000	24.2	0.42H	7.8H
Birches	6	5	2	2,200	64.4	3.7H	10.9H

v_m (m/s) = mean minimum leeward wind speed
v_0 (m/s) = mean undisturbed windward wind speed
L_m (m) = Distance of mean minimum wind speed from the edge of shelterbelt
L_{70} (m) = Distance from shelterbelt where the wind speed does not exceed 70% of v_0

Table 8.2 Mean minimum wind speed v_m/v_0 and distance L_m of mean minimum wind speed behind the windbreaks at different sites. Increasing v_m/v_0 means decreasing wind speed reduction

Trees	Full-height porosity (%)	Bottom-half porosity (%)	v_m/v_0 (%)	L_m (m)
Norway spruce (five sites)	16.5–21.4	0.8–5	39.3–45	4H
Cedar (three sites)	10.6–20.4	0–0.1	31.5–34.6	3H–4H
Scots pine (four sites)	18.1–36.9	12.4–13	60.6–70.8	3H–4H

The mean minimum wind speed v_m/v_0 could be significantly related to the optical bottom-half porosity BHP of the windbreaks, (Fig. 8.35).

The relationship between windbreak structures and their function has been summarised as follows (Heisler and Dewalle 1988):

- The horizontal extent of wind protection is generally proportional to windbreak height.
- The wind speed reduction is related to the open area of the windbreak.
- Very dense barriers are less effective than medium porous barriers for wind speed reduction of 10–30 % at larger distances.
- Height growth of a natural windbreak may be more important than density when areas as large as possible have to be protected. But shading should be considered for the wind protection of greenhouses.
- Natural barriers with width less than height and a steep side produce a larger wind reduction over a greater distance than very wide windbreaks or streamlined windbreaks in cross-section.
- Tree windbreaks lose less effectiveness in oblique winds than thin artificial wind screens
- Turbulent wind flow decreases with increasing open area of the windbreak.

Chapter 9
Ventilation

Ventilation is the exchange of air between the inside and outside atmosphere.

The purpose of greenhouse ventilation is:

- Exchange of carbon dioxide and oxygen
- Dissipation of surplus heat and temperature control
- Humidity control

Ventilation is a major factor of climate control for crop production with good yield and quality. Good ventilation efficiency is very important for the production of quality throughout the year.

There are two possibilities to ventilate greenhouses:

- Natural ventilation by opening ventilators at side walls, gables ridge or roof area.
- Forced ventilation by fans, which depends on electricity supply and cost for electricity.

Important criteria for ventilation systems are:

- The ventilation rate.
- The temperature difference inside to outside.
- The homogeneity of temperature distribution inside the greenhouse.
- The air velocity near and inside the crop canopy.

Air movement by ventilation has influence on the uniformity of greenhouse climate factors, and consequently influences crop growth and quality (Sase 2006).

9.1 Natural Ventilation

Natural ventilation means an air exchange through ventilator openings. Physical pressure differences are necessary for free air exchange and air movement. The pressure differences are caused by

C. von Zabeltitz, *Integrated Greenhouse Systems for Mild Climates*,
DOI 10.1007/978-3-642-14582-7_9, © Springer-Verlag Berlin Heidelberg 2011

- Temperature and humidity differences (stack effect)
- Wind influence on different surfaces of the greenhouse (wind effect)

If the wind is blowing against the greenhouse, different pressure and suction zones occur at the outside surface of the greenhouse, and the air exchange occurs due to those pressure differences. The position of the pressure and suction zones depend on the wind direction and roof shape. Small roof slopes have suction zones over the whole roof surface. Steeper roof slopes have suction zones on leeward and pressure zones on windward sides. Wind pressure zones for different roof surfaces, round-arched or pitched roofs, are given in the standards for wind loads, for example the European greenhouse standard EN 13031-1:2001.

The operation of natural ventilation by opening ventilators at side walls, gables and ridge or roof area is cheaper than forced ventilation by fans.

Ridge ventilation is necessary if the mean maximum outside temperature exceeds 27°C and if the total width of the multi span greenhouse unit exceeds 18 m, but ridge or roof ventilation is very costly (see Chap. 5).

Ventilation openings with insect screens reduce ventilation efficiency (see Chap. 10).

9.1.1 Air Exchange

Characteristics for air exchange are (von Zabeltitz 1986a, von Zabeltitz 1999):

1. The ventilation rate V_E [m^3/m^2 h] is the ventilation flow rate V_v exchanged in relation to the greenhouse floor area A_G.

$$V_E = V_V / A_G [m^3/(m^2 h)] \tag{9.1}$$

   V_V (m^3/h; m^3/s) Ventilation flow rate.
   A_G (m^2) Greenhouse floor area.
2. The air exchange number N [1/h] is the ventilation flow rate V_v exchanged in relation to the greenhouse volume V_G.

$$N = V_V / V_G \left[m^3/m^3 h = 1/h\right] \tag{9.2}$$

With the mean greenhouse height (h) given, the relation between the two characteristics can be calculated

$$V_E = h \times N \tag{9.3}$$

3. Another characteristic figure for evaluating the ventilation is ventilation efficiency as the relation of ventilation flow rate V_v to the area of the ventilation opening A_V (Bartzanas et al. 2004):

$$V_{eff} = V_V/A_V\left[m^3/(m^2h)\right] \quad \text{or} \quad V_{eff} = N/A_V\left[1/(m^2h)\right]. \tag{9.4}$$

9.1.2 Ventilation Opening Area

The question is: how large should the ratio of ventilator opening to greenhouse floor area be for sufficient ventilation efficiency?

The ventilation rate V_E necessary to keep a definite temperature difference inside outside can be calculated by an energy balance (von Zabeltitz 1986a, b; Bailey 1998; Willits 2006; ANSI/ASAE EP 406.4, Jan 2003) Neglecting the energy flux into the soil and for photosynthesis, the energy balance for a ventilated greenhouse can be written as

$$\tau q_O = u\frac{A_C}{A_G}\Delta T + V_E c_p \rho \Delta T + Ef\tau q_O. \tag{9.5}$$

τ (–) is transmittance of greenhouse
q_O (W/m_2) outside global radiation
c_p [(Wh)/(kg K)] specific heat of air
ρ (kg/m^3) density of air
ΔT (°C) temperature difference inside outside the greenhouse
u (W/m^2 K) overall heat transfer coefficient
A_C (m^2) greenhouse surface area
E (–) evaporation coefficient; ratio of energy used to evaporate water from the canopy to incoming solar energy.
f (–) area factor, area of greenhouse covered by crop.

$$\Delta T = \frac{\tau q_o(1 - Ef)}{V_E \rho c_p + u A_C/A_G} \, (°C) \tag{9.6}$$

$$V_E = \frac{\tau q_o(1 - Ef)}{c_p \rho \Delta T} - \frac{u A_C}{c_p \rho A_G}\left[m^3/(m^2h)\right]. \tag{9.7}$$

The *evaporation coefficient E* is difficult to calculate (Willits 2006). As long as the incoming air does not participate in the water evaporation from the canopy, E has values between 0 and 1. But if the incoming air in arid regions has very low humidity, then this dry air participates in the evapotranspiration and additional energy will be taken from the air, and E can have values >1. If $E > 1$, then transpiration requires more energy than the net radiation above the canopy can supply. The air temperature decreases from the inlet to the outlet in this case (Willits 2003, 2006).

As a first assessment for design purposes, the following values can be used if outside humidity is not too low:

$E = 0$ for empty greenhouses, $E = 0.5$ half cropped, $E = 0.8$–1.0 fully cropped greenhouse.

The area factor f can be set

$f = 0.8$ for vegetables on ground beds, $f = 0.8$ for cut flowers on ground beds, $f = 0.6$–0.8 for pot plants on tables.

Assumption for Fig. 9.1:

$q_O = 700$ W/m^2; $\tau = 0.6$ for single cladding; $c_p = 0.28$ Wh/kg; $\rho = 1.2$ kg/m^3; $A_C/A_G = 1.5$; $u = 7$ W/(m^2 K); $f = 0.8$.

One conclusion of Fig. 9.1 is: temperature differences of less than 3°C can be achieved with a ventilation rate of about 200 m^3/m^2 h in greenhouses with crop, whereas the temperature difference is more than 5°C in an empty greenhouse.

The ventilation flow rate can be calculated by the equation (Bailey 2000a, b):

$$V_V = G_{(\alpha)} A_V v_w (\mathrm{m}^3/\mathrm{s}) \tag{9.8}$$

A_V (m^2): Ventilator area, v_w (m/s): Wind speed.

$G_{(\alpha)}$: Non-dimensional number depending on dimensions of ventilator opening and opening angle. The non-dimensional parameter $G_{(\alpha)}$ has been used to characterise the air flow through a ventilator opening (Bot 1983; De Jong 1990; Bailey 2000a, b).

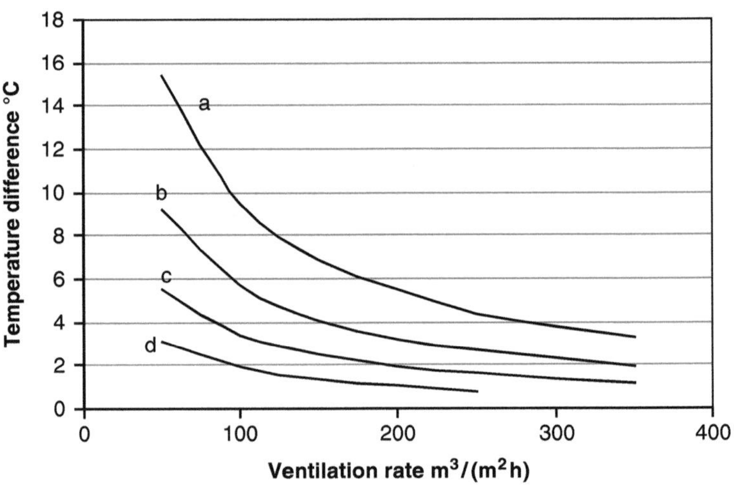

Fig. 9.1 Example of the temperature difference depending on ventilation flow rate. Curve a: $E = 0$; b: $E = 0.5$; c: $E = 0.8$; d: $E = 1$.

For flap ventilators with flap opening angle of 40°–50°, $G_{(\alpha)}$ has values of 0.08–0.1 (Boulard and Draoui 1995).

One can get the ratio of ventilator opening to greenhouse floor area by combining equations (9.1) and (9.8)

$$\frac{A_V}{A_G} = \frac{V_E}{G_{(\alpha)} v_w} \tag{9.9}$$

Example: $V_{Emin} = 200\ m^3/m^2h = 0.056\ m^3/m^2s$
$G_{(\alpha)} = 0.08–0.1$ for open ventilators (Bailey 2000a).
$v_w = 3$ m/s
$A_V/A_G = 0.23$

Another method to calculate the ventilation flow rate has been developed by using the Bernoulli equation. Boulard and Baille (1995) derived the following equation for the combined wind- and temperature-driven ventilation by roof ventilators (Boulard 2006; Perez-Parra et al. 2006; Bailey 2000a):

$$V_V = \frac{A_V}{2} C_d \sqrt{2g(h/4)\Delta T/T_m + C_W v_w^2} (m^3/s). \tag{9.10}$$

A_V (m_2) area of vent opening, g (m/s^2) acceleration of gravity, h (m) vertical distance between the centres of the regions of inflow and outflow, ΔT (°C) temperature difference, T_m (°C) mean outside temperature, v_w (m/s) wind speed, C_d (–) discharge coefficient, C_w (–) global wind pressure coefficient.

The first part below the radical sign represents the temperature-driven (stack effect), the second part the wind-driven ventilation rate.

The discharge coefficient C_d can be calculated (Bailey et al. 2003; Perez-Parra et al. 2004):

$$C_d = 1 \Big/ \sqrt{1.9 + 0.7 \exp\{-L_0/[32.5 H_0 \sin\alpha]\}} \tag{9.11}$$

where L_0 and H_0 are the length and width of the ventilation opening in the plane of surface (Fig. 9.2), and α is the angle of the flap opening.

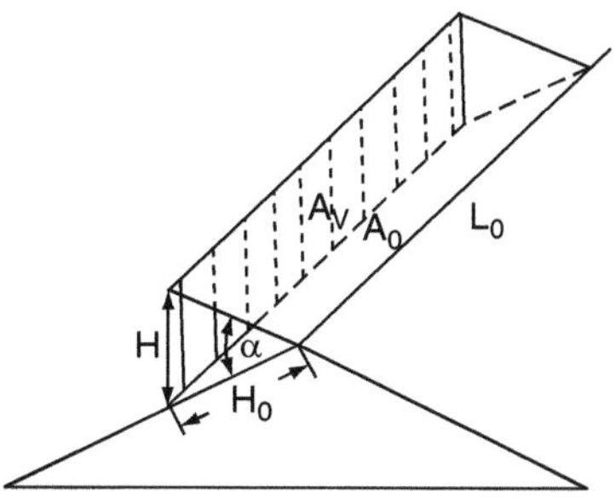

Fig. 9.2 Dimensions of flap ventilator openings

When a greenhouse has ventilation openings in the roof and side walls, the ventilation flow rate can be expressed by (Kittas et al. 1996; Perez-Parra et al. 2004):

$$V_V = C_d \left[\left(\frac{A_r A_s}{\sqrt{A_r^2 + A_s^2}} \right)^2 \left(2g \frac{T_i - T_o}{T_m} h' \right) + \left(\frac{A_r + A_s}{2} \right)^2 C_w v_w \right]^{0.5} \tag{9.12}$$

A_r and A_s (m^2) are the areas of the roof and side wall ventilators and h' (m) is the distance between the midpoints of the side wall and roof openings. Neglecting the temperature-driven ventilation effect, (9.12) has the same value as in (9.10) with $A_r + A_s = A_V$.

Temperature-driven ventilation can be significant with side wall and roof ventilation, and when there is a large internal/external temperature difference (Bailey 2000a). In multi-span greenhouses with only roof ventilation, the temperature-driven ventilation rate is smaller because of the small vertical distance h'.

The temperature-driven part can be ignored if:

- Wind speed $v_w > 2$ m/s (Papadakis et al. 1996),
- $v_w/\Delta T^{0.5} > 1$ (Perez-Parra et al. 2006, Boulard 2006, Kittas et al. 1996).

The ventilation flow rate can be calculated with the simplified assumption neglecting the temperature-driven ventilation

$$\frac{V_V}{A_G} = V_E = \frac{A_V}{2A_G} C_d v_w \sqrt{C_w} \ [m^3/(m^2 s)] \tag{9.13}$$

The factor 2 was introduced because half of the vent area is used to provide half inflow and half outflow. The discharge coefficient C_d and the wind pressure coefficient C_w have been quantified by several researchers and summarised by Bailey (2000b) and Boulard (2006).

The discharge coefficient C_d has more or less the same value for different vents. The overall wind pressure coefficient C_w depends on the wind speed, and decreases with wind speed. The coefficient is $C_w = 0.1$–0.11 for larger greenhouse areas. Perez-Parra et al. (2006) got lower C_w-values for Parral-type greenhouses with flap and roll-up ventilation (Tables 9.1 and 9.2).

The product $C_d \times C_w^{0.5}$ depends on wind speed also, and has values of 0.2–0.27 (Bailey 2000b) or 0.185–0.207 (Boulard and Baille 1995).

Table 9.1 Discharge coefficients C_d

Discharge coefficient	Conditions	Source
0.65–0.7	Greenhouse roof vents	Bot (1983)
0.65–0.75	Greenhouse roof vents	De Jong (1990)
0.644	Greenhouse continuous vents	Boulard and Baille (1995)

Table 9.2 Wind pressure coefficients C_w

Wind effect coefficient	Greenhouse area (m^2)	Source
0.10	416 (two-span)	Boulard and Baille (1995)
0.14	179 (one-span)	Kittas et al. (1995)
0.071	900 (tunnel)	Kittas et al. (1996)
0.13	416 (two-span)	Papadakis et al. (1996)
0.09	204 (four-span)	Baptista et al. (1999)
0.11	38,700 (60-span)	Bailey (2000b)

Combining the two equations (9.7) and (9.13) for the ventilation rate gives the possibility to calculate the ratio A_V/A_G for the ventilation area A_V:

$$\frac{A_V}{A_G} = \frac{2}{c_p \rho v_w C_d \sqrt{C_w}} \left[\frac{\tau q_O (1 - Ef)}{\Delta T} - \frac{A_C}{A_G} u \right] \tag{9.14}$$

With the values used for Fig. 9.1 and $E = 0.5$, $C_d \times C_w{}^{0.5} = 0.22$ derived by Bailey (2000b), $v_w = 3$ m/s $= 10{,}800$ m/h, $\Delta T = 3$°C the ratio A_V/A_G becomes

$A_V/A_G = 0.184$ for $\Delta T = 3$°C; $A_V/A_G = 0.289$ for $\Delta T = 2$°C.

That means a ventilator area opening of 18–29% should be designed for sufficient ventilation.

Practical experience suggests a ventilator area of 18–25% related to greenhouse floor area.

The ANSI/ASAE EP 406.4 standard (Jan 2003) gives a figure of 15–25%.

Ventilators covered with insect screens have to be enlarged for sufficient ventilation (see Chap. 10).

The consequences are:

- The small openings in round-arched tunnels, which can be found in many countries, are not sufficient (Sect. 5.1).
- If round-arched tunnels are built, through side wall ventilators should be installed.
- Multi-span greenhouses with more than three spans should have roof ventilators.

9.1.3 Ventilation Efficiency

Not only the ventilation opening area is very important, but also the arrangement of the ventilation openings, roof or side wall, windward or leeward openings. The ventilation efficiency V_{eff} in relation to the type of the vents was investigated for a single-span arched greenhouse by experiments and calculation (Bartzanas et al. 2004). The measurements of the greenhouse were: floor area 8 by 20 m, eaves height 2.4 m, ridge height 4.1 m.

The following configurations and vent arrangements were investigated:

A. Roll-up side wall vent only. $A_V = 36\ m^2$; $A_V/A_G = 22.5\%$.
B. Flap side wall vents: $A_V = 27.5\ m^2$; $A_V/A_G = 17\%$.
C. Roof flap vent windward: $A_V = 18\ m^2$; $A_V/A_G = 11\%$.
D. Roof and roll-up side wall vents: $A_V = 54\ m^2$; $A_V/A_G = 33.75\%$.

The calculated values validated by experimental data for the vent efficiency and the converted data for the ventilation rate are:

Configuration	Vent efficiency	Ventilation rate converted with $A_G = 160\ m^2$
	V_V/A_V	V_V/A_G
A	540	122
B	468	80.4
C	252	28
D	576	194

The conclusion of these results are:

The vent efficiency of configurations A and D are not very different, although the combined side and roof vents (D) has the highest ventilation rate. The roll-up side wall is better than the flap side wall vent because of higher ventilation opening. Roof vent only is not sufficient, whereas the side wall roll-up vent with $A_V/A_G = 22.5$ may be sufficient for single-span greenhouses. The measured temperature homogeneity was the best with combined side wall and roof vent (D).

9.2 Types of Natural Ventilation Openings

9.2.1 Side Wall Ventilation

The most used vent openings are the roll-up and flap ventilation.

Roll-up ventilation at side walls has proved very efficient in many countries, if it is installed correctly with resistance to wind forces.

Figure 9.3 shows a roll-up ventilation system at a side wall. The plastic film is fixed at the gutter or eaves and rolled up on a steel tube. It is very important to fix the film exactly on the horizontal tube at the beginning, so that the tube remains horizontal during operation. The horizontal tube with the film has to fit tightly at the vertical stanchions, and must not be moved away from the structure by wind forces (Figs. 9.4–9.10).

To prevent the rolling-up tube and plastic film from moving to the side, ropes or steel rods have to be stretched close to the film from eaves to bottom. Steel ropes or wires should be protected by plastic to prevent the plastic film from being destroyed by friction.

At the gable ends of the greenhouse, a plastic film about 1.5–2 m wide should be stretched and fixed from eaves to bottom to avoid leakage. This overlapping of roll-up ventilation and fixed film is absolutely necessary.

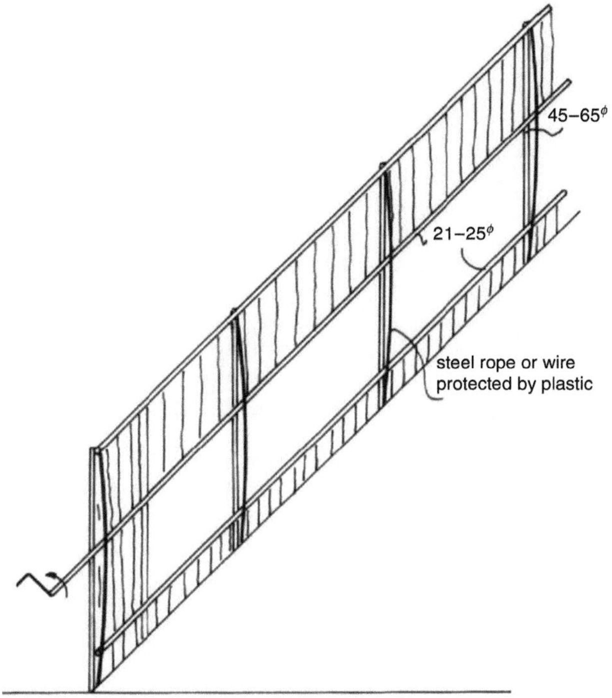

Fig. 9.3 Roll-up ventilation at side wall

Fig. 9.4 Very well-designed roll-up ventilator with sufficient overlapping at the gable end to prevent leaks and destruction by wind forces. The horizontal roll-up tubes are nearly straight at a length of about 30 m

9.2.2 Roof Ventilation

Roof ventilation should be included in the design if the mean maximum outside temperature exceeds 27°C and if the width of the greenhouse unit is wider than 18 m (Fig. 9.11).

Fig. 9.5 Although a plastic film is fixed over 1.5 m from eave to bottom, there is no overlapping with the roll-up vent, with the consequence of leaks, danger of destruction and cold air infiltration

Fig. 9.6 No overlapping and fixed plastic film at the gable end

9.2.2.1 Examples

See Figs. 9.12–9.19.

Fig. 9.7 The roll-up vents are moving and fluttering at the side wall due to wind forces, causing leakage and danger of destruction. The roll-up vents are not stretched close enough to the construction by ropes or steel rods

Fig. 9.8 Ropes and plastic-covered steel ropes as holder for the rollup vent against the structure

9.2.2.2 Greenhouses with Flap Ventilation

The mean wind speed varies in different countries and climates. Continental climates have lower mean wind speeds than maritime climates. The ratio of ventilator

Fig. 9.9 The vertical steel tube holder for the rollup vent is not close enough to the roll-up vent, and causes leakage

Fig. 9.10 Roll-up vents at side walls, 40–60 m long, where the plastic film of the roll-up side wall vent as well as the plastic film of the roof are fixed by rolling them up on one tube only. This method is very recommendable, because the roof plastic film can be stretched in summer

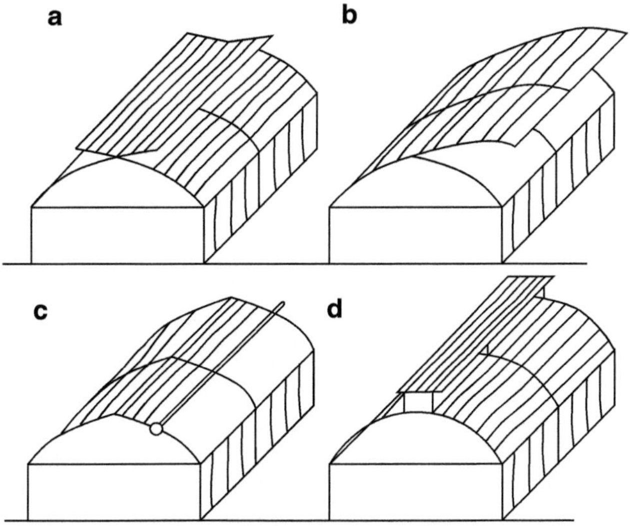

Fig. 9.11 Different types of roof or ridge ventilation. The flap ventilation (**a**) at the ridge is used both for rigid plastic and glass greenhouses and for plastic-film greenhouses. It can be opened optionally to leeward, to windward or on both sides. The ventilation opening (**b**) where half of the roof is opened, and the roll-up ventilation (**c**), are suitable for plastic-film greenhouses. The ventilation efficiency of type (**d**) is not as efficient as with the other types (see Chap. 5). The external flow passes directly through the ventilator, and has little effect on the internal flow (Montero et al. 2001)

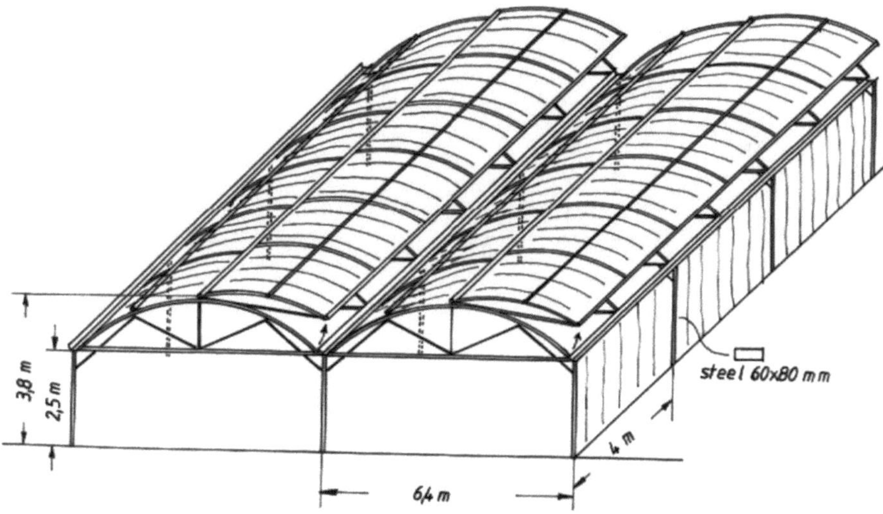

Fig. 9.12 Roof vent hinged to the ridge. Half of the roof of each span can be opened for ventilation

Fig. 9.13 Rolling-up through ventilation in a round-arched multi-span greenhouse, where both roof sides can be opened for ventilation at the gutter (Iran). Both plastic films of the roof as well as the rollup-vent are fixed by one tube even here. The forgotten clamps in the gutter must be removed

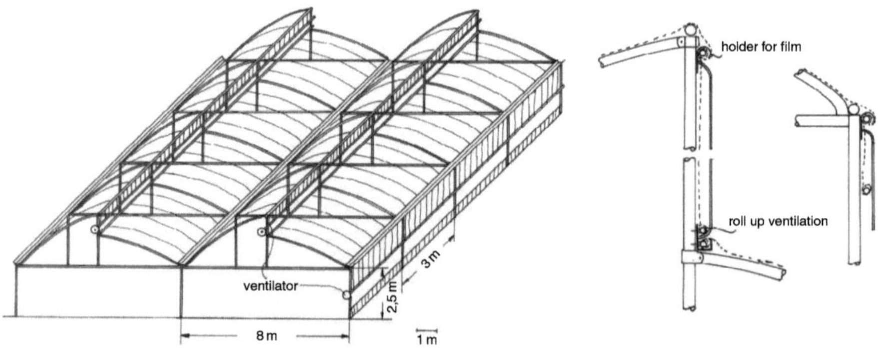

Fig. 9.14 A vertical roll-up vent at the ridge. This kind of ventilation is more influenced by the wind direction. It can be compared with a one side open roof flap vent. The details show the possibilities for stretching and fastening the plastic film at the ridge vent and at the side wall if there is no gutter at the eave

opening to greenhouse floor area has to be larger in continental climates than in maritime climates. It is therefore not recommendable to transfer typical greenhouse structures from maritime to continental climates, as happened in the past.

Fig. 9.15 Roll-up ventilation at the ridge of a round-arched greenhouse that has fan and pad cooling at the same time (Iran). The natural ventilation is for transit times, and has to be closed when fan and pad cooling is on

Fig. 9.16 Vertical ridge vents for a saddle-roof structure (Israel). Each roll up tube of the vents has a counterweight to pull and hold the roll-up tube down

Example:
Mean wind speeds:
Antalya, Turkey (July, August): $v_w = 2.1–2.6$ m/s
Netherlands (July, August): $v_w = 2.9–6$ m/s
Necessary ventilator opening calculated by (9.9): $V_E = 200 m^3/m^2$ h $= 0.56$ m^3/m^2 s, $G_{(\alpha)} = 0.08$.
Antalya Turkey $A_V/A_G = 0.33–0.27$
Netherlands $A_V/A_G = 0.24–0.11$
The ventilation opening in Antalya should be larger than in the Netherlands.

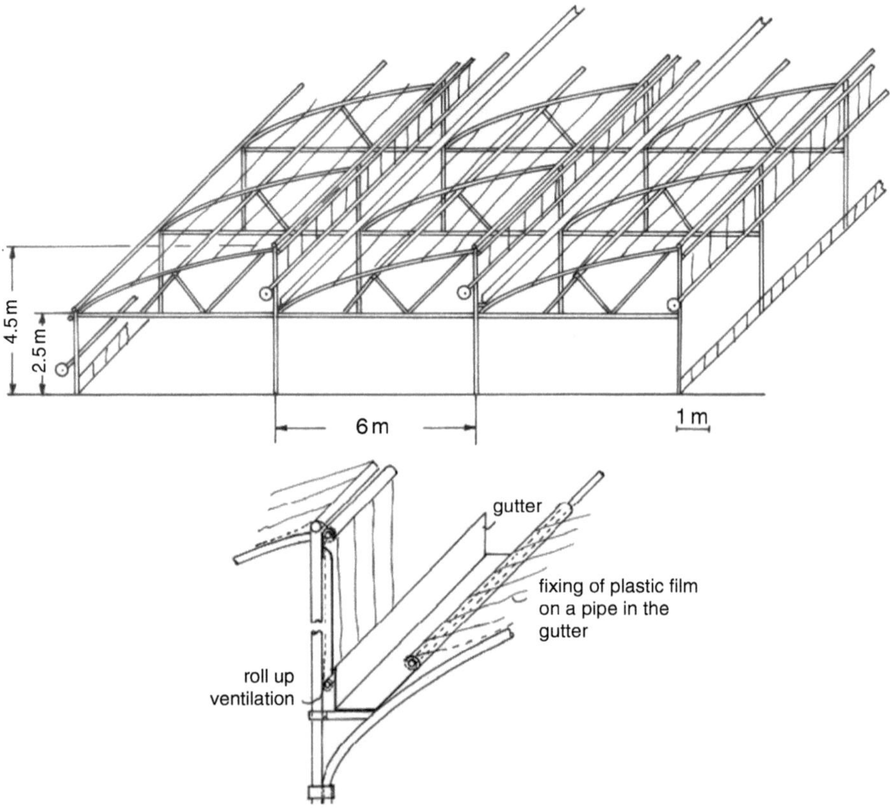

Fig. 9.17 Vertical roll-up ridge vent in a shed-roof greenhouse with the details for fastening and stretching the plastic film

Fig. 9.18 Vertical roll-up ridge vent in a shed roof (Tunisia). The roll-up vent will be operated by a rope and driving disc. Leakage at the gable end should be avoided by overlapping with a fixed part of film

The wind speed must be higher than 5 m/s for a ventilation opening of 18% and for a sufficient ventilation rate of more than 170 m^3/m^2 h. A wind speed of 2 m/s causes only a ventilation rate of less than 100 m^3/m^2 h (Figs. 9.20). Examples for flap vents area Figs. 9.21–9.23.

Fig. 9.19 Growers in India found that ventilation efficiency can be improved if the vertical ventilation openings at the ridge are directed in alternate directions. In most cases, the ridge vent of the last span is directed alternately. The openings of the ridge vents are open, while the side wall vents have roll-up vents

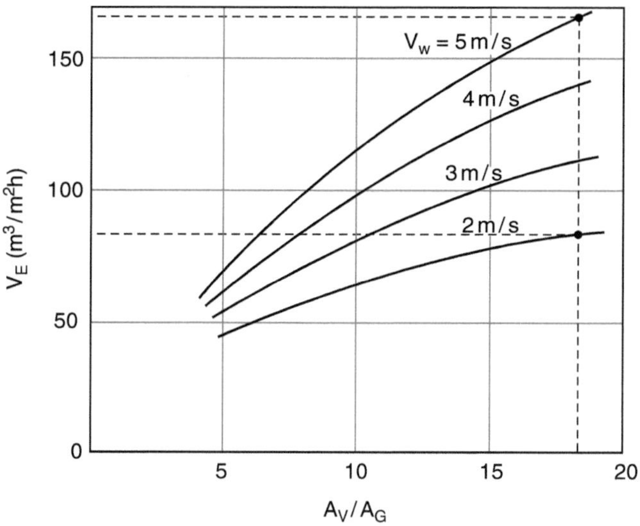

Fig. 9.20 Ventilation rate versus ventilation opening for single-span glass greenhouse with continuous flap roof vents only, both sides open (Baytorun 1986; von Zabeltitz 1997)

Waaijenberg (In Bakker et al. 1995) gives ratios A_V/A_G for different Venlo-type greenhouses and also for wide-span greenhouses with continuous vent opening on both sides and a pane length of 1.4 m.

Greenhouse span (m)	Ratio A_V/A_G
8	33.6
9.6	28
12.8	21

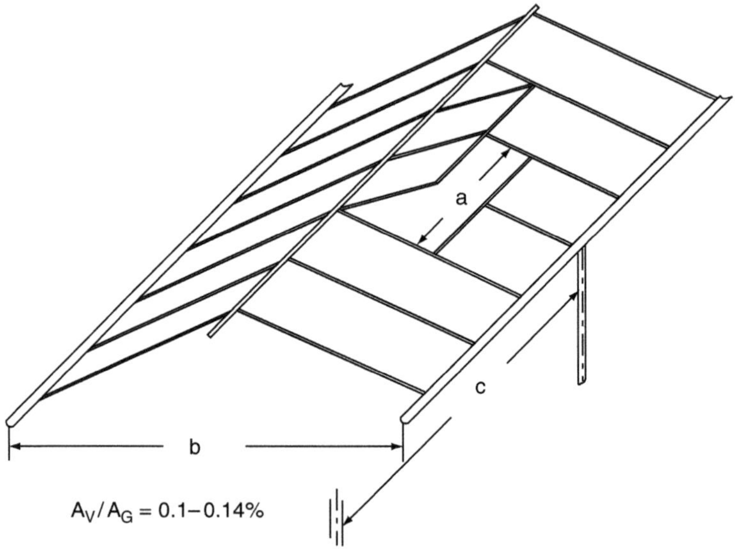

Fig. 9.21 Venlo-type greenhouse with a "two half glass pane" ventilation window installed alternately on both sides of the ridge, ratio A_V/A_G = 10–14%. The ventilation efficiency is not sufficient for continental climates

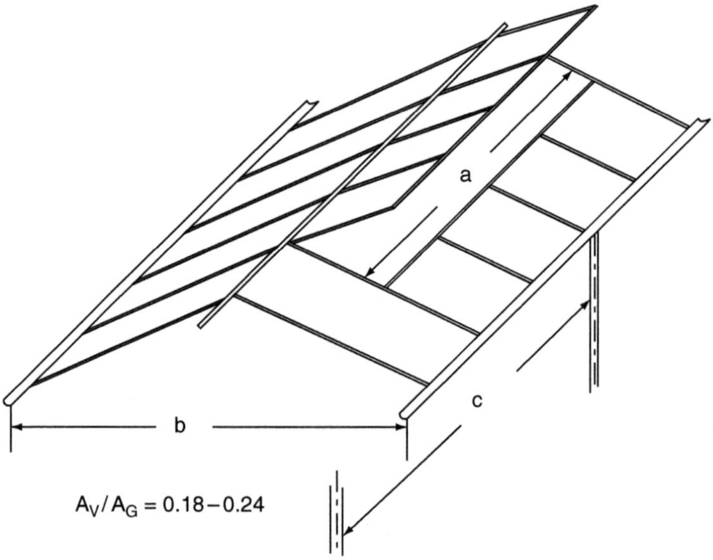

Fig. 9.22 "Four half glass pane" installed alternately on both sides of the ridge. Ratio A_V/A_G = 18–24%

Different configurations of continuous flap vents have been investigated by Bournet and Ould Khaoua (2008) and by Ould Khaoua et al. (2006) in a four-span glass house with 2,600 m^2 (4 × 9.6 × 68 m) floor area, 3.9 m height to the

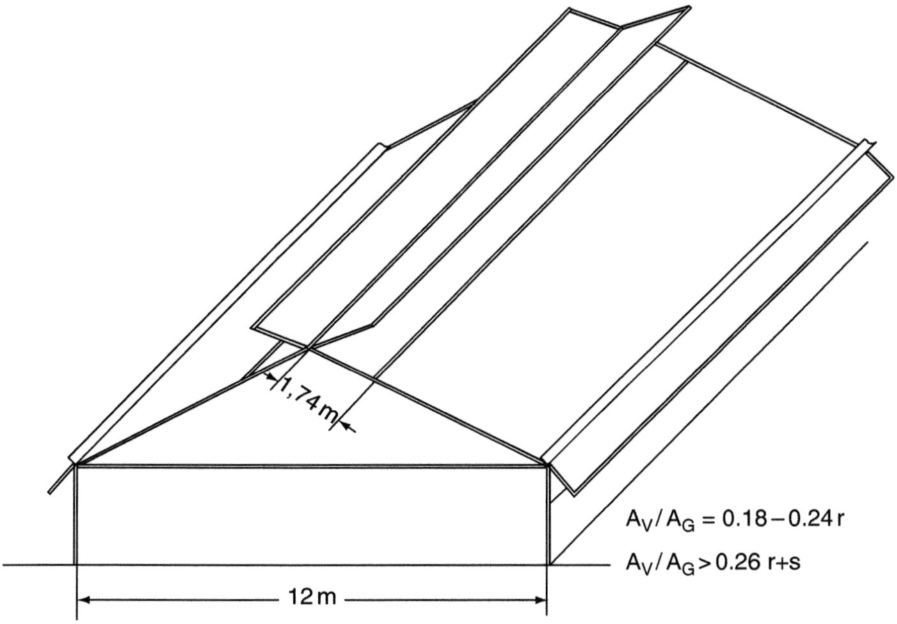

Fig. 9.23 Wide-span glass house with continuous flap ridge vents on both sides. Ratio $A_V/A_G = 18$–24% for ridge vent only. Ratio $A_V/A_G > 26.5$ for ridge and side wall ventilation in single-span greenhouse

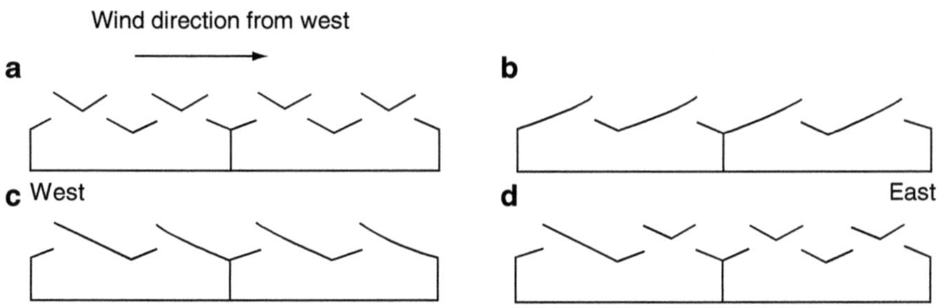

Fig. 9.24 Different combinations of flap vents, investigated by Bournet and Ould Khaoua (2008)

eave and 5.9 m ridge height (Fig. 9.24). Pot plants were grown on benches. The greenhouse unit was divided into two compartments, two spans each, in a westerly and easterly direction and with wind direction from the west.

Bournet and Ould Khaoua (2008) studied four roof vent combinations with wind speeds of 1.3–2.6 m/s:

Case A: Continuous roof vents, both sides open.
Case B: Continuous roof vents, leeward open.

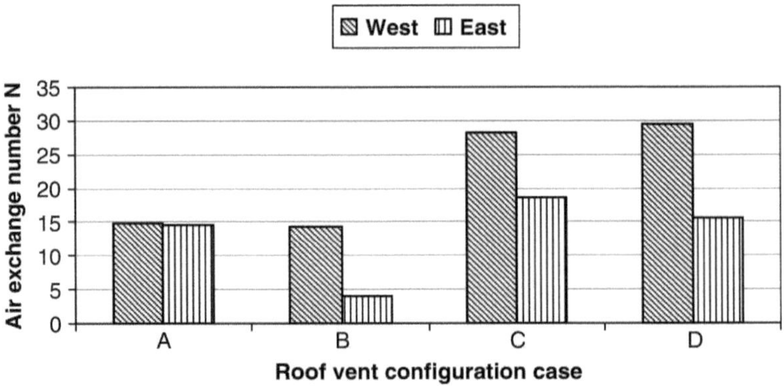

Fig. 9.25 Air exchange number N (1/h) for different roof vent configurations A–D (Bournet and Ould Khaoua 2008)

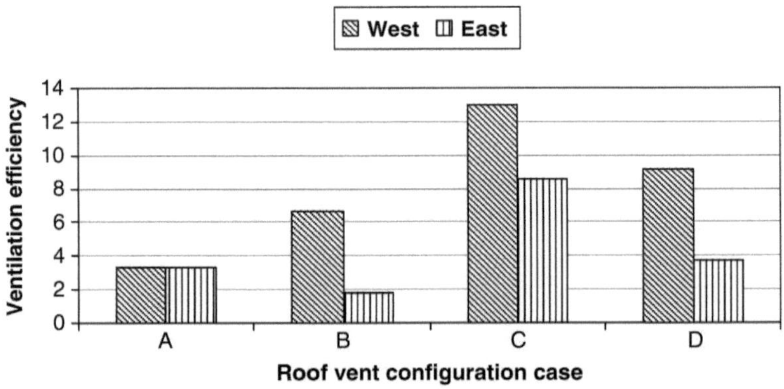

Fig. 9.26 Ventilation efficiency V_{eff} for different roof vent configurations A–D (Bournet and Ould Khaoua 2008)

Case C: Continuous roof vents, windward open.
Case D: First span windward open, the others both sides open.

Figure 9.25 shows the calculated air exchange numbers N(1/h) for the different vent configurations in the western and eastern compartments.

Figure 9.26 shows the ventilation efficiencies $V_{eff} = N/A_V$ (1/m^2 h) for the different vent configurations in the western and eastern compartments.

The air exchange is highest for case D, followed by cases C, A and B respectively. But the ventilation efficiency is highest in case C, followed by case D. The temperature difference inside at 1-m height above the crop and outside was 4°–6°C for cases C and D in the west compartment and 6°–11° for cases A and B respectively. The cases C and D seem the best with regard to low temperature differences.

The homogeneity of the temperature distribution at plant level is a very important factor. It is defined by the standard deviation of temperature difference related

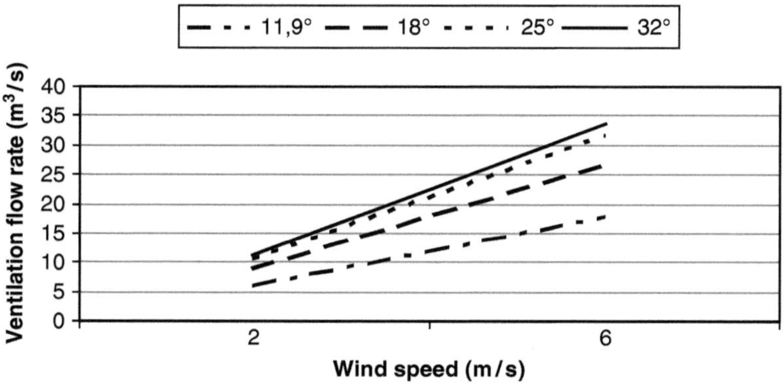

Fig. 9.27 Ventilation flow rate V_V (m^3/s) for different roof slopes and wind speeds (Baeza 2007)

to the outside temperature. Cases C and D had better homogeneity than cases A and B in the western compartment. There were no significant differences in the eastern compartment.

Considering the ventilation rates, temperature differences and the homogeneity, case D is a good compromise for efficient ventilation and climate conditions at crop level.

Ould Khaoua et al. (2006) found similar results for cases A, B, and C.

Bailey et al. (2004) tested models of a five-span Venlo-type glass house. They found the following values for $G_{(\alpha)} = V_V/(A_V \times v_w)$ at different opening angles of the leeward and windward flap vents.

Windward 42°, leeward 42° $G_{(\alpha)} = 0.096$
Windward 42°, leeward 0° $G_{(\alpha)} = 0.04$
Windward 0°, Leeward 42° $G_{(\alpha)} = 0.026$

Figure 9.27 shows the ventilation flow rate V_V (m^3/s) based on different roof slopes of 11.9°, 18°, 25°, 32°, and wind speed (Baeza 2007). The best ventilation flow rate is given with roof slopes of more than 18° (see Sect. 9.2.4). Even light transmittance is better with steeper roof slopes.

Figure 9.28 shows the air flux around and inside greenhouses with windward and leeward roof ventilators calculated by CFD (computational fluid dynamics) (Montero et al. 2009). The air velocity distribution and air flow rate through ventilators is more intensive with windward ventilators, but more equal inside greenhouses with leeward ventilators.

The Use of Deflectors or Baffles

Ventilation efficiency is reduced if roof ventilators are open to both sides leeward and windward, because the air stream is flowing directly through the ventilator openings and the air movement in the canopy area is poor, creating hot zones. This

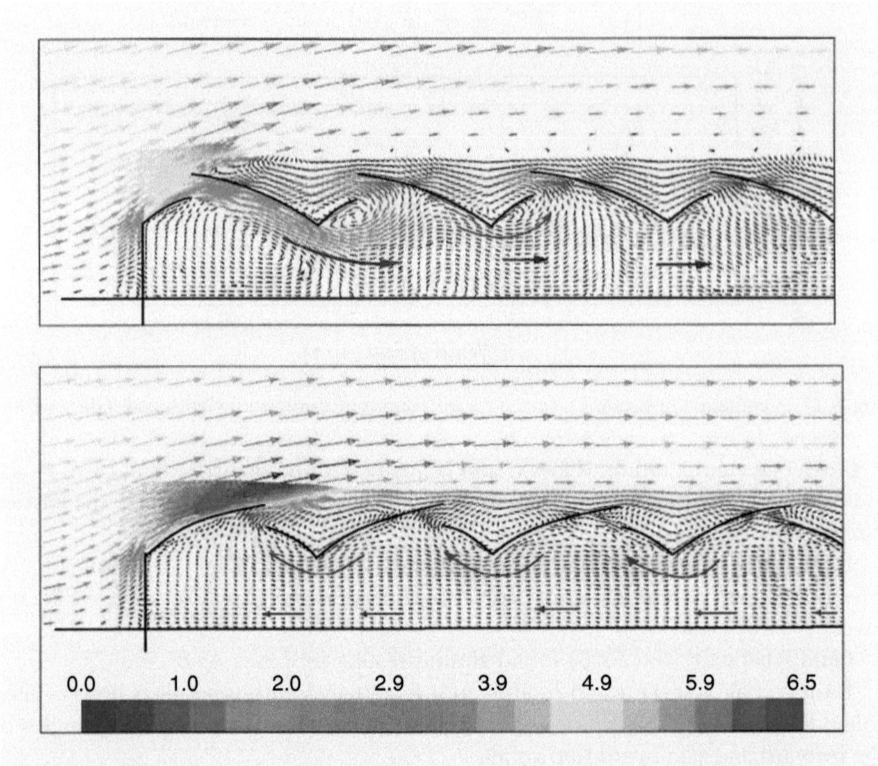

Fig. 9.28 Calculated air flux inside and outside of greenhouses with windward and leeward vents (Montero et al. 2009)

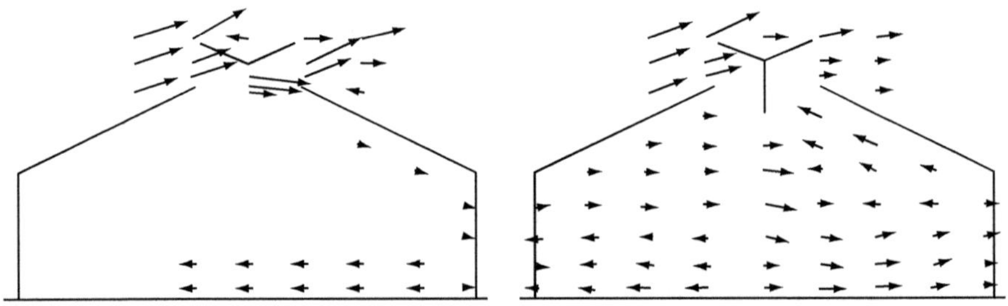

Fig. 9.29 The use of deflectors at the continuous roof flap ventilator in the first windward span can improve homogeneity considerably (Nilsen 2002, Baeza 2007, Baeza et al. 2008a, b, Montero et al. 2009)

negative effect can be considerably reduced by so-called deflectors or air-flow baffles at the first ventilator to the windward side (Fig. 9.29) (Nilsen 2002; Baeza 2007; Baeza et al. 2008a, b; Montero et al. 2009). The air stream will be distributed much better inside the whole greenhouse volume. Nilsen (2002) mounted a 1 m-high

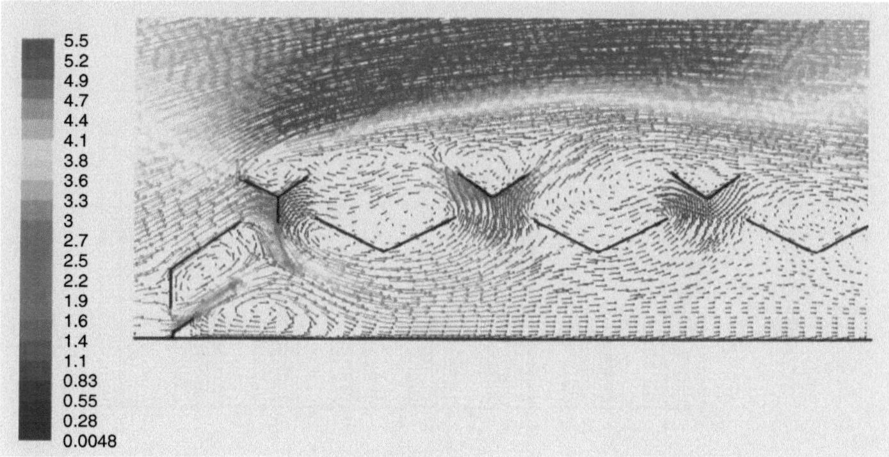

Fig. 9.30 A further improvement is the opening of the side ventilator flap inwards and distribution of the air flux, as shown in this figure (Baeza 2007)

vertical screen to direct the air flow into the crop canopy, and achieved 50% improvements of air exchange. The installation of these baffles forces the incoming air to move down into the crop area. Due to the baffles, a more homogenous temperature distribution with lower temperatures in the crop area can be obtained, although the total ventilation rate will be decreased a little bit by the baffles. It is also recommended to install these baffles in the first and last span of multi-span greenhouses (Fig. 9.30).

9.2.3 Side Wall and Roof Ventilation

Kacira et al. (2004b) investigated the effect of different ventilator configurations on ventilation rate V_E in a two-span greenhouse with a tomato crop, where the plant rows were parallel to the side vents. A special ventilation rate for the plant canopy has been defined and compared with the greenhouse ventilation rate (Sase 2006). Eight vent configurations were investigated, Fig. 9.31, Table 9.3.

The best ventilation rate was achieved with both roof flap vents open and roll-up side wall vents (configuration 3). This configuration (3) was much better than with flap side wall vents (1). The best ventilation rate for the plant canopy was achieved with roll-up side wall vents and leeward roof flap vents (8). The roll-up side vents significantly improved the ventilation rate in the plant canopy, configuration (3) through (8). The incoming air moves more into the plant canopy with side vent roll-up vents.

The effect of side wall vents in relation to the number of gothic-arched greenhouse spans, in combination with a continuous leeward roof vent at each ridge,

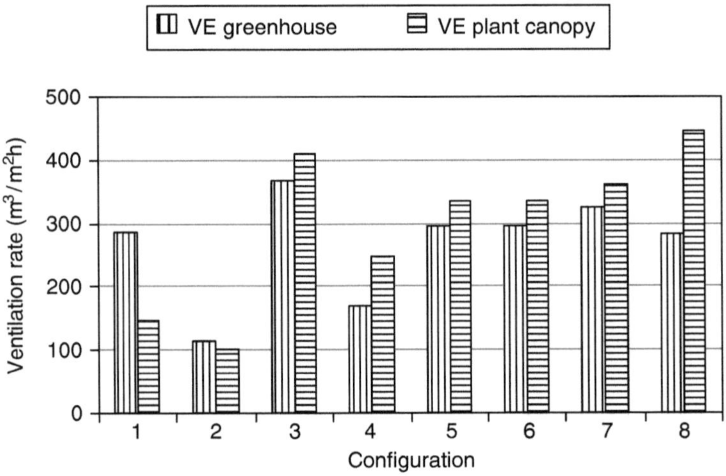

Fig. 9.31 Effect of different ventilator configurations on the ventilation rate V_E (m^3/m^2 h) in the greenhouse and in the plant canopy (Kacira et al. 2004b, Sase 2006)

Table 9.3 Eight configurations of different vent openings

Configuration	Type of ventilators	Symbol
1	Roof flap vents both sides open and flap side wall vents open	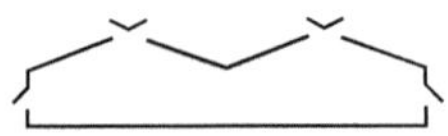
2	Side wall flap vents open	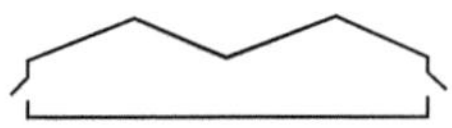
3	Roof flap vents both sides open and rolling-up side wall vents	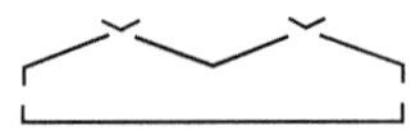
4	Side wall rolling-up vents only	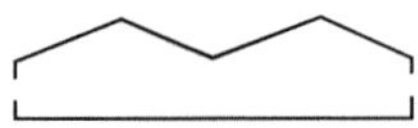
5	Roof flap vents windward open and rolling-up side wall vents	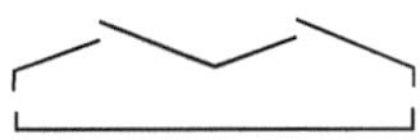
6	First roof flap windward, second leeward open and rolling-up side wall vents	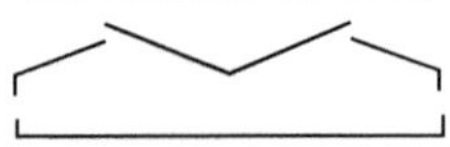
7	First roof flap vent windward, second both sides open and rolling-up side wall vents	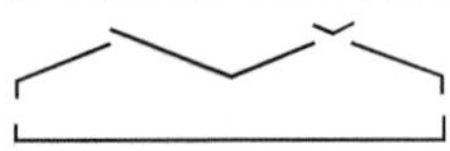
8	Both roof flap vents leeward open and rolling-up side wall vents	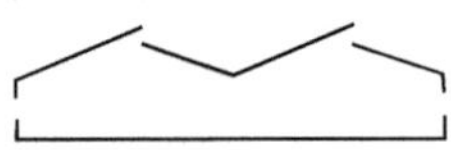

investigated by Kacira et al. (2004a), is described by Sase (2006). The ventilation rate increased when both side wall vents were opened but decreased with increasing number of spans, because the ratio A_V/A_G decreases with constant side wall opening area and increasing number of spans.

Bournet et al. (2007) investigated a four-span glasshouse (4 × 9.6 m wide, 68 m long), N–S-oriented with continuous flap roof ventilators and flap side wall ventilators at both sides. The greenhouse was the same as described by Ould Khaoua et al. (2006), with 3.9 m height to the eave and 5.9 m ridge height and roof vents only (see Sect. 9.2.2). Nine vent configurations of roof and side wall ventilators have been tested. The combinations of configurations were:

A. Roof vents, both sides open
B. Windward roof vent open
C. Leeward roof vent open
1. Side vents at the base of the side wall
2. Side vents in the middle of the side wall
3. Side vents at the top of the side wall near the eave

Considering only the side vents in the middle of the side wall (2) as the location found to be best, the following air exchange rates N (1/h) and temperature differences ΔT have been derived:

Configuration	A2	B2	C2
N (1/h)	32.9	40.5	12.7
ΔT (°C)	2.95	2.64	4.58

The roof vent orientation plays an important role. The windward open roof vents have the highest air exchange rate and the lowest temperature difference. The configuration with both roof vents open is less efficient. The temperature homogeneity is the best in combination B2.

Multi-span greenhouses with side wall and roof flap ventilation have very good ventilation efficiency when windward roof vents and side wall vents that are installed not too close to the soil are opened.

Montero et al. (2001) investigated the predominant wind-driven ventilation for different configurations of ventilators at side wall and roof in a round-arched single-span greenhouse with vertical side walls and a roof vent type as shown in Figs. 9.11 and 9.32. The greenhouse has two side walls and a roof vent, both sides open. If two side walls were open, the wind-driven air passed the greenhouse from windward to leeward side wall opening, and only a small exchange of flow was observed through the roof opening. A greenhouse with roof vents only had the lowest ventilation rate. The external air flow passed directly through the vents, and had very little effect on the internal air exchange.

Montero et al. (2008b) reported that windward ventilation openings have better ventilation rates but are more dangerous for crop and structure damage. They investigated the following configurations of a 15-span, 90 m long saddle-roof greenhouse with 26° roof slope by CFD study:

Configuration 1: Leeward roof vents open, 0.8 m vent opening. Side wall vent closed.

Configuration 2: Leeward roof vents open, windward side vents closed, leeward side vents open. Additional 1 m wide vents on the roof planes of spans 1 and 2.

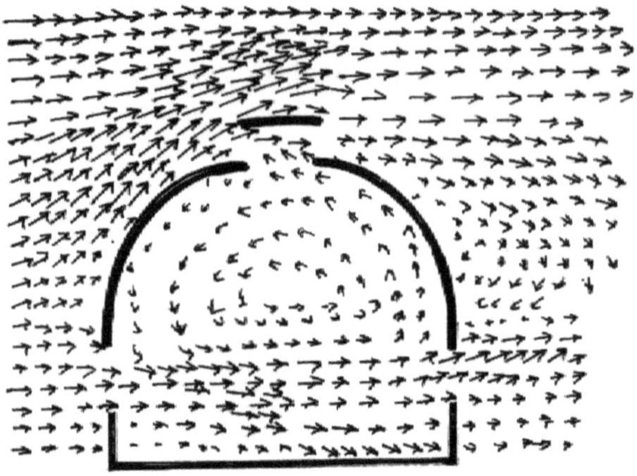

Fig. 9.32 Velocity field of air flow in a single-span greenhouse with side wall and roof ventilators (Montero et al. 2001)

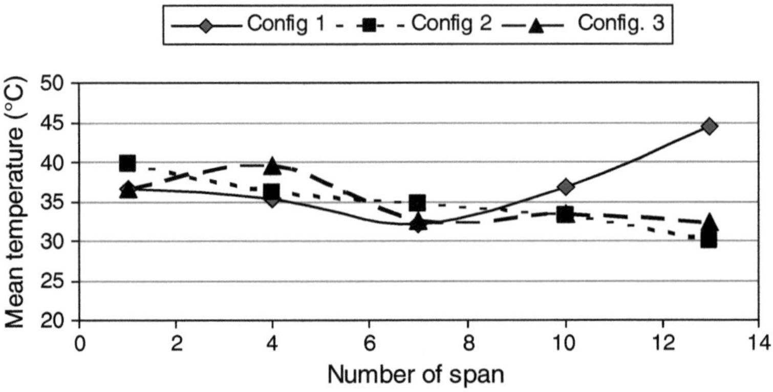

Fig. 9.33 Mean temperature in different spans of a 15-span greenhouse (Montero et al. 2008)

Configuration 3. As configuration 2, plus larger vent surface at spans 5 and 10. The plastic film on the roofs of span 5 was replaced by insect-proof screens.

Figure 9.33 shows the mean temperatures in the different spans.

Configuration 3 has the best uniformity of the temperature distribution in all spans. Configuration 1 has a good uniformity in the first seven spans, and then a steep increase in temperature. Configuration 2 has higher temperatures in the first half of the greenhouse unit, and the lowest at the leeward side.

Lee and Short (2000) investigated a round-arched four-span, north–south-oriented, plastic-film greenhouse, 7.32 m span width, 3.5 m side wall height, 5.34 m ridge height by CFD modelling. The greenhouse unit had a side wall vent at the west side only, and a half roof flap vent at each span with opening to the east

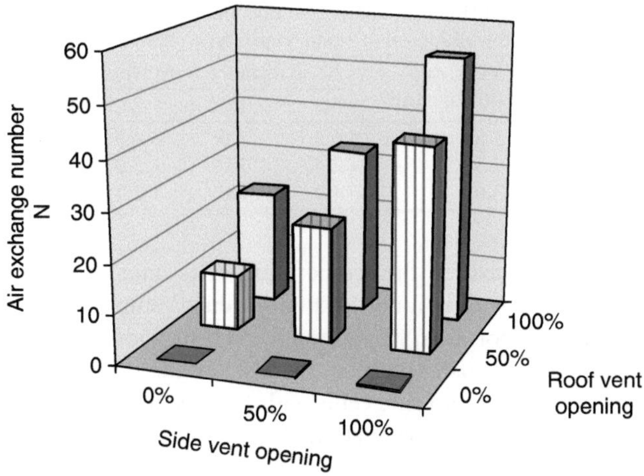

Fig. 9.34 The air exchange number N (1/h) for *leeward* wind direction, and different combinations of side wall and roof vent openings (Lee and Short 2000)

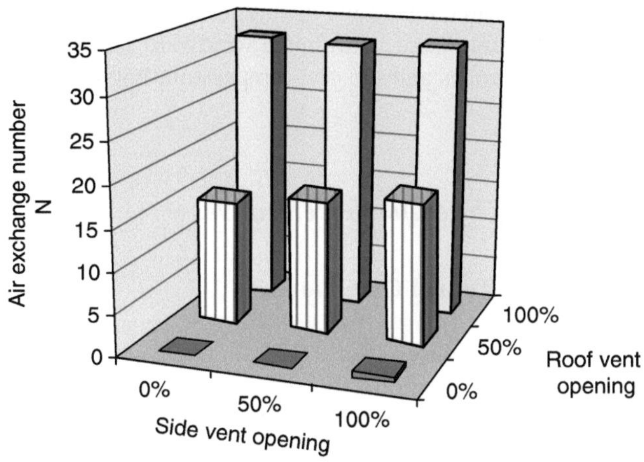

Fig. 9.35 The air exchange number N (1/h) for *windward* wind direction, and different combinations of side wall and roof vent openings (Lee and Short 2000)

side, as shown in Figs. 9.11b and 9.12. No side wall vent was installed at the east side wall. Figs. 9.34 and 9.35 show the predicted air exchange number N (1/h) of different combinations of roof and side wall openings for leeward and windward wind direction. The highest air exchange occured with leeward wind direction and all vents fully open. The air exchange with windward direction and open leeward side wall vent was lower, and more or less equal for all leeward side vent openings. The air distribution inside the greenhouse with crop was better with leeward wind direction than with windward wind direction.

Table 9.4 Mean temperature difference, ventilation rate and temperature differences at different distances from windward side wall in a multi-span Venlo-type greenhouse (Ishii et al. 2008)

	Mean ΔT (°C)	Vent. rate (m^3/m^2 min)	ΔT at distance from windward side wall				
			5.2 m	16.2 m	27.2 m	38.2 m	49.2 m
Case 1	5.3	1.83	6.6	5.6	4.9	5.5	4.2
Case 2	5.0	2.33	2.4	3.9	5.0	6.6	6.4
Case 3	2.9	2.91	1.9	1.6	2.7	3.7	4.1

Ishii et al. (2008) measured the temperature difference at different distances from the windward side wall in a north–south-oriented empty 17-span Venlo-type greenhouse, 1,741 m^2 floor area, 32 m length, 54.5 m width, 4.0 m eaves height, 4.7 m ridge height, 3.2 m span width, at 35°6′N latitude. Each span had two roof flap vents. Two side wall roll-up side vents with 1 m opening width were installed at both sides. The following cases were studied:

Case 1: Roof vent fully open, side vents closed.
Case 2: Roof vent fully open, western side vent closed, windward eastern side vent open.
Case 3: Roof vents fully open, both side vents open.

Table 9.4 shows the mean temperature differences, ventilation rates and temperature differences at different distances from the windward side wall. Case 3 has the lowest temperature differences and the best temperature homogeneity.

9.2.4 *Ventilation of Specific Greenhouses*

9.2.4.1 Round-Arched Tunnel Greenhouse

Round-arched tunnel greenhouses (Sect. 5.1) should have continuous side wall ventilators for sufficient ventilation efficiency. Kittas et al. (1996) investigated the ventilation of a tunnel greenhouse, 9 × 30 m, $A_G = 270$ m^2, equipped with a continuous rolling-up side wall ventilator. The ventilation opening was $A_V/A_G = 0.177$.

A non-dimensional parameter $G_{(\alpha)}$ of the ventilation flow rate related to the greenhouse floor area was identified:

$$G_{(a)} = \frac{V_V}{A_g \times v_w} = 0.1\frac{A_V}{A_g} + 0.0006$$

With the non-dimensional number $G_{(\alpha)}$ related to the ventilation opening

$$G_{(\alpha)} = \frac{V_V}{A_V \times v_w}$$

is

$$G_{(\alpha)} = \frac{A_g}{A_V} G_{(a)}$$

For the tunnel greenhouse, it becomes: $G_{(\alpha)} = 0.103$.

In comparison to other greenhouses with continuous side wall flap vents, the rolling-up ventilators are more effective.

The expression $C_d \times {C_w}^{0.5} = 0.2$ has been found.

The crop in a greenhouse has remarkable influence on the ventilation efficiency. The effect of tomato crop rows in a tunnel-type greenhouse on the ventilation rate has been studied by Fatnassi et al. (2008). The tunnel greenhouse, 128 m^2 floor area, 717 m^2 cladding area, 1,500 m^3 air volume, was N–S-oriented and had two continuous side vents, net area 28.8 m^2, equipped with insect screens. Tomatoes were grown in five N–S-oriented rows with 0.5 m inter-row space.

The ventilation rate is according to (9.13)

$$V_E = \frac{A_V}{2A_G} C_{ds} \sqrt{C_w} v_w$$

C_{ds} = discharge coefficient of side wall opening with screen.

C_w = wind pressure coefficient.

V_w = wind speed.

The product $C_{ds}\sqrt{C_w}$ is a measure for the ventilation rate if the other factors do not change. Three cases were studied:

Case 1: Wind parallel to the side openings.

Case 2: Wind perpendicular to the side openings.

Case 3: Wind 45° to the side openings.

The product $C_{ds}\sqrt{C_w}$ is given in Table 9.5.

The ventilation rate is 2.03 times higher with perpendicular wind (case 2) than with parallel wind in an empty greenhouse, and nearly identical with a 2.13 m-high tomato crop. The ventilation rate with 1 m high and 2.13 m high tomato crops is 74% and 45% respectively compared to the empty tunnel.

Table 9.5 The product $C_{ds}\sqrt{C_w}$ as measure for the ventilation rate for a round-arched tunnel greenhouse with insect screens and a tomato crop of different heights $C_{ds}\sqrt{C_w}$ (Fatnassi et al. 2008)

	$C_{ds}\sqrt{C_w}$ Case 1	$C_{ds}\sqrt{C_w}$ Case 2	$C_{ds}\sqrt{C_w}$ Case 3
No crop	0.26	0.53	0.46
1 m-high tomato crop	0.22	0.39	0.33
2.13 m-high tomato crop	0.21	0.24	0.22

9.2.4.2 Parral-Type Greenhouses

Too high temperatures due to insufficient ventilation efficiency, with the consequence of negative effects on yield and quality, are a serious problem of many greenhouse structures in subtropical climates. Insufficient ventilation creates high humidity, with the risk of fungal diseases.

The Parral-type greenhouse with 84% of the whole greenhouse area is the most used structure in Spain. The original Parral greenhouses have flat roofs and side wall ventilators (see Sect. 5.7.1). Improved Parral greenhouses have multiple pitched roofs with a small roof slope or asymmetrical saddle roofs. They are equipped with continuous roof ventilators placed on one side of the ridge. Twenty-three percent of the Parral greenhouse have roll-up ventilators, and 14% flap ventilators (Castilla and Hernandez 2005).

Measurements and calculations were made in a five-span Parral greenhouse, 38 m long, 23.2 m wide and 4.4 m ridge height. The roof inclination was 12°. Three types of ventilators were investigated, Fig. 9.36, (Perez-Parra et al. 2004):

A. Roof flap ventilator, linked to the ridge, 30.5 m^2 ventilator area, $A_V/A_G = 0.035$.
B. Rolling up roof ventilator, 92.3 m^2 ventilator area, $A_V/A_G = 0.105$.
C. Side wall rolling-up ventilator, 98.25 m^2 ventilator area, $A_V/A_G = 0.111$.

Measurements were made without and with insect screens, porosity 0.39, thread diameter 0.28 mm (see Chap. 10).

The regression equations for the air flow through the ventilators, depending on the wind speed, are:

Rolling-up roof vents, windward $V_V = 5.31 + 1.376 \times v_w$
Rolling-up roof vents, leeward $V_V = 5.03 + 1.403 \times v_w$
Flap roof vents, windward $V_V = 2.72 + 2.485 \times v_w$

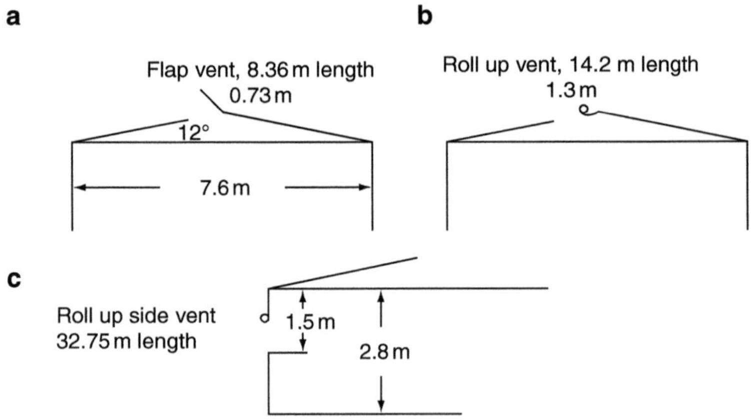

Fig. 9.36 Parral-type greenhouse with different roof ventilations (Perez-Parra et al 2004)

Flap roof vents, leeward $V_V = 2.87 + 1.428 \times v_w$
Side wall vent, leeward and windward $V_V = 11.96 + 1.305 \times v_w$
Rolling-up roof and side wall vents $V_V = 19.5 + 2.339 \times v_w$

One has to take into consideration differences of vent opening areas.

The side wall vents are statistically equal for leeward and windward ventilators. They have an important influence on the ventilation rates. The roof flap vents have lower ventilation flow rates than the rolling-up vents, because the vent opening area is much smaller. There is a clear difference between windward and leeward flap roof vents.

The non-dimensional parameter $G_{(\alpha)}$ was obtained for fully opened roof flap vents:

Windward $G_{(\alpha)} = 0.1$
Leeward $G_{(\alpha)} = 0.071$

Taking into consideration the difference in the ventilation opening areas, one has to compare the ventilation efficiencies $V_{eff} = V_V/A_V$, Fig. 9.37

The ventilation efficiency of the rolling-up ventilation on the roofs with low inclination is lower than for the flap ventilators. The flap vent is a barrier to the air flowing over the low-sloped roof.

A ranking of the different ventilation systems for Parral-type greenhouses with small roof slopes with regard to decreasing ventilation efficiency is:

Windward flap ventilator 100%
Leeward flap ventilator 67%
Rolling-up side wall ventilator 42%
Rolling side wall and rolling roof ventilators 37%
Rolling roof ventilator 28%

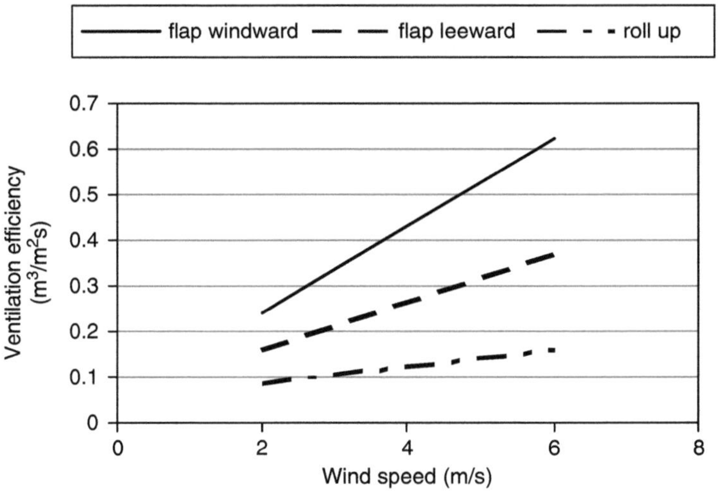

Fig. 9.37 Ventilation efficiency V_V/A_V for roof ventilators of Parral greenhouses

That means that roof flap ventilators of the same size as rolling-up roof ventilators have a better ventilation efficiency on a greenhouse with low roof inclination of about 12°.

The calculated discharge coefficients C_d and global wind pressure coefficients C_w are:

	C_d	C_w	$C_d C_w^{0.5}$
Roof flap windward	0.649	0.1	0.205
Roof flap leeward	0.649	0.037	0.125
Rolling-up roof	0.645	0.0058	0.046
Rolling-up side wall	0.666	0.006	0.047
Rolling-up roof and side wall	0.656	0.0017	0.025

The global wind pressure coefficient depends on the wind speed. The following functions were developed by Perez-Parra et al. (2006)

Flap vent windward	$C_w = 0.1866 \times v_w{}^{-0.3872}$
Flap vent leeward	$C_w = 0.1667 \times v_w{}^{-0.1896}$
Rolling-up side wall	$C_w = 0.1122 \times v_w{}^{-1.4737}$
Rolling-up roof vent	$C_w = 0.0301 \times v_w{}^{-0.8176}$

The following published values for C_d and C_w are given by Perez-Parra et al. (2004)

	L/H	C_d	C_w	$C_d C_w^{0.5}$	Source
Flap vent, curved roof	32	0.705	0.079	0.199	Boulard and Baille (1995)
Flap vent, curved roof	33	0.707	0.121	0.246	Papadakis et al. (1996)
Rollup, side wall tunnel	37	0.712	0.709	0.2	Kittas et al. (1996)
Flap, glasshouse roof	3	0.63	0.117	0.15	Bailey (2000b)

Baeza et al. (2005) and Baeza (2007) simulated multi-span Parral-type greenhouses with different vent configurations. First investigation was a five-span greenhouse, the same type as Perez-Parra et al. (2004) with windward flap vents of various opening sizes from 0.4 to 1.9 m. The ventilation rate increases with increasing opening width of the vents; the air movement is enhanced, and the temperature gradients are more uniform.

The second investigation concerns a ten-span greenhouse with the following configurations and ventilation rates:

C1: One windward flap vent on each span, 0.8 m opening width. The ventilation rate is slightly higher than C2.

C2: Two windward flap vents on the first two spans and two leeward flap vents on the last two spans, 0.8 m vent opening. The vents in the middle were closed. This configuration had the lowest ventilation rate.

C3:The same configuration as C2, but 1.3 m vent opening size of the flap. The ventilation rate was slightly higher than C1.

C4:The same configuration as C3 but rolling up vents installed on the middle spans. This configuration had the highest ventilation rate.

The combination of flap vents and rolling-up vents, which are cheaper, provides a cheap and effective vent configuration.

Baeza et al. (2006) investigated the influence of the number of spans on ventilation rate in a multi-span Parral-type greenhouse with low roof slope, span width of 7.6 m and one vent on each span. The roof ventilation configurations were:

- Single windward flap vents.
- Single leeward flap vents.
- Flap vents with alternate orientation on the spans (windward–leeward–windward–leeward).
- Two flap vents per span, both sides open.

A simple linear equation was found for the air exchange number N (1/h) depending on wind speed v_w(m/s).

$$N = m \times v_w$$

m is the slope of the regression line and a measure for the ventilation rate.

The ratio A_v/A_g remains constant for all numbers of span.

Figure 9.38 shows the slope m of the regression lines depending on the number of spans. The ventilation rate decreases with increasing number of spans for the windward and leeward single-flap vents. The ventilation rate for the double-flap vents and alternate flap vents is larger than for the single-flap vents, in this case for low roof slopes. If single-flap vents should be installed, alternate flap vents may be favourable. The mean temperatures were nearly the same in all spans, even with various ventilation rates.

Baeza et al. (2008a, b) studied the natural ventilation at very low wind speeds (only buoyancy-driven ventilation) in a large Parral-type greenhouse with different numbers of spans, gutter height 3.6 m, ridge height 4.4 m, span width 7.6 m, 23.2 m length. The greenhouse unit had one roof flap vent per span, and rolling-up vents, 1.2 m wide, at the two side walls in longitudinal direction. The results of a CFD simulation with and without open side wall ventilation are given in Fig. 9.39 for the

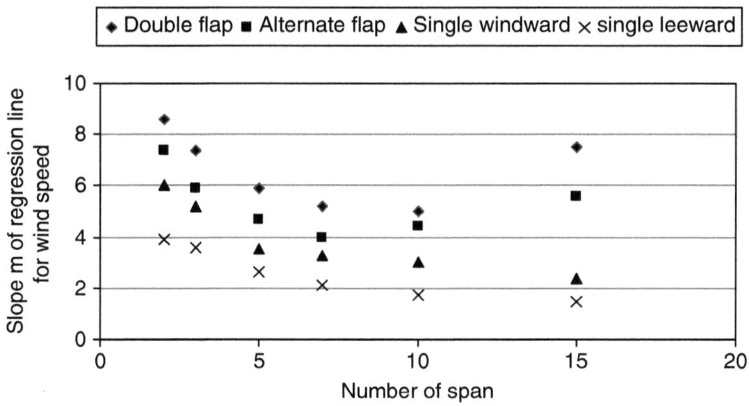

Fig. 9.38 Slopes m for the regression lines $N = m \times v_w$ depending on the number of spans (Baeza et al. 2006)

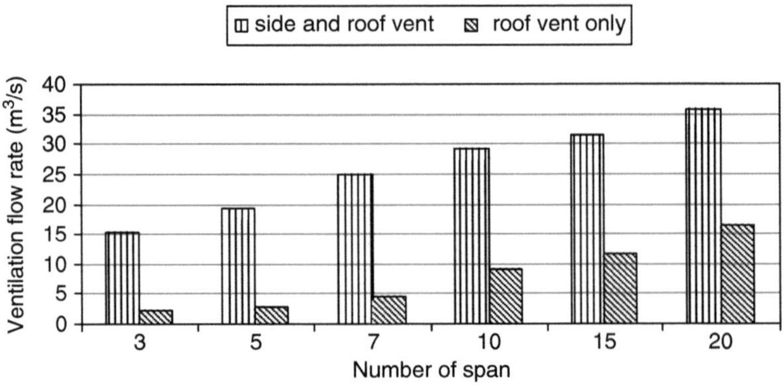

Fig. 9.39 Ventilation flow rate V_V (m^3/s) in various spans of a multi-span Parral-type greenhouse with and without side wall ventilation (Baeza et al. 2008a, b)

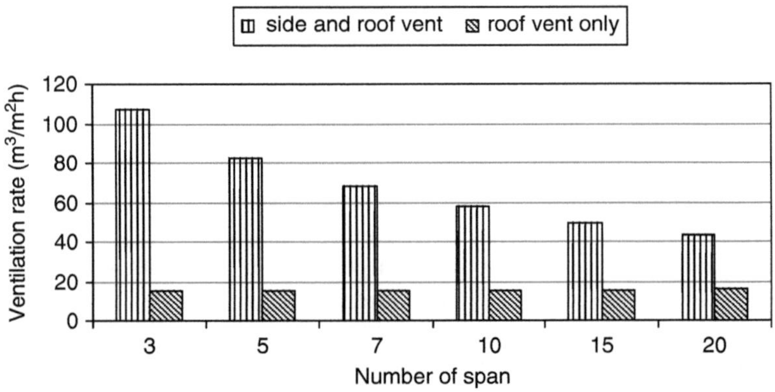

Fig. 9.40 Ventilation rate V_E (m^3/m^2 h) in various spans of a multi-span Parral-type greenhouse with and without side wall ventilation (Baeza et al. 2008a, b)

ventilation flow rate V_V(m^3/s) and in Fig. 9.40 for the ventilation rate V_E (m^3/m^2h) respectively.

It is clear that side wall vents are absolutely necessary for sufficient ventilation at low wind speeds, when buoyancy forces are active only. The temperature differences >4°C up to 9°C maximum were large with roof ventilation only. The ventilation rate V_E with roof and side wall ventilation decreases with the number of spans, because the ratio of side wall vent opening to floor area decreases with increasing number of spans. With a large number of spans, the ventilation rates with roof and side wall ventilation approximate the ventilation rate without side wall ventilation, which is nearly constant over the number of spans. The recommendation is to design greenhouses with side wall vents against prevailing wind direction and maximum distance of 50 m between the side vents to guarantee a relatively sufficient ventilation efficiency.

The discharge coefficient and the wind-pressure coefficient have been investigated for a five-span saddle-roof Parral-type greenhouse in Almeria (36°50′N) with 3.2 m ridge height, 2.6 m gutter height, 44 m length, and 38.5 m width. Roof and side wall vents were covered by insect screens, 10 × 20 threads/cm^2, 0.34 porosity, 0.272 mm thread diameter and 0.233 × 0.741 pore width and height respectively (Molina-Aiz et al. 2009). The following discharge coefficients have been calculated:

With insect screen: $C_{ds} = 0.228$ for wind speeds $v_w = 7–9$ m/s
$C_{ds}=0.158$ for wind speeds $v_w = 3–4$ m/s

Mean value $C_{ds} = 0.193$

Without screen $C_d = 0.668$

The reduction of the discharge coefficient is 65–76% for insect screens in comparison to vents without screen. The mean values for the wind pressure coefficients have been calculated: $C_w = 0.049–0.017$ considering wind speed and stack effect.

The mean value for $C_d \times Cw^{0.5}$ was 0.05 with screen and 0.173 without screen. That means a reduction of 71% for the main influence factor for the ventilation rate besides the wind speed (see 9.13).

9.2.4.3 Banana Greenhouse

Banana greenhouses are large units with heights up to 6–7 m (Sect. 5.7.2). Demrati et al. (2001) investigated a 32-span round-arched plastic-film banana greenhouse, 100 by 100 m (1 ha) with 6–7 m ridge height in Morocco. Roof and roll-up side wall ventilators were installed, in total 874 m^2 ventilator opening area, $A_V/A_G = 0.087$. They identified $C_d \times C_w^{0.5} = 0.27$ and measured 3–4°C temperature difference.

The non-dimensional parameter $G_{(\alpha)}$ gets:

$$G_{(\alpha)} = \frac{V_V}{A_V v_w} = \frac{1}{2} C_d \sqrt{C_w} = 0.135$$

Using (9.13), the ventilation rate becomes $V_E = 126$ (m^3/m^2 h).

9.2.4.4 Summary of Recommendations for Natural Ventilation

The following recommendations in particular can be summarised for the natural ventilation of multi-span greenhouses (Montero et al. 2009; Montero 2009; Baeza 2007).

- Windward ventilation can cause damage to structure and crop.
- Windward ventilation creates high ventilation rates, in particular at the first ventilators, but the incoming air mainly follows the inner surface of the roof

and does not mix with the air in the crop area. Deflectors or baffles can avoid this disadvantage and create better homogeneity.

- The greenhouse units should not be wider than 50 m, with windward ventilation to avoid excessive temperatures.
- The ventilation efficiency and temperature homogeneity with leeward ventilation in multi-span greenhouses can be improved by increasing the vent opening at intervals of every five spans.
- It is recommended to close the windward and open the leeward side wall vents with leeward roof vents.
- Increasing the roof slope improves the air flow to the crop area. The ventilation increases sharply with roof slopes up to 25°.
- A vent opening in the roof near the gutter of the first span can improve the homogeneity of air mixture.
- Increasing the ventilator size in the first two and last two spans increases the ventilation rate and air movement into the crop in multi-span greenhouses. The temperature distribution is more homogeneous.
- The ventilation rate increases with both side wall vents open, but decreases with the number of spans.

9.3 Forced Ventilation

Forced ventilation by fans is an effective method of ventilation, but it needs electricity and is more expensive than natural ventilation: it will be used for fan and pad cooling systems (Chap. 11). Fans suck air out on one side, and openings on the other side let air in.

The total ventilation flow rate by the fans can be calculated directly with the ventilation rate V_E.

$$V_V = V_E \times A_G \ [m^3/h]$$

The ventilation flow rate V_E can be calculated by an energy balance, (9.7) and Fig. 9.1.

Axial fans will be used for forced ventilation.

Some essential design criteria are (von Zabeltitz 1999):

1. The fans must exhaust air out of the greenhouse, because the temperature distribution on the suction side is more even.
2. The distance between two fans should not exceed 8–10 m.
3. Ventilation fans should develop a capacity of more than 30–50 Pascal (3–5 mm water gauge) with regard to fan and pad cooling and the use of insect screens.
4. If possible, the fans should be located on the leeward side of the greenhouse.
5. A space of at least 1.5 times the fan diameter should be left between the fan discharge and the nearest obstruction.

6. The inlet opening should be at least 1.25 times the total area of the fan openings.
7. The velocity of the incoming air must not be too high. In the plant area, the air speed should not exceed 0.5 m/s.
8. The openings must close automatically when the fans are not in operation.

Forced ventilation produces gradients in temperature and humidity between air inlet and outlet. The distance should be limited to 30–40 m (Bailey 2006). Short (2004) reported a distance of 50 m or less (Fig. 9.41).

Fans should have low speeds, because the power consumption and noise increase with increasing speed. The necessary high air volume can only be produced by fans with low speed if the diameter is big enough. Low-pressure and low-speed axial fans with diameters of 750–1,250 mm are suitable for greenhouses. Table 9.6 shows some data of those possible fans for greenhouses.

Fig. 9.41 Fans for forced ventilation (*left side*) and air inlet flaps (*right side*) combined with roof flaps for natural ventilation (USA)

Table 9.6 Data of axial fans for forced ventilation in greenhouses (von Zabeltitz et al. 1989)

Diameter (mm)	750–1,250
Speed (1/min)	380–725
Air flow (m^3/h)	12,000–40,000
Power consumption P (kW)	0.4–2.2
Electricity consumption (Wh/100 m^3 airflow)	2.9–6.0

The installation power P_i is ventilation rate $V_E \times$ power consumption P.

$P_I = V_E \times P$

For a ventilation rate of 150 m^3/m^2h, the installation power is 4.4–9.0 W/m^2.

The hours of operation for forced ventilation in the climate of Hannover, Germany has been calculated, and is shown in Table 9.7 (von Zabeltitz et al. 1989).

The influence of the ventilation rate V_E is smaller than of ventilation set point.

The electricity consumption per m^2 of greenhouse floor area for different configurations has been calculated for given fan data in Table 9.8.

Combined forced and natural ventilation, where roof vents are partially open, can improve the temperature control and homogeneity of temperature distribution inside the greenhouse (Baeza et al. 2005). Experiments were carried out in a three-span round-arched greenhouse in southern Spain, Almeria, with through roof vents, one exhaust fan in the eastern gable of each span and a rectangular air inlet in the western gable (Florez-Velasquez et al. 2009). All vents were covered by insect screens. The greenhouse was 22.5 m wide, 28 m long, 3 m gutter height, and 4.7 m ridge height. Three cases of ventilation were investigated, and the air exchange number N (1/h) was measured:

- Forced ventilation only, roof vents closed: $N = 21.2$ (1/h).
- Forced ventilation and 30% open roof vents: $N = 24.5$ (1/h).
- Forced ventilation and 100% open roof vents: $N = 23.4$ (1/h).

Table 9.7 Hours of operation for a forced ventilation system in the climate of Hannover, Germany (von Zabeltitz et al. 1989)

Configuration	Ventilation set point (°C)	Ventilation flow rate V_E ($m^3/(m^2$ h)	Hours of operation (h/a)
A. single glass	24	125	830
B. single glass	24	150	777
C. single glass	24	175	729
D. single glass	24	200	694
E. single glass	22	150	1,030
B. single glass	24	150	777
B. double glass	24	150	773
F. single glass	26	150	570
G. single glass	28	150	408

Table 9.8 Electricity consumption per m^2 floor area (kWh/m^2 a)

Configuration	Installation power (W/m^2)	Operation hours (h/a)	Electricity consumption (kWh/m^2a)
A: 24/125	5.5	830	4.6
B: 24/150	6.6	777	5.1
C: 24/175	7.7	729	5.6
D: 24/200	8.8	693	6.1

Fan data: diameter: 1 m; speed: 725 1/min; air flow rate: 25,000 m^3/h; power consumption: 1.1 kW; electricity consumption: 4.4 Wh/(100 m^3); static pressure: 30 Pa (von Zabeltitz et al. 1989)

There is a temperature gradient with forced ventilation only from air inlet to exhaust fan as well as a vertical stratification if the roof vents are closed. A better mixing of air stream near the plant canopy can be obtained with open roof vents, and as a result a reduction of vertical temperature stratification and longitudinal temperature gradient. The distance between forced air inlet and outlet can be extended when additional roof vents are opened.

Chapter 10
Insect Screening

Vegetable production under protected cultivation in humid tropics and generally in warm climates is vulnerable to climate stresses (temperature, humidity), biotic stresses by insects, and plant virus diseases vectored by insects (Kumar and Poehling 2006). Pests penetrating the greenhouse damage the crop by feeding and by transmit of phytopathogenic organisms. Therefore, the exclusion of pests by insect screens is an important factor for an integrated approach of pest management (Teitel 2007). Fighting against insect pests in greenhouses becomes more and more important, because many chemicals are not allowed anymore, and pest insects become resistant against special chemicals. Physical and optical methods should be used therefore for integrated production and protection (IPP). The IPP aims in finding alternative solutions for reducing pesticide application (Fig 10.1).

Insect screens are used in front of the ventilation openings and doors to keep useful insects inside, and to prevent pest insects from penetrating the greenhouse. They have different mesh openings because the insects are of various sizes.

Criteria for the choice of insect screens are:

- The species of insects to be screened out
- The influence on the greenhouse climate
- The UV stability and the mechanical durability (thickness of threads)
- The cost in comparison to the economical value of the crop

The efficiency of insect screens depends on the mesh size of the screens, the cross-section of the screens, and the colour of the screens, as well as on the tightness of the greenhouse structure and doors, and on the tight fastening of screens. Insect screens can prevent the penetration of insects only if the mesh size is smaller than the widest part of the insect.

The following insect pests can be screened out by the given mesh sizes (Bethke and Paine 1991; Bethke 1994; Bethke et al. 1994; Teitel 2006, 2007):

C. von Zabeltitz, *Integrated Greenhouse Systems for Mild Climates*,
DOI 10.1007/978-3-642-14582-7_10, © Springer-Verlag Berlin Heidelberg 2011

Fig. 10.1 Greenhouses with insect screens (Turkey and Malta)

Insect pest	Hole size (mm)	
	Bethke	Teitel
Leaf miners (*Lyriomyza trifoli*)	0.64	0.61
White fly (*Bemisia tabaci*)	0.462	0.46
Aphid (*Aphis gossipii*)	0.341	0.34
Greenhouse white fly (*Trialeurodes vaporariorum*)		0.29
Silverleaf white fly (*Bemisia argentifolii*)		0.24
Thrips (*Frankliniella occidentalis*)	0.192	0.19

Insect screens have to be installed without gaps in the structure. Insecure insect screens where insects can penetrate are useless (Fig. 10.2).

A *disadvantage* of insect screens is the *reduction of ventilation efficiency* with influence on temperature and humidity, as well as reduction of *light transmittance*.

Fig. 10.2 Secure and insecure insect screens

Sufficient ventilation efficiency has to be guaranteed even in screened greenhouses. To guarantee similar climate conditions to those in unscreened houses, enlarged ventilation openings should be designed. Different principles of greenhouse construction, cladding materials and climate control measures are available to meet the requirements.

The main factors of characterisation of insect screens are:

- The porosity, the ratio of open area to total area of the screen
- The mesh or hole size
- The thread dimension (woven or knitted)
- The light transmittance
- The colour and its influence on pest behaviour.

The expression "mesh" means the number of open spaces per inch in each direction. The characterisation of insect screens only by the expression "mesh" is not sufficient, because it does not give information about the thread diameter and thereby about the hole size. Data about insect screens should contain thread diameter, hole size in both directions and porosity.

10.1 Enlargement of Screened Vent Openings

The ratio of ventilator opening to greenhouse floor area in unscreened houses should be 18–29% (see Chap. 9, Ventilation).

The ANSI/ASAE (2003) gives values of 15–25%.

There are two possibilities when designing screened openings:

- The enlargement of vent openings and direct screening by nets
- Enlargement of the screened openings in front of the existing vent opening

New greenhouses can be designed according to the requirement for enlarged vent openings. If the vent openings of existing greenhouses cannot be enlarged easily, enlarged screened components have to be installed in front of the vents, Fig. 10.3.

Figures 10.3–10.5 show the possible enlargement of existing vent openings by insect screens at side walls with flap ventilation, with roll-up ventilation and with ridge ventilation. If the flap ventilation opens, a tube in a longitudinal direction

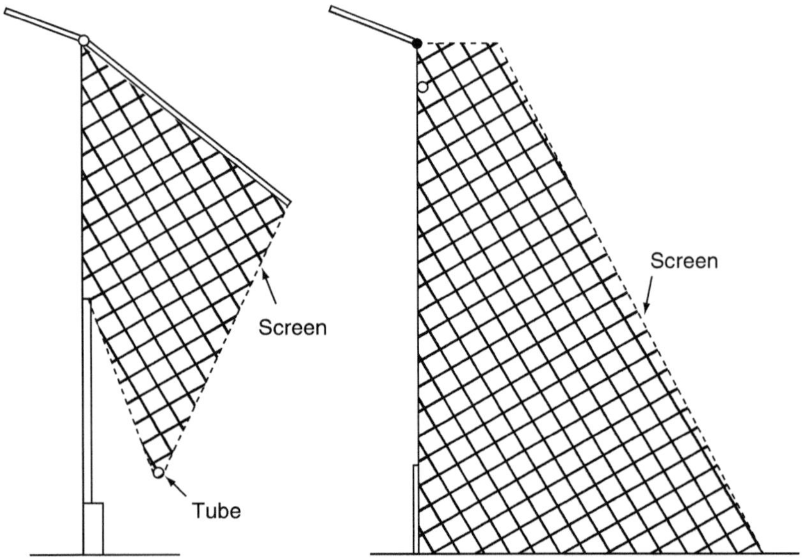

Fig. 10.3 Enlargement of existing vent openings covered by insect screens

Fig. 10.4 Screened side wall vent openings (USA)

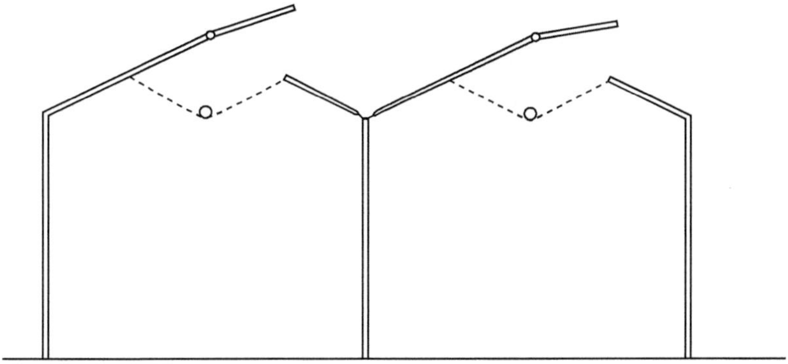

Fig. 10.5 Screened ridge vent openings

inside the net will be raised up and the net is stretched. Screened porches can be installed in front of smaller roll-up ventilation openings.

The inclination of screens relative to the horizontal air stream influences the flow rate through the screen (Teitel et al. 2008a; Teitel et al. 2008b; Teitel et al. 2009). Experiments were carried out with screens of different porosities, and with inclinations of 45°, 90°, and 135° to the horizontal air flow direction. A forty-five degree inclination means that the upper edge of the screen is directed towards the air flow. The flow rate is highest with an inclination of 45°, followed by 90° and 135°. The air flow with 45° inclination is 10 and 6% higher than with 90° for porosities of 0.4 and 0.52 respectively. The air flow through a 135° inclined screen is 14 and 6% lower than with 90° for 0.4 and 0.52% porosity.

10.2 Necessary Enlargement of Screened Vent Openings

The question is:

How much have the vent openings to be enlarged to guarantee nearly the same climate conditions as in unscreened greenhouses?

Harmanto (2006) investigated insect screens under a hot climate in Thailand (Table 10.1).

10.2.1 Method of Sase and Christiansen (1990)

Sase and Christiansen (1990) presented a diagram that gives the relation of the ratio screened vent opening to unscreened opening (A_S/A_V) versus the discharge coefficient (C_d) for screened vents to achieve the same ventilation rate.

Table 10.1 Insect screens investigated by Harmanto (2006), Harmanto et al. (2006a, b)

	Econet M 40 mesh Anti leaf miners and larger	Econet SF 52 mesh Anti white flies and larger	Econet T 78 mesh Anti thrips and larger
Hole size (mm)	0.4 × 0.45	0.25 × 0.8	0.18 × 0.29
Thread diameter (mm)	0.25	0.31	0.19
Discharge coefficient C_d	0.31	0.28	0.21
Screen porosity ε	0.41	0.38	0.3

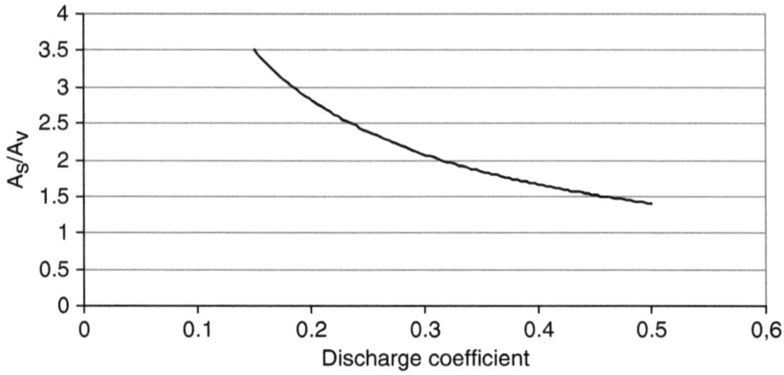

Fig. 10.6 Relation of screened vent opening to unscreened opening depending on discharge coefficient (Sase and Christianson 1990)

Figure 10.6 shows the function for 50° flap opening (global radiation: 500 W/m^2; wind speed: 1 m/s). This function approximately can be taken for open ventilators. The function can be represented by the equation:

$$\frac{A_S}{A_V} = 0.8294 \times C_d^{-0.7608} \tag{10.1}$$

The enlargement of the screened vent opening A_S in comparison to the unscreened opening A_V becomes for the three screens:

	Econet M	Econet SF	Econet T
C_d (Harmanto 2006):	0.31	0.28	0.21
A_S/A_V	2.02	2.18	2.72

10.2.2 Calculation of A_S/A_V by using methods of Bailey et al. (2003) and Teitel (2006)

The pressure loss through ventilation openings without insect screen is:

$$\Delta p_0 = \frac{1}{2} k_0 \rho \left(\frac{V_V}{A_V}\right)^2 \tag{10.2}$$

For a rectangular opening without flap is:

$$k_O = [1.9 + 0.7\exp\{-(L_O/32.5 \times H_O)\}] \tag{10.3}$$

L_O = length of vent opening
H_O = height of vent opening
With $L_O/H_O = 40$ is
$k_O = 2.64$
The pressure loss through a screen is:

$$\Delta p_S = \frac{1}{2} k_S \rho \left(\frac{V_V}{A_S}\right)^2 \tag{10.4}$$

with

$$k_S = \left[\frac{1-\varepsilon^2}{\varepsilon^2}\right]\left\{\frac{18}{Re} + \frac{0.75}{\log\,(Re + 1.25)} + 0.055\log Re\right\} \tag{10.5}$$

The Reynolds number was calculated with the thread diameter

$$Re = \frac{d \times v_w}{\nu}$$

The pressure loss coefficient k_S for different porosities ε becomes with $Re = 20$:

ε	0.5	0.41	0.38	0.3
k_S	3.07	7.59	9.1	15.5

The total pressure loss through the ventilator opening with insect screen Δp_{Sco} is the sum of the pressure losses Δp_O and Δp_S (Teitel 2006; Bailey et al. 2003; Fatnassi et al. 2002):

$$\Delta p_{Sco} = \Delta p_O + \Delta p_S = \frac{1}{2}\rho\left(\frac{V_V}{A_S}\right)^2 (k_S + k_O) \tag{10.6}$$

If the ventilation openings for unscreened and screened greenhouses have the same size and if the air flow should be the same to keep the same temperature difference, then the pressure loss through the screened opening is much higher than through the unscreened opening, because $k_S > k_O$. Higher pressure loss reduces the ventilation rate. Assuming that the pressure losses should not be too different to guarantee a reasonable ventilation efficiency, one can assume

$$\Delta p_O = \Delta p_{Sco}$$

Combining equation (10.2) and (10.6):

$$k_O \left(\frac{V_V}{A_V}\right)^2 = (k_S + k_O)\left(\frac{V_V}{A_S}\right)^2 \tag{10.7}$$

The discharge coefficient C_d is by definition

$$C_d = \frac{1}{\sqrt{k}}$$

If V_V should not change, (10.7) becomes

$$\frac{A_S}{A_V} = \sqrt{\frac{k_S + k_O}{k_O}} \tag{10.8}$$

The enlargements of the ventilation openings for different porosities ε are:

Screen		Econet M	Econet SF	Econet T
Porosity ε	0.5	0.41	0.38	0.30
Enlargement A_S/A_V	1.47	1.96	2.1	2.62

The values correspond well with the data of Sase and Christiansen.

The ventilation openings covered by different insect screens for various insects have to be enlarged by the following factors to get the same climate conditions compared to unscreened openings:

Econet M against leaf miners $A_S/A_V = 1.9–2.03$

Econet SF against white flies $A_S/A_V = 2.1–2.2$

Econet T against thrips $A_S/A_V = 2.6–2.7$

10.3 Climate Conditions and Pest Control

The climate conditions of screened greenhouses have been investigated by several authors.

Fig. 10.7 shows a screened greenhouse for tropical regions. Harmanto (2006) measured the climate conditions in this greenhouse with different ratios of vent opening (Figs. 10.8 and 10.9).

The functions for the air exchange rate for fully cropped and empty greenhouse can be expressed by:

Fully cropped greenhouse:

$$N = -24.65\left(\frac{A_V}{A_G}\right)^2 + 46.5\frac{A_V}{A_G} + 11.16 \tag{10.9}$$

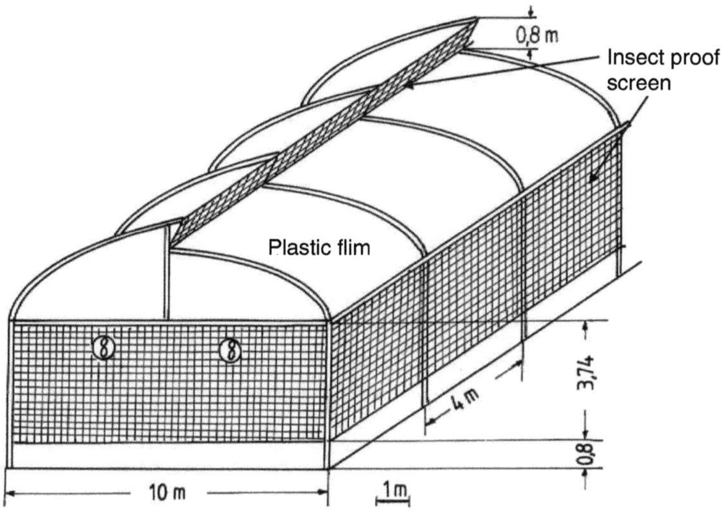

Fig. 10.7 Screened greenhouse for tropical regions (manufactured in India, built and investigated in Thailand). The relation A_S/A_G is 1.05 with fully opened ventilators. The vent openings were covered by a screen against white fly (52-mesh), porosity $\varepsilon = 0.38$ (Harmanto 2006; Harmanto et al. 2006a, b)

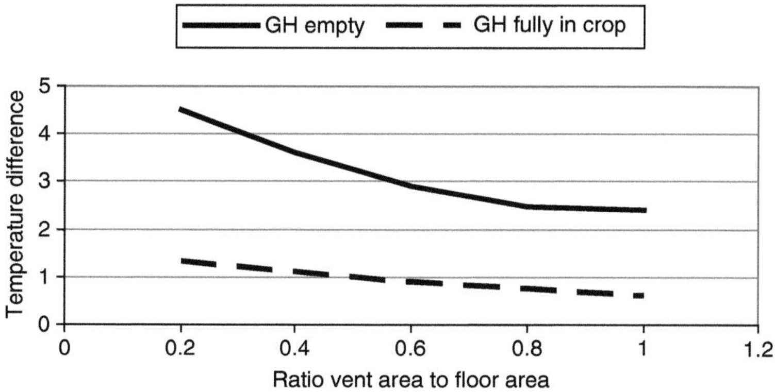

Fig. 10.8 The temperature rise in the greenhouse, shown in Fig. 10.7, depending on the ratio A_S/A_G for fully cropped and empty greenhouse

Empty greenhouse:

$$N = -\left(\frac{A_V}{A_G}\right)^2 + 46.5\frac{A_V}{A_G} + 7.21 \tag{10.10}$$

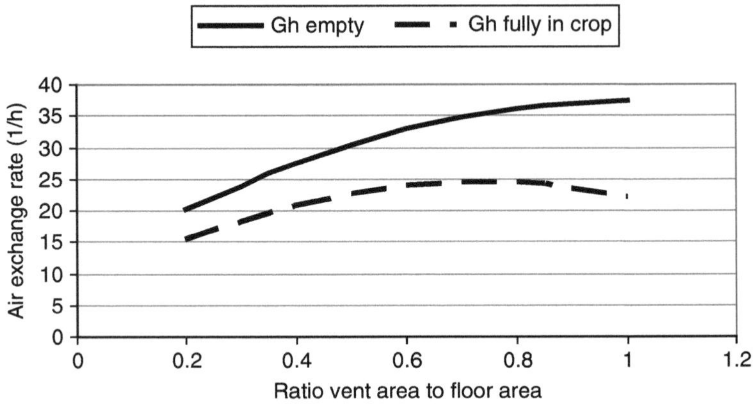

Fig. 10.9 Air exchange rate N (1/h) in the greenhouse, shown in Fig. 10.7, depending on the ratio of vent opening to floor area, vent opening covered by 52-mesh insect screen (see Table 10.1)

Table 10.2 Mean values for air exchange N, air temperature T and relative humidity φ (Harmanto et al. 2006b)

Screen	N (1/h)	V_E (m^3/m^2h)	T (°C)	φ (%)
Anti leaf miners (40-mesh)	52	250	30.8	69.7
Anti white fly (52-mesh)	33.6	161	31.1	70.3
Anti thrips (78-mesh)	26.1	125	31.9	74

A minimum vent area of 60% is necessary to maintain temperature rise and air exchange favourable for growing tomatoes in tropical greenhouses. The temperature difference was below 3°C in the empty greenhouse or in a greenhouse with small plants. The air exchange rate was not significantly changed with vent ratios above 60%.

The combination of side wall and roof ventilation plays a significant role in the greenhouse shown in Fig. 10.7.

If the ventilator opening in unscreened greenhouses is 25–29%, the vent opening in screened greenhouses should be about twice the size of that in the unscreened greenhouse.

Several scientists (reported in Teitel 2006) found a factor of about two for the temperature rise between screened and unscreened greenhouses for an anti white fly screen.

All those results confirm the necessary enlargement of the vent opening in screened greenhouses by about twice and more.

The climate conditions and air exchange were measured for the different screens (Table 10.1.) in three identical greenhouses (Fig. 10.7), with $A_S/A_G = 1.05$ (Table 10.2) (Harmanto et al. 2006b).

The temperature differences under the different insect screens with high rate of vent to floor area are not very different, although the mean air exchanges vary considerably.

Blue and yellow coloured sticky traps (10 × 12 cm) were mounted at various randomly selected locations in the greenhouses to capture white flies and thrips (Table 10.3).

The 40-mesh screen (anti leaf miners) was unable to exclude white flies. Only little thrips were found under 52- and 78-mesh screens on young plants, but many insects were found on mature plants. Bethke (1994) recommended much smaller mesh sizes to exclude thrips, but it is not possible to use these screens under tropical conditions, because of the reduction of ventilation rate and rise of inside temperature. The compromise of 78-mesh screen was unable to exclude thrips entirely, but it can reduce the infestation considerably.

The 52-mesh and 78-mesh screens cause similar climate conditions with a vent ratio of 1.05 and similar exclusion of insects on young plants. Thus, the 52-mesh screen can be taken for tropical regions.

Crop infestation can be reduced by UV-blocking greenhouse-covering materials, because the orientation of white flies, thrips and aphids are dependent on UV light (see Sect. 7.3). Material hindering insect invasion but permitting effective ventilation is desirable for humid tropics. Net or screen houses are not suitable for the humid tropics, because of heavy rain falls. Plastic film covering on the roofs, and suitable insect screens for the ventilation openings, should be used. Several combinations of UV-blocking PE film for roof covering and UV-blocking insect screens were investigated in eight small greenhouses with vertical side walls (Kumar and Poehling 2006).

The covering materials were:

- UV-blocking Bionet screen (50 mesh)
- UV-transmitting screen
- UV-blocking PE film
- UV-transmitting PR film

The four combinations were:

A: UV-blocking screen and UV-blocking film BN + BP

B: UV-transmitting screen and UV-blocking film TrN + BP

C: UV-blocking screen and UV-transmitting film BN + TrP

D: UV-transmitting screen and UV-transmitting film TrN + TrP

Each greenhouse had doors that were opened every morning from 6.00 to 10.00 to let in the insects during their peak activity time. White flies and aphids were captured by yellow sticky traps, YST. Figures 10.10 and 10.11 show the results.

Table 10.3 Mean values of captured insects in the greenhouses covered by three different insect screens

Screen	White flies/trap		Thrips/plant	
	Young	mature plants	Young	mature plants
Anti leaf miners (40-mesh)	0.43	1.46	2.08	77
Anti white fly (52-mesh)	0	0.08	0.63	42
Anti thrips (78-mesh)	0	0.04	0.55	21

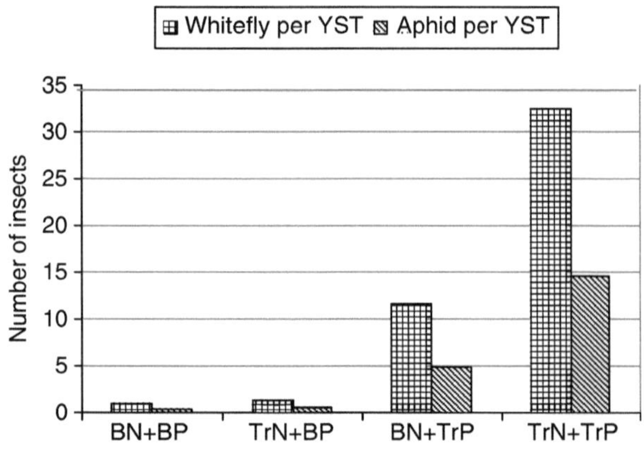

Fig. 10.10 Weekly mean number of white flies and aphids on yellow sticky traps YST 28 days after transplanting (Kumar and Poehling 2006). UV-blocking plastic film has the best effect on keeping out the pest insects, but UV-blocking insect screen keeps out many insects even if the plastic film is UV-transmitting, BN + TrP

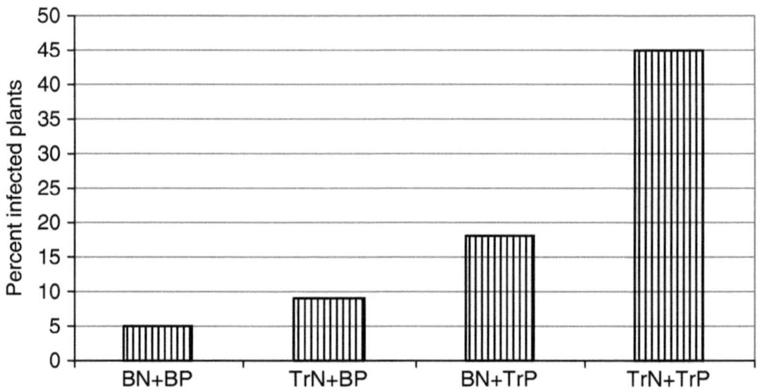

Fig. 10.11 Percentage of virus-infected tomato plants 35 days after transplanting (Kumar and Poehling 2006). UV-blocking insect screens protect the plants additionally from infestation by virus.

A three-span round-arched plastic-film greenhouse, 6.4 m width of span, 12 m length, 2.5 m height to the eave, 4 m ridge height, 230 m^2 floor area, has been investigated with two ventilation openings (Munoz et al. 1999):

G1: Half through roof flap vent on each span, linked to the ridge and opened 0.6 m at the gutter, maximum vent opening $A_V = 21.6\ m^2$, $A_V/A_G = 0.094$.

G2: The flap vent was replaced by a rolling-up roof vent so that each span has vent opening of half roof area. $A_V = 128.5\ m^2$, $A_V/A_G = 0.56$.

An anti-aphid insect screen of 0.4 × 0.4 mm mesh size and a porosity $\varepsilon = 0.45$ was installed on all vent openings.

The ventilation flow rate V_V was calculated for the ventilation openings:

Opening	Ventilation flow rate (m^3/s)	V_V for 2 m/s wind speed
G1 with anti-aphid screen	$V_V = 1.64v_w - 0.77$	2.51 (m^3/s)
G1 without anti-aphid screen	$V_V = 3.44v_w + 3.79$	10.67 (m^3/s)
G2 with anti-aphid screen	$V_V = 4.34v_w - 0.34$	8.34 (m^3/s)

The enlargement of the vent opening from G1 to G2 by five times causes an increase in ventilation flow rate by the factor 3.3 for screened openings at a wind speed of 2 m/s. The ventilation flow rate of G1 with screen is 76% lower than for G1 without screen. The ventilation flow rate for G2 with screen is not very different from G1 without screen.

The discharge coefficients for the vent openings were different at the central and lateral spans. The mean values for the discharge coefficients were found:

G1 with screen $C_d = 0.419$

G2 with screen $C_d = 0.121$

The global wind pressure coefficient C_w depends on the type of vent, type of screen, wind speed, wind direction and surroundings of the greenhouse. The mean wind-pressure coefficients were found for wind speeds $2 < V_W < 4$ m/s:

G1 with anti-aphid screen $C_w = 0.089$

G1 without screen $C_w = 0.32$

G2 with anti-aphid screen $C_w = 0.31$

Montero et al. (1999) evaluated the same greenhouse with an anti-thrips screen (0.18×0.18 mm mesh size) and found less than half of the ventilation flow rates in comparison to the anti-aphid screen. They got for $V_W = 2$ m/s:

G1 with anti-thrips screen $V_V = 1.3\ m^3/s$

G2 with anti-thrips screen $V_V = 3.5\ m^3/s$

They concluded that the climate conditions are not suitable enough by natural ventilation under anti-thrips screen.

The pest control in the greenhouses G1 without screen and G2 with anti-aphid screen gave the following results of pest presence:

	Greenhouse G1	Greenhouse G2
White flies	Very serious attack	Light attack
Thrips	Very serious attack	Very serious attack
Aphids	Light attack	No attack

Katsoulas et al. (2006) investigated a single-span round-arched greenhouse with vertical side walls, 8 m width, 2.4 m height to the eave, 4.1 m height to the ridge, 160 m^2 floor area, and 572 m^3 greenhouse volume. The greenhouse was equipped with rolling-up side walls, $2 \times 0.9 \times 15\ m = 27\ m^2 = 117\%$ vent opening area, and a flap roof vent linked to the ridge, 2 m long, 0.9 m maximum opening height equal to 18 m^2 opening area. An anti-aphid insect screen with 50% porosity was installed.

The calculation resulted in the following values:

Vent	Screen	$C_dC_w^{0.5}$
Side wall	no	0.078
Side wall	yes	0.052
Roof flap	yes	0.028
Side wall and roof	yes	0.096

Neglecting the stack effect and considering the equation for the wind-driven ventilation only, the ventilation flow rate becomes (see Chap. 9):

$$V_V = \frac{A_S}{2} v_w C_d \sqrt{C_w}$$

In order to obtain the same ventilation flow rate for the screened and unscreened side wall ventilation and with the same wind speed, one can get the relation of screened and unscreened vent opening.

$$\frac{A_S}{A_V} = \frac{C_d\sqrt{C_W}}{C_{dS}\sqrt{C_{wS}}} = \frac{0.078}{0.052} = 1.5$$

This is nearly the same enlargement factor of 50% as given above for a porosity of $\varepsilon = 0.5$.

With regard to ventilator openings with screen, the most effective vent configuration was the combined roof and side wall, while the roof vent was the least effective one.

Fatnassi et al. (2006) carried out experiments and CFD calculations (computational fluid dynamics) in a four-span round-arched plastic-film greenhouse, 922 m^2 floor area (4 × 9.6 m width, 24 m long and 5.9 m maximum height) in Southern France. The greenhouse was equipped with roof flap vents, maximum opening height 1.5 m, and rolling-up side wall vents. Two types of insect screens were considered:

	Hole size (mm)	Thread diameter (mm)	Porosity ε
Anti white fly	0.78 × 0.25	0.22	0.41
Anti thrips	0.18 × 0.18	0.22	0.2

The parameter $C_dC_w^{0.5}$ for the roof flap vents with screen was 0.03. This value corresponds very well with the figure of 0.028 obtained by Katsoulas et al. (2006) for roof flap vents with an insect screen.

The mean temperature difference in the greenhouse with and without screens for the configuration open windward roof and side wall vent was:

Screen	Temperature difference (°C)
Without	2.4
Anti white flies	5.1
Anti thrips	7.1

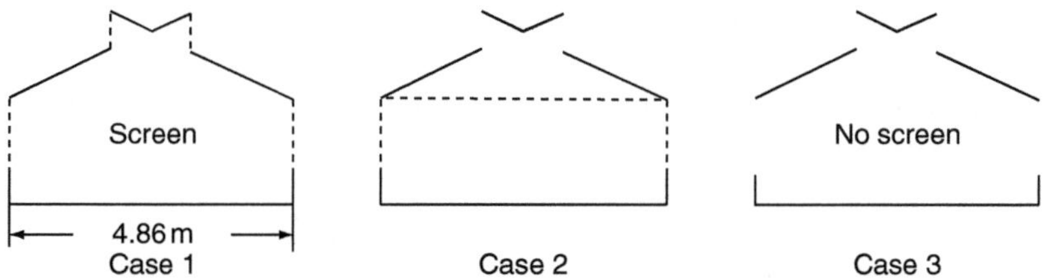

Fig. 10.12 Greenhouses with different screen installations (Sase et al. 2008)

The temperature rise under the white fly screen is twice the temperature difference without screen, and nearly three times that under thrips screen. The only possibility of getting appropriate climate conditions is the enlargement of the screen openings. Climate control under anti-thrips screen is very difficult by natural ventilation.

Sase et al. (2008) investigated three identical small greenhouses with continuous ridge vents and sliding door type side vents with different types of screen installation (Fig. 10.12). The dimensions of the greenhouses were: 4.86 m width, 7.44 m length. 1.74 m eaves height, and 2.98 m ridge height. The ratios of the vent area were: for the roof vent $A_V/A_G = 0.23$; for the side vent $A_V/A_G = 0.25$. Screen porosity 55%, thread diameter 0.15 mm, hole size 0.41 × 0.45 mm. Three cases were investigated:

- Case 1: Screens at side wall and ridge vent opening
- Case 2: Horizontally installed and side vent screen, about 10% light loss.
- Case 3: Control without screen.

The mean natural ventilation rate on a clear day with 782 W/m^2 radiation peak was 40 and 77% for cases 1 and 2 respectively in comparison to case 3. The horizontal installation of a screen might be a less limiting factor to the ventilation rate compared to the screen installation at the vent openings, but with the consequence of about 10% light loss.

The natural ventilation rate increases with increasing side vent opening, and the temperature difference decreases correspondingly:

Side vent opening (%)	Ventilation rate N (1/h)		
	Case 1	Case 2	Case 3
0	17.3	23.5	29
50	25.5	39	42.8
100	27.5	48.3	84.5

Ben-Yakir et al. (2008a, b) published the proposal to cover the vents only when and where risks of pest invasions are imminent. They made studies about the behaviour of white flies in a non-screened four-span greenhouse in Israel, north–south-oriented,960 m^2 floor area, 240 m^2 side vents, and four vertical roof vents with 84 m^2 total open area.

The number of white flies trapped in the greenhouse with closed side vents and open roof vents was:

Hour of the day	Night	7–8	8–9	9–10	10–11	11–12	12–13
% trapped white flies	0	2	20	41	28	8	1

There is a majority of white fly penetration in the morning hours between 8.00 and12.00. The number of white flies trapped at the eastern vents was two to three times higher than those trapped at the other vents.

The greenhouses for the studies of thrips behaviour were east–west-oriented round-arched tunnels, 6–8 m wide and 250–1,000 m^2 floor area, with open doors at the ends for ventilation. Two poles with blue sticky traps were positioned at the entrances of the tunnels and two in the open field nearby. Sixty to eighty percent of the thrips were caught below 1.0 m above ground in the open field: 85% were caught in the morning between 7.00 and 11.00, and 10% at dusk, when the wind speed was below 9 km/h (1.05 m/s). The prevailing wind speed exceeded 10 km/h between 11.00 and 18.00. The prevailing wind direction was south–west. The number of thrips caught at the eastern entrances was twice the number caught at the western entrance. This means that most of the thrips were caught at the leeward side, upwind.

Both white flies and thrips are relatively weak flyers, and therefore their penetration is affected by the prevailing wind direction and the air movement around the crop. The risk of penetration of the pests was significantly higher at the leeward side.

All these results indicate that greenhouses should be designed and positioned to minimise the risk of pest entry by wind flow. Ventilation openings and rolling-up screens should be correspondingly controlled.

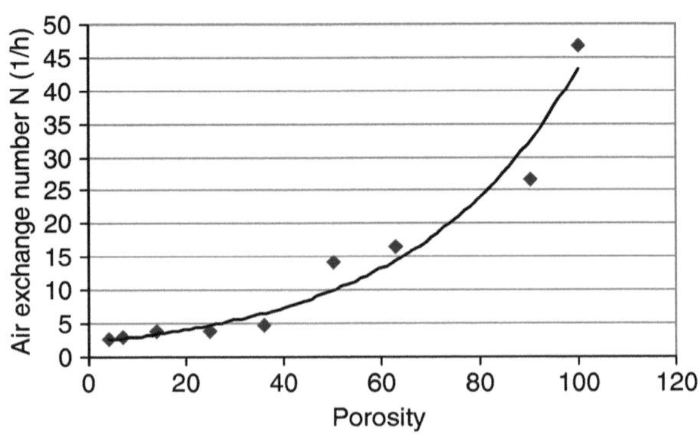

Fig. 10.13 Air exchange number N (1/h) depending on the screen porosity ε (Bartzanas et al. 2009)

A proportionality between screen porosity and air exchange number N (1/h) has been evaluated by Bartzanas et al. (2009) in a round-arched plastic-film-covered greenhouse with vertical side walls at 39°44′N latitude in Greece. The greenhouse, 2.4 m gutter height, 4.1 m ridge height, 8 m width, and 280 m length, had two side wall roll-up vents of 27 m^2 vent area (2 vents 0. 9 m opening, 15 m length). The ratio of vent area to floor area was 17%. The vents were covered by eight different

Fig. 10.14 Very well-installed and fixed insect screens at side walls and ridge ventilators

Fig. 10.15 The insect screens are not fixed tightly enough to the structure, so that leaks occur and pest insects can penetrate

insect screens with porosities from 4 to 90%. Figure 10.13 shows the air exchange number N (1/h) depending on the porosity of the tested screens. The temperature difference inside to outside the greenhouse changed from 4.14°C for 4% porosity to 3.12°C for 90% porosity, and 2.08°C for open vents without screens.

Figure 10.14 and 10.15 show examples for well-installed and insecure screens.

Chapter 11
Cooling

The plant temperature in greenhouses can exceed the air temperature by 5–10°C in case of high global radiation in hot seasons and in arid regions, even with good ventilation. Efficient ventilation provides insufficient cooling for many plants. Even the air temperature is too high in many climates. It is also necessary to increase the inside humidity for appropriate crop growth in arid and semi-arid regions with very low outside humidity. For additional cooling and increasing humidity of greenhouses, evaporative cooling by the following systems may be used (von Zabeltitz 1999, 1986a)

- The fan and pad system
- The fog system
- The spray cooling system

Evaporative cooling can compensate for the lack of ventilation in greenhouses with insect screens, and is necessary in arid and semi-arid climates. The possibility and necessity of evaporative cooling in subtropical climates has to be checked before installation, because in some cases (crop and crop cycle) no advantage of evaporative cooling may be found.

Different methods of cooling were evaluated in southern Spain (36°41′N), for example in a greenhouse with a pepper crop (Gazquez et al. 2006):

1. Forced ventilation by fans with roof vents opened 30% of maximum aperture.
2. Natural ventilation and high pressure fog cooling system.
3. Natural ventilation plus shading by whitening, 30% reduction of solar radiation transmittance during the first two months of crop cycle before September and natural ventilation only after removing the whitewash in September.

The summary of results was:

- The forced ventilation was the least effective strategy. It was not possible to reduce the temperature sufficiently during the first weeks of cropping before September.
- The temporary reduction of 30% solar radiation did not affect the total and marketable yield.

C. von Zabeltitz, *Integrated Greenhouse Systems for Mild Climates*,
DOI 10.1007/978-3-642-14582-7_11,

- More fruits were infested by blossom end rot under the fogging system because of higher humidity.
- The temporary use of whitening was the most effective system with regard to yield and water use efficiency in the subtropical climate of southern Spain.

Further experiments were carried out under the same conditions (greenhouse, location and climate) with the following cooling configurations (Gazques et al. 2008):

A. Natural ventilation plus whitening by calcium carbonate, and natural ventilation only after washing off the whitening in September.
B. Natural ventilation with high pressure fog cooling system, and whitening with reduced concentration of calcium carbonate.

The transmittance was 12.5% lower in A than in B during the shading period. The air temperatures were similar in both configurations during the whole growing cycle, but the canopy temperature was lower under the whitening. The total and marketable yield was higher under the whitening (A). Thus, natural ventilation plus temporary whitening of the cladding material before September was the most appropriate strategy for a pepper crop under the subtropical climate of southern Spain.

However, evaporative cooling may be the only possibility for successive crop growing under arid and semi-arid climate conditions.

The precondition for sufficient evaporative cooling is a low outside air humidity.

The rates of air and water flow necessary for air cooling inside the greenhouse depend on:

- The radiation inside the greenhouse, influenced by the cladding material and shading
- The heat transfer through the cladding material
- The evaporation of the cooling system
- The efficiency of the system

Before installing a system, the air and water flow rates must be calculated. Fresh water is a scarce resource, therefore brackish or sea water should be used for evaporative cooling if possible. The water use efficiency of evaporative cooling systems, investigated by Sabeh (2007) will be presented in Chap. 13.

11.1 Fan and Pad Cooling

The fan and pad cooling system is most commonly used in horticulture. Air from outside is blown or sucked through pads with a large surface. The pads are kept permanently wet by sprinkling water that evaporates on the surface of the pad and

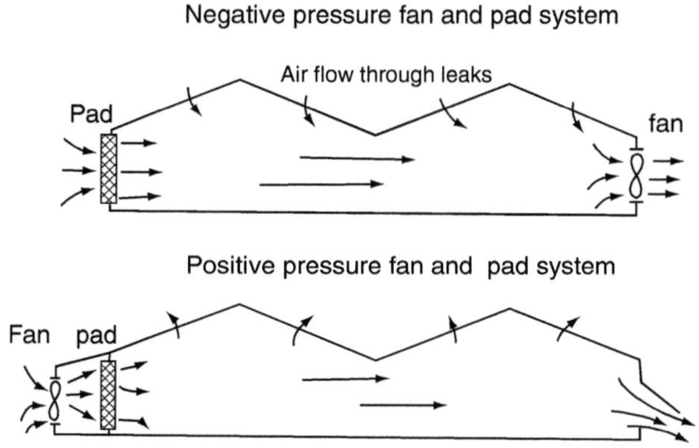

Fig. 11.1 Negative and positive fan and pad systems

cools the air down. For optimal cooling the greenhouse should be shaded. There are basically two systems of fan and pad cooling (Fig. 11.1)

The fans suck the air through the pad and greenhouse in the negative pressure system. The disadvantage is that air and dust can get into the greenhouse through leakage.

The fans and pads are located on one side and vents on the other side in the positive pressure system. The fans blow the air through the greenhouse so that an overpressure occurs. The advantage is that dust cannot penetrate the greenhouse, because of the overpressure (Fig. 11.2).

Advantages of fan and pad cooling systems are:

- Simple systems and low water pressure for the distribution of water over the pads.
- Possibility of self-made pads from various materials that are locally available.
- The use of untreated water if simple operation rules are kept.

The *disadvantages* of fan and pad cooling systems are:

- Forced ventilation by fans is a prerequisite.
- Temperature and humidity gradients between the pad and air outlet along the greenhouse.
- Good maintenance to prevent clogging of the pads.

Fan and pad cooling systems should fulfil the following demands:

- Sufficient water use efficiency
- The use of brackish water without clogging of the pad material.
- Efficiency even with strong wind
- Low temperature gradients in the greenhouse
- No dust penetration in desert regions.

The pad itself, the water distribution system, the water and air flow rates, and the pump capacity have to be designed and calculated carefully to provide sufficient

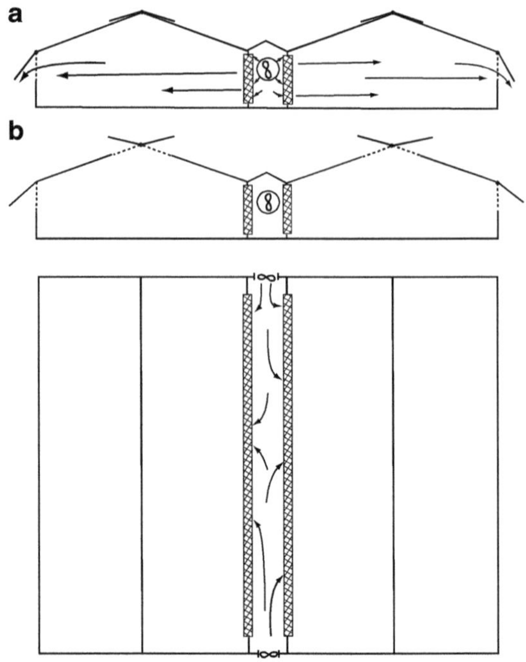

Fig. 11.2 One can combine the positive pressure system with natural ventilation

cooling efficiency and wetting of the pad, as well as to avoid the clogging of the pad by deposition of materials. The following *general rules* should be considered when designing a fan and pad cooling system (ANSI/ASAE 2003; Von Zabeltitz 1986a).

11.2 Pad Location

There are two possibilities for installing the pads, horizontal and vertical pads.

Vertical pads are supplied with water from a perforated tube along the top edge.

Horizontal pads are supplied with water by spraying it over the upper surface. The air flows upwards through the pad. They are more suitable for saline water and for dusty air in desert regions (Fig. 11.3).

11.3 Vertical Pad System and Water Circulation

The vertical height of the pads should be 0.6 m minimum and 2.4 m maximum to guarantee uniform wetting of the pad. The pad must not dry out during operation.

If saline water is used, a high amount of water has to be circulated.

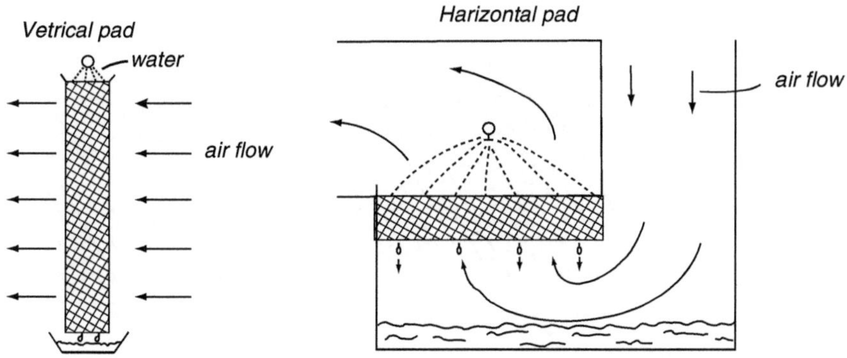

Fig. 11.3 Vertical and horizontal pads

To avoid salt concentration, some of the recirculating water needs to be discharged continuously (bleed-off). The continuous bleed-off should be 0.002 l/min per m^3/s air flow for mineral concentration below 700 ppm in the water and evaporation rate less than 0.012 l/min per m^3/s air flow. In areas with up to 1,500 ppm mineral concentration (the upper limit for irrigation water), the bleed-off rate should be 0.006 l/min per m^3/s air flow (ANSI/ASAE 2003) (Figs. 11.4–11.6).

11.4 Horizontal Pads

Horizontal pads are suitable for arid climates and desert conditions. A high rate of water flow can prevent clogging by sand and dust, which will be washed out (Fig. 11.7–11.8).

11.5 Pad Efficiency

The cooling efficiency of the pad can be calculated by the following equation:

$$\eta = \frac{t_o - t_{in}}{t_o - t_{wb}}$$

t_o = outside air temperature
t_{in} = air temperature just behind the pad
t_{wb} = wet bulb air temperature of outside air
t_{in} can be calculated by: $t_{in} = t_o - \eta(t_o - t_{wb})$

The cooling efficiency should provide an inside humidity of a maximum of 85% at the outlet. A higher humidity slows down the transpiration rate of the plants. The plant temperature can increase above air temperature.

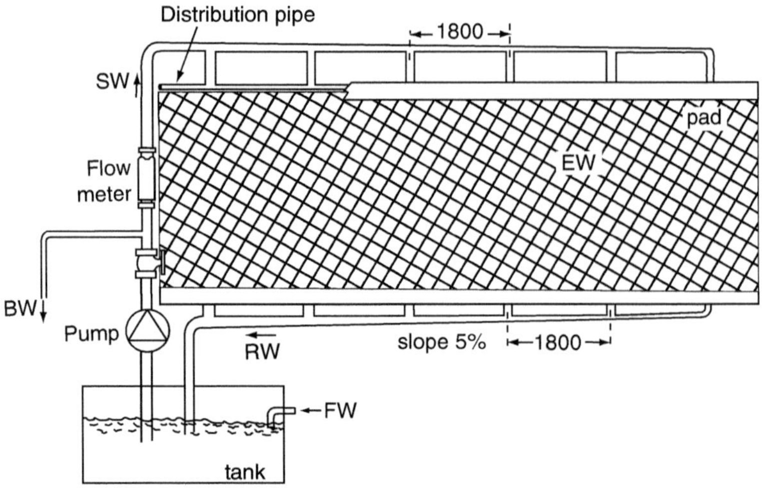

Fig. 11.4 Pad system. Water supply to the pad: SW >6 times EW (Evaporated water). Bleed-off ratio: BW/EW depends on water quality. BW/EW = 0.1–4. Fresh water supply: FW = BW + EW

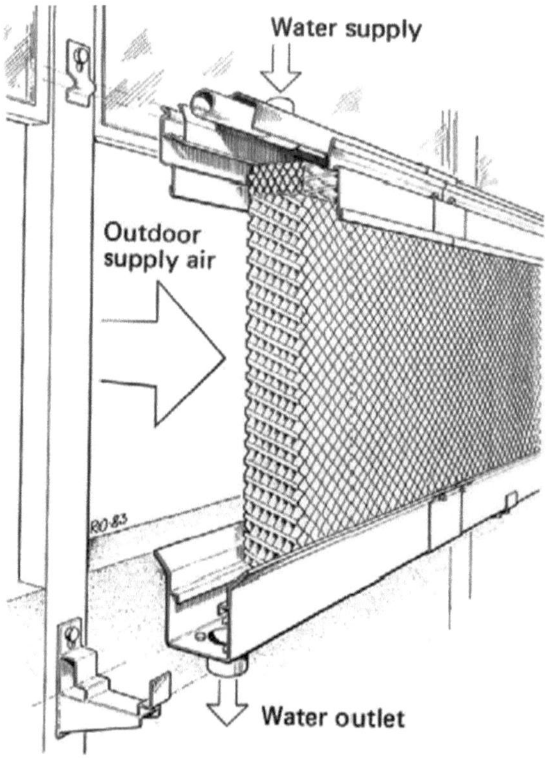

Fig. 11.5 Mounting of pads. The pads must be sealed tightly at the upper and lower edges so that the total air flows through the pad and not through leaks (Celdec company)

Fig. 11.6 Well-installed, operated, and maintained corrugated cellulose pads

Fig. 11.7 Double horizontal pads

11.6 Pad Material and Thickness

Various materials are available for cooling pads, sometimes made from local available materials. Requirements for pad materials are:

- High surface area for air water contact
- Good wetting properties
- High cooling efficiency
- Little pressure loss
- High durability
- No or little shrinking. Shrinking materials have to be filled up if leaks occur.

Fig. 11.8 Horizontal pad, view from below and intensive water flow rate

The pad thickness is about 100–200 mm.

Possible pad materials are corrugated cellulose (commercial material CELdec), wood excelsior, aspen fibre, expanded clay or other locally available materials with high surface area (Figs. 11.9–11.14).

Gunham et al. (2007) reported a literature review about various pad materials and their efficiency (shown in parenthesis), for example wood shavings (69%), discarded clay brick, charcoal, coir fibre (89–90%), fine fabric (47–85%), and coarse fabric (64–86%). Von Zabeltitz (1986a) presented values for wood excelsior (90–95%), coke (70–90%), and expanded clay (85–95%).

Gunham et al. (2007) compared pumice stones, volcanic tuff, shading material and corrugated cellulose as pad materials. A pad efficiency of 70–85% was measured for pumice stones and volcanic tuff, with a water flow rate of 3.5 l/min per m pad length, air velocities of 0.5–1.5 m/s and 150 mm pad thickness. The corrugated cellulose pad had an efficiency of 75–80% and the shading net only 35–50 %.

The pressure losses of pumice stones and volcanic tuff are relatively high compared to the corrugated cellulose material CELdek. The pressure loss of the CELdek material is below 30 Pa for thickness up to 150 mm and air velocity up to 1.6 m/s. The pressure losses of pumice stones and volcanic tuff are below 30 Pa only for 0.6 m/s up to 150 mm pad thickness. Commercial cellulose pads fulfil the conditions with regard to efficiency and pressure loss, but they are very expensive. Volcanic tuff is an alternative, with efficiencies from 63 to 81%, but air flow velocities should be limited to 0.6 m/s.

Another comparison of three pad materials has been reported by Al-Massoum et al. (1998). The pad materials, with pad thickness 100 mm each, were:

A: Mats of date palm fronds weaved from 25 mm-wide date frond leaves.

B: Chopped date fronds, 20 mm chip size.

C: Corrugated cellulose CELdek (Table 11.1).

The cooling efficiency of chopped date fronds is sufficient with high air face velocity through the pad and high water flow rate, but decreases with decreasing air velocity and water flow rate. The efficiency of the corrugated cellulose pad does not vary as much with changing water flow rate at high air velocity, but decreases with decreasing air velocity. The date palm fronds are cheap and

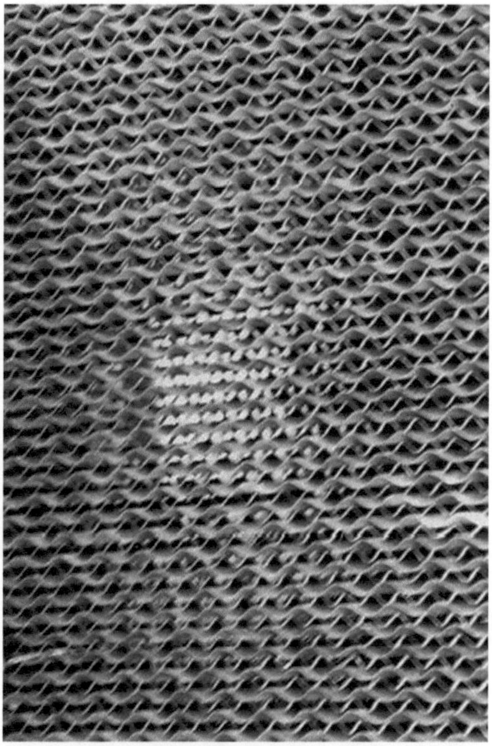

Fig. 11.9 New corrugated cellulose pad

Fig. 11.10 Pads from natural plants and wood excelsior

Fig. 11.11 A pad made from vertical ropes

Fig. 11.12 A single row rope pad is not sufficient because of free entrance for air and low efficiency

available in several countries. They can be an alternative material for cooling pads if the service life is sufficient and the pressure loss through the pad not too high.

11.6.1 Pad Area

The pad area depends on the air flow rate necessary for the cooling system and the maximum approach velocity at the pad. Average approach velocities are

Fig. 11.13 Double- or multi-layer rope pads are recommended

0.75–1.5 m/s. Excessively high velocities cause too great a pressure loss across the pad. Excessive velocities may cause problems with drop penetration into the greenhouse, wetting the plants. Recommended air velocities through various pad materials are (ANSI/ASAE 2003):

Aspen fiber, mounted vertically	0.76 m/s
Aspen fiber, mounted horizontally	1.00 m/s
Corrugated cellulose (100 mm thick)	1.27 m/s
Corrugated cellulose (150 mm thick)	1.78 m/s

Fig. 11.14 If rope pads are clogged, they can easily be cleaned by shaking and knocking

Table 11.1 Cooling efficiencies of the different pad materials

Air velocity (m/s)	Water flow rate (l/min m^2 pad area)	Pad efficiency (%)		
		Pad A	Pad B	Pad C
1.8	5	71.7	90.1	86.3
1.8	2	73.9	60.9	76.9
1.0	5	62.8	71.2	67.8
1.0	2	59.6	63.9	52.2

The pad area should be about 1 m^2 for 20–30 m^2 of greenhouse floor area. The maximum fan to pad distance should be 35–40 m. For very long or wide greenhouses, the installation of fans in the roof or walls at the midpoints and pads on both ends may be possible.

11.6.2 Pad Installation and Maintenance

It is very important that there are no leaks in the pad where the air can pass through without making contact with the pad. It should not be possible to look through a pad (Fig. 11.15).

Fig. 11.15 Leaks in pad systems should be avoided. Shrinking of natural material has to be avoided by continuous refilling

The fan and pad cooling system is sensitive to sand and dust in desert regions, because the pads clog very easy by a mixture of salt and sand. This clogging can be like concrete (Fig. 11.16).

The pads should be *protected from sandstorm* and dust (Fig. 11.17).

The pads have to be *protected from direct sunlight* to prevent local drying out and clogging.

The pads should be located on the side facing the prevailing wind (Figs. 11.18–11.19). Distance between inlet and outlet must be wide enough, Fig. 11.20. The pad has to be sealed Fig. 11.21.

11.6.3 Operation of Fan and Pad System

The pad must not dry out during fan operation, to avoid clogging by sand, dust and calcium (Fig. 11.22).

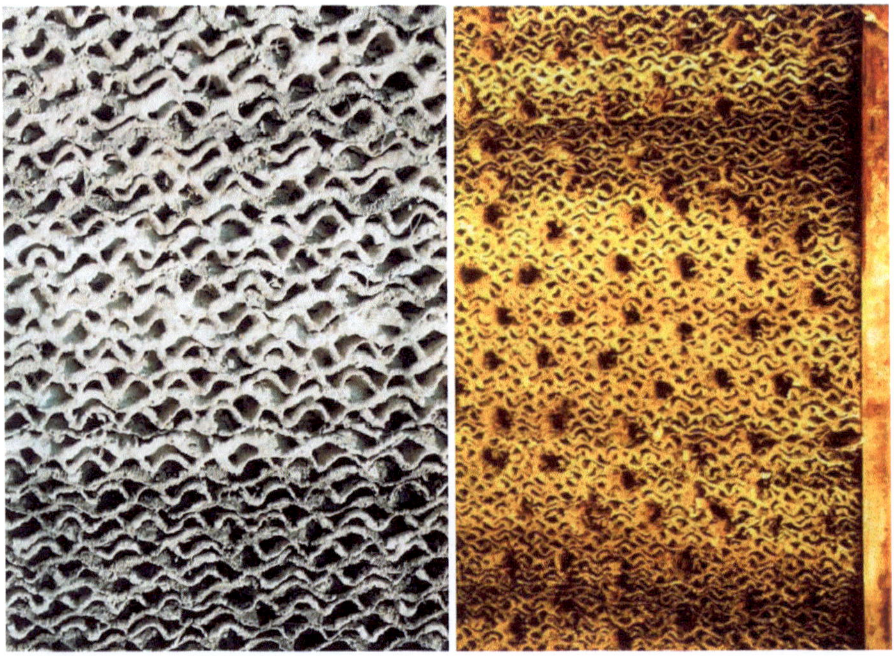

Fig. 11.16 Totally clogged pad. Pads must not be clogged by sand and salt from saline water. Drilled holes in a clogged pad as first aid but without great success

Starting the operation

- Turn on water supply first to prevent pad from clogging.
- Start the fan after the whole pad is completely wetted.

End of the operation

- Turn off the fan first.
- Then turn off the water supply.

11.7 Fans

Belt-driven or directly driven propeller fans are used. Directly driven fans are better to maintain. The overall pressure loss of the fans is about 30 Pa (3 mm H_2O). The fans should be placed on the leeward side of the greenhouse. The distance between the fans should not exceed 7.5–10 m, and the fans should not discharge towards the pads of an adjacent greenhouse less than 15 m away. All exhaust fans should be equipped with automatic shutters to prevent air exchange when the fans are not operating, and also to prevent back draughts when some of the fans are not in operation.

Fig. 11.17 A wall of bricks protects the pad made from aspen fibre, as well as palm fronds in front of the pad

Fig. 11.18 The pads are partly shaded, but one part is under direct sun radiation. Pads must not be located under direct sunlight

11.7.1 Air and Water Flow Rate

The *air flow rate* depends on the solar radiation inside the greenhouse and on evapotranspiration from plants and soil. The air flow rate can be calculated by an energy balance (Chap. 9). Generally, a ventilation rate of 120–150 $m^3/(m^2\ h)$ permits satisfactory operation.

Fig. 11.19 Pads properly shaded by special roof construction or shading cloths

Fig. 11.20 The pad is installed nearby the pad outlet of the next greenhouse, so that the warm and humid air directly enters the next greenhouse and reduces the cooling efficiency remarkably

Fig. 11.21 The pad is not sealed enough at the lower edge, so the air can penetrate without touching the pad surface and being cooled

Fig. 11.22 Partly dried out pads reduce the cooling efficiency

The following water flow rates and water reservoir capacities can be assumed for vertical pads (ANSI/ASAE 2003)

	Water flow rate per m of pad (l/min m)	Reservoir per m^2 pad area (l/m^2)
Aspen fiber, 50–100 mm thick	3.7	20
Aspen fiber, desert conditions	5.0	20
Corrugated cellulose, 100 mm	6.2	33
Corrugated cellulose, 150 mm	9.9	40

The pipe system should include control valves and a gauge for measuring the water flow rate.

Salty or brackish water may be used for fan and pad cooling. The high amount of water flow needs a recirculation system. To avoid a concentration of salt, some of the recirculation water needs to be discharged and replaced by fresh water. The relation of discharged water (bleed-off) to evaporated water depends on the water quality, and has to be calculated according to the manufacturer's guidelines.

Horizontal pads can be supplied by water flow rates of maximum 0.12 l/s per m^2 of pad area.

Willits (2003, 2006) developed a model for forced ventilation and cooling in greenhouses with and without fan and pad cooling. The outside climate conditions were set to:

Solar radiation: 955 W/m^2
Temperature: 36.8°C
Relative humidity: 8.6; 22.8; 37; 54.3; 74.9% (Fig. 11.23).

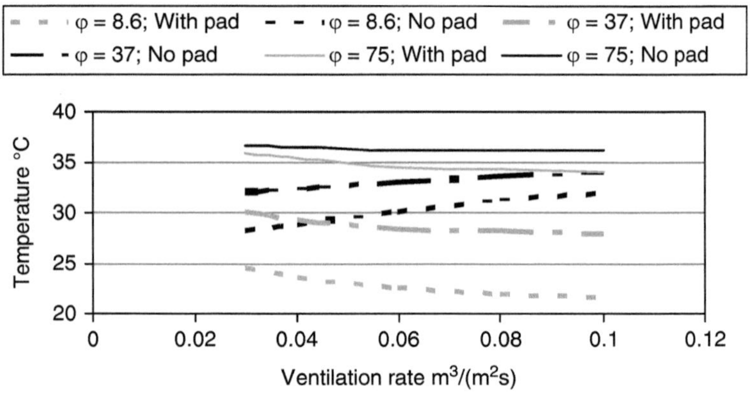

Fig. 11.23 Shows the canopy air temperature at the air exit of the greenhouse depending on ventilation rate V_E and relative humidity φ. The leaf area index was 3.0 m^2/m^2, the outside temperature 36.8°C

Results

1. The air temperatures at the outlet of forced ventilated greenhouses (without pad) are below the outside temperature of 36.8°C for low humidity in arid regions, but the cooling effect decreases with increasing ventilation rate. For high relative humidity (75%), the air temperature remains near the outside temperature of 36.8°C.
2. The canopy air temperature at the outlet increases with increasing ventilation rate V_E for forced ventilation without fan and pad cooling up to an outside relative humidity of 37% (arid conditions). That means there is an increasing amount of incoming dry warm air, and the plant canopy can not cool down this air by transpiration.
3. The canopy air temperature at the outlet decreases with increasing ventilation rate in greenhouses with fan and pad cooling, but the decrease is lower above a ventilation rate of 0.05 m^3/m^2 s or 180 m^3/m^2 h. There is a cooling effect with forced ventilation in arid regions with outside low humidity, but the cooling effect increases remarkably with fan and pad cooling.

Crop production in greenhouses in arid regions can be improved and extended by evaporative cooling, but the cooling systems need a relative high amount of water, which is a rare commodity in arid regions. Sabeh et al. (2006) and Sabeh (2007) investigated two single-span greenhouses with fan and pad cooling in a semi-arid climate with 39°C outside temperature, 15% relative humidity, and 845 W/m^2 solar radiation. The greenhouses had 278 m^2 floor area each, 4 m gutter height, and 6.3 m ridge height. They had a double-inflated PE roof cladding and a 40% shading curtain. The pad had dimensions of 1.1 m height by 8.5 m length, and was designed of 150 mm-thick corrugated cellulose. A tomato crop with 2.3 plants per m^2 was cultivated in the greenhouses.

Figure 11.24 shows the cooling efficiency and water consumption of the fan and pad system depending on ventilation flow rate. The cooling efficiency reduced with increasing ventilation flow rate V_E from 85 to 73.5%. The water consumption per m^2 pad area increased linearly from 504 to 1,404 kg/(m^2 h) for ventilation flow rates per m^2 floor area of 61.2–284.4 m^3/m^2 h.

The mean air temperature decreased slightly with increasing ventilation rate of about 180 m^3/m^2 h, and was then constant, Fig. 11.25. The mean relative humidity decreased with increasing ventilation flow rate in the fully cropped greenhouse.

If the minimum acceptable relative humidity for a tomato crop is 50%, then the ventilation rate should be lower than 180–200 m^3/m^2 h. If the relative humidity for the crop should be near 70%, then the lowest ventilation rate of about 60 m^3/m^2 h is optimal. If water saving is the priority, then low ventilation rates are to be preferred.

The temperatures increase from pad to fan, but the increase is lower in the fully cropped greenhouse. The differences of air temperatures from pad to fan decrease with increasing ventilation rate from 6.8°C for $V_E = 61$ m^3/m^2 h to 2°C for $V_E = 284$ m^3/m^2 h, but there is no significant decrease beyond 180 m^3/m^2 h.

There are no significant changes in mean air temperature, relative humidity and air temperature differences from pad to fan at ventilation rates above 180 m^3/m^2 h. Increasing ventilation rate above 180 m^3/m^2 h did not produce an overall better climate for growing tomatoes (Fig. 11.25).

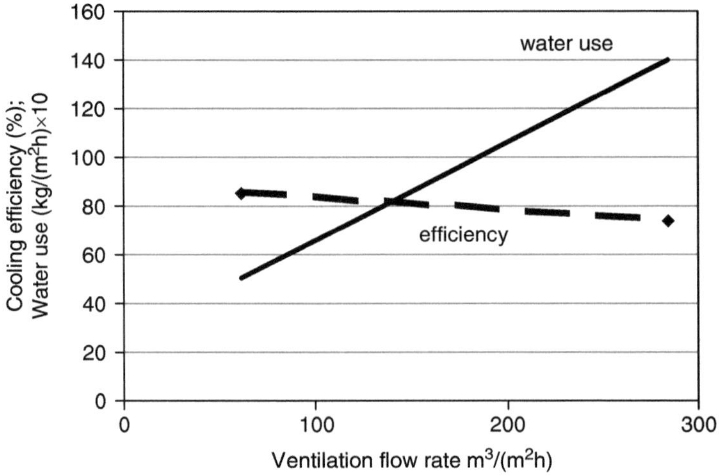

Fig. 11.24 Cooling efficiency and water consumption of fan and pad cooling system (Sabeh et al. 2006)

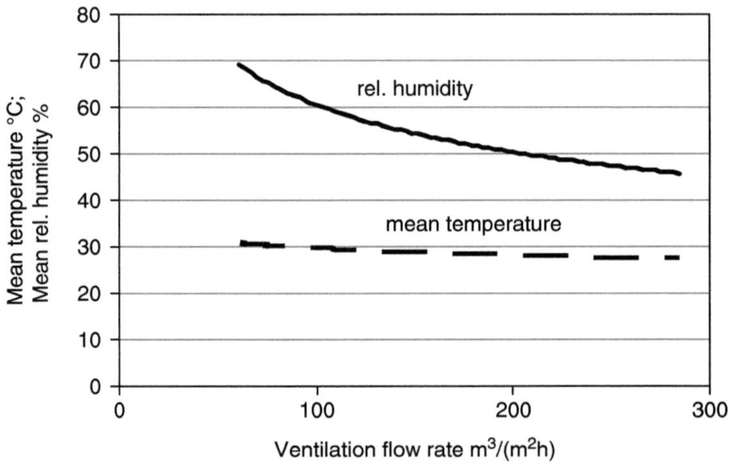

Fig. 11.25 Mean temperature and mean relative humidity in evaporative cooled greenhouse with a tomato crop (Sabeh et al. 2006)

11.8 Fog System

The fog cooling system is based on spraying very small water droplets inside the greenhouse above the crop area. The drops in the fog should have small diameters to create a large surface for evaporation, and the drops should not fall down into the crop area but should evaporate during free fall in the air stream inside the greenhouse. The fog droplets can be generated by the following techniques (Arbel et al. 1999; Li and Willits 2008):

- Twin fluid nozzles combining an air flow of 2–2.5 bar and a water flow of 2–5 bar. The pressures are relatively low, to generate the necessary droplet sizes, but the system needs an air compressor and double tubes for air and water flow to the nozzles.
- Low pressure nozzles work at water pressure of 3–5 bar, and create droplet sizes of 50–100 μm.
- High pressure nozzles need water pressure of 35–70 bar and very clean pre-treated water, because the orifice diameters of the nozzles are very small. The droplet diameters are 2–60 μm.

The *advantage* of fog cooling systems is the more uniform temperature and humidity distribution, and the possible operation with forced ventilation as well as with natural ventilation.

A fog cooling system consists of a water softener, good filters to prevent nozzle clogging, a water reservoir, pumps, pressure regulation valves, and tubes with nozzles installed above the crop. The distances between nozzle tubes and nozzles on the tubes have to be designed according to the rules of manufacturers. The water and air flow rates depend on the desired inside climate conditions, the ambient air temperature and humidity.

The cooling efficiency of fogging systems is defined as (Li et al. 2006; Li and Willits 2008):

$$\eta_{\text{cool}} = \frac{t_{\text{unfog}} - t_{\text{fog}}}{t_{\text{unfog}} - t_{\text{wbfog}}}$$

t_{unfog} = air temperature in the unfogged greenhouse.
t_{fog} = air temperature in the fogged greenhouse.
t_{wbfog} = wet bulb air temperature in the fogged greenhouse.

The fog evaporation efficiency η_{evap} is defined as the ratio of fog evaporation rate to spray rate. It indicates the ratio of fog that contributes to the cooling. The rest of the sprayed fog is shifted out of the greenhouse by ventilation, or sinks down to the ground.

$$\eta_{\text{evap}} = \dot{m}_{\text{e}}/\dot{m}$$

$\dot{m}_{\text{e}}$ = fog evaporation rate per m^2 floor area $(\text{g/m}^2\ \text{s})$
$\dot{m}$ = spray rate per m^2 floor area $(\text{g/m}^2\text{s})$

Montero (2006) presented a comparison of fully cropped greenhouses (LAI = 3), shaded, non-shaded, and with evaporative fog cooling. The outside conditions were 900 W/m^2 solar radiation, 30°C air temperature, and 40% relative humidity.

Figure 11.26 shows the temperature differences inside–outside depending on the air exchange number N (1/h). Shading has more effect on cooling when the air exchange is below 20 (1/h). The air temperature can be decreased to 6°C below outside temperature by fog cooling in this case, but above air exchange numbers of about 45 the cooling effect is more or less constant.

The effect of natural ventilation on the air and foliage temperature is different for low and high outside relative humidity. The foliage temperature can sink below inside and outside temperature in warm and dry climates ($t = 35°C$, $\varphi = 25\%$), but is higher in warm and humid climates ($t = 35°C$, $\varphi = 75\%$).

Sase et al. (2006) and Sase (2008) studied a high-pressure fog system in a greenhouse, 9.6 m wide, 28.2 m long, 4.0 m gutter height, 6.3 m ridge height, covered by an air-inflated double PE film. The greenhouse had a roll-up side vent, 1.48 m wide and a ridge vent, hinged to the gutter, 1.3 m opening width, both covered by an insect screen, porosity 0.25.

The high-pressure system consisted of a nozzle tube along the centre of the greenhouse at gutter height with 68 nozzles, 0.2 mm orifice size, 15 μm droplet size and 10 MPa (100 bar) pressure. The water output per nozzle was 103.3 g/(min nozzle) equal to 25.95 g/m^2 min. The outside radiation was 1,008 W/m^2.

The regression curves of the experiments resulted in the following conclusion:

The relative humidity decreased from 80 to 65% when the ventilation rate increased from 60 to 210 m^3/m^2 h. The water consumption per m^2 floor area increased from 1,080 to 1,260 g/m^2 h for ventilation rates from 60 to 210 m^3/m^2 h.

Ishii et al. (2006) investigated the same greenhouse with a tomato crop, 3.3 plants per m^2 in hydroponics. Two vent configurations were studied with the following mean values

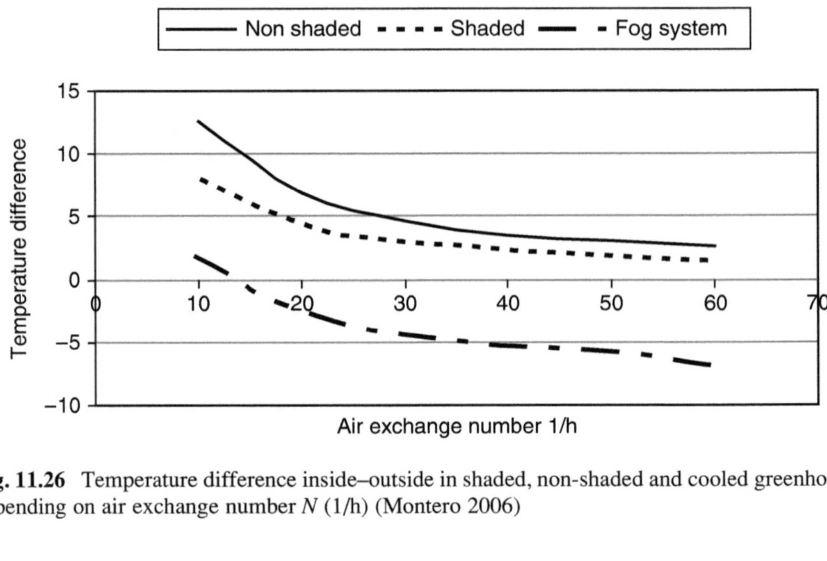

Fig. 11.26 Temperature difference inside–outside in shaded, non-shaded and cooled greenhouses depending on air exchange number N (1/h) (Montero 2006)

	Case 1	Case 2
Roof vent	100% open	100% open
Side vent	Closed	25% open
Ventilation rate (m^3/m^2 h)	72	132
Inside temperature (°C)	24	25.4
Relative humidity (%)	89.7	72.7
Water consumption (g/m^2 h)	792	909

The set point for the inside temperature was 24°C, and the system could keep the temperature in the range of 23–26°C.

With increasing ventilation rate, the inside temperature and the relative humidity increased, and leaf temperature decreased. The water consumption was reduced with decreasing ventilation rate and increasing relative humidity.

A good controlled restricted ventilation rate can decrease the water consumption and keep the inside temperature in the range of set point temperature.

Fog cooling systems have to be combined with appropriate control systems for natural ventilation and temperature to keep the set point temperature and to reduce the water consumption.

Kim et al. (2007) evaluated the performance of a fog cooling system in a three-span glass house without plants in Korea, 7.5 m span width, 2.3 m gutter height, and 28° roof slope. The fog nozzles had a water output of 1.56 g/s nozzle at a pressure of 6.3 MPa (63 bar). The droplet size was 20 μm. Thirty two nozzles with 1 m spacing were installed on two tubes per span.

Different nozzle tube configurations were simulated with various nozzle tube distances from the left windward side wall of the greenhouse. The fog was sprayed upwards, in order to maximize the falling length of the droplets.

The most uniform temperature and relative humidity distribution was achieved with the following configuration:

Nozzle tube height above ground: 2.3 m

Nozzle locations from left side wall: 1.9; 5.6; 9.4; 13.1; 16.9; 20.6 m.

The nozzle tube should be installed as close as possible to the side walls but the spray should not reach the side wall. An air entry speed of 0.5 m/s resulted in the best uniformity in temperature and relative humidity.

Li and Willits (2008) compared a low- and high-pressure cooling system in two identical greenhouse types, 6.4 × 11 m, 3.45 m gutter height, 5.36 m ridge height, vent opening $A_v/A_g = 0.35$. One greenhouse was fogged, and the other one was not fogged.

Low-pressure system. Twenty four nozzles on tubes in three rows parallel to the ridge. Nozzles on the middle row 5 m above ground, those on the side rows 4 m above ground, 405 kPa (4.05 bar) water pressure, 3.79 l/h nozzle flow rate.

High pressure system. Twenty four nozzles at a height of 2.36 m on seven nozzle tubes normal to the ridge, 6.89 MPa (68.9 bar) water pressure, 5.49 l/h nozzle flow rate. Water filtered and decalcified (Table 11.2).

The evaporation efficiency η_{evap} in the low-pressure fogging system is lower than in the high-pressure system. The very small droplets of the high pressure

Table 11.2 Shows the mean values of the experiments with the two fogging systems

	Low pressure	High pressure
Outside air temperature (°C)	31.7	28.7
Solar radiation (W/m^2)	570	630
Relative humidity (%)	55	48
Spray rate (g/m^2 s)	0.336	0.53
t_{unfog}–t_{fog}	4.3	5.8
η_{cool}	0.42	0.53
η_{evap}	0.39	0.71

system can evaporate better. A higher amount of droplets of the low-pressure system do not evaporate in due time. Even the spray rate is higher in the low-pressure system. The cooling efficiency of the high-pressure system is higher. But one has to take into consideration the higher operation and investment costs for the high-pressure pump, as well as the absolute necessity of very clean water.

Ishigami et al. (2009) compared fog systems of single fluid nozzles and twin fluid nozzles with compressed air in a greenhouse of 26.4 m^2 floor area with tomatoes grown in it. The single fluid nozzle system worked with eight horizontally spraying nozzles in two lines, alternately 30 s spraying and 30 s not spraying. The twin fluid nozzles worked with six horizontally spraying nozzles in two lines. The inside temperature could be decreased from 30.5°C outside to 28.7 and 29°C with single fluid and twin fluid nozzles respectively. The twin fluid nozzle system used up to 50% less water than the single fluid nozzle system. No wetting of plants was observed under the twin nozzle fluid system, while plants were wetted by the single nozzle system. The reason for that may be the bigger droplets produced by the single fluid nozzle system, which do not evaporate during the state of suspense in the air.

11.8.1 Upward Fogging

The installation of fogging systems may be less expensive than fan and pad cooling systems, but there are some problems that have to be avoided:

- The dripping of water from the nozzles after stopping the spraying interval.
- The clogging of nozzles by chemicals and dust in the water.
- The mechanical damage of connections by a water hammer effect.

Fog cooling systems normally are operated intermittently to prevent wetting of plants, but the intermittent operation mode causes fluctuations of air temperature and humidity.

The greenhouse air will be cooled by evaporation of the small droplets, and the cooled air moves down by free convection with a velocity of 0.5–1.0 m/s, which is higher than the terminal velocity of the small droplets. This downward air stream by convection contains droplets which have not evaporated yet, and which fall on the plants. This phenomenon may be the reason for the plants becoming wet under fog systems. The downward movement can be slowed down or prevented by an upward air stream that can be generated by a small fan below the nozzles, or by upward-directed nozzles operated by twin fluid nozzles with compressed air. Thus, the dwell time of the fog droplets in the air is longer, and there is more time for evaporation (Toida et al. 2006; Ohyama et al. 2008).

Two small fans with a flow rate of 1.1 m^3/min were installed 15 cm below an upright spraying nozzle 1 m above ground. In comparison to that, a horizontally slightly upward-spraying nozzle was installed. The pressure of the nozzle operation was 2 MPa (20 bar) (Toida et al. 2006). No downward air stream could be observed by visualisation under the nozzles with fan, but a downward air stream was observed with upward nozzles without fan support and with horizontal nozzles.

The evaporation ratio η_f as relation of evaporated fog to generated fog was measured

Upward nozzle with fans	$\eta_f = 0.96$
Upward nozzle without fan	$\eta_f = 0.6$
Horizontal nozzle without fan	$\eta_f = 0.65$

The fog evaporation was about 1.6 times higher than without fans. The upright nozzles with fan support spread the fog over a wider area, and resulted in a more uniform temperature distribution than the others.

Further experiments were carried out in a single-span glass house, 4.8 m wide, 5.6 m long, 3.4 m ridge height in Japan (35°50′N). Two roof vents (5.6 × 0.6 m) and two side wall vents (1.4 × 0.9 m) were installed. Two nozzle lines with three upward-directed nozzles each and two small fans (2 W power, 1.1 m^3/min flow rate, 90 mm diameter) were installed below the nozzles and 1.8 m above ground (Ohyama et al. 2008). The system was operated continuously with 0.22 g/m^2 s water. In comparison to the upright nozzles, a conventional system (two lines with four horizontally spraying nozzles each, 2.4 m above ground) was operated in cycles of 1 min fogging with 0.24 g/m^2 s water and 2 min no fogging. No crop wetting was observed in the continuous spraying system with upright nozzles, while the crop was wetted in the conventional system. The mean temperature decrease was constantly about 2.4°C below outside temperature during fog operation in the upright fog system, and fluctuated with mean temperature difference of 0.5°C in the conventional system. The temperature decrease in the new system is caused by evaporation of all droplets in the air before touching the crop.

Continuous fogging with upright nozzles reduces the wetting of the crop, decreases the temperature difference inside–outside, and lowers the fluctuations of temperature and humidity.

11.9 Spray Cooling

The fan and pad cooling system is sensitive to sand and dust in desert regions, because the pads clog very easy by a mixture of salt and sand. This clogging can be like concrete (see Fig. 11.16). The fog cooling system needs clean water and is cost-intensive.

Learning from experiences in various countries, a simple spray cooling system has been designed which fulfils the demands for cooling systems, Fig.11.27, (Lüchow and von Zabeltitz 1992). It is relatively easy to build, and can be added to existing greenhouses as an alternative to pad cooling systems.

Figure 11.27 shows the cross-section of the first experimental design. There are two chambers for concurrent and counter current flow. The width of the whole cooler is 1 m and the width of each chamber 0.5 m. A spray tube with 15 hollow cone nozzles is installed at the top of each chamber, with a distance between the nozzles of 0.6 m. The cooler is constructed of timber, covered with plastic film and insulated by a reflecting plastic material. This insulation is very necessary, because the direct penetration of solar radiation influences the cooling efficiency remarkably. The height of the cooler is about 2 m. The air is sucked through the greenhouse by axial fans, and enters the cooler from the outer upper side. First, the air moves with the current and then against the current with the sprayed water. A problem was

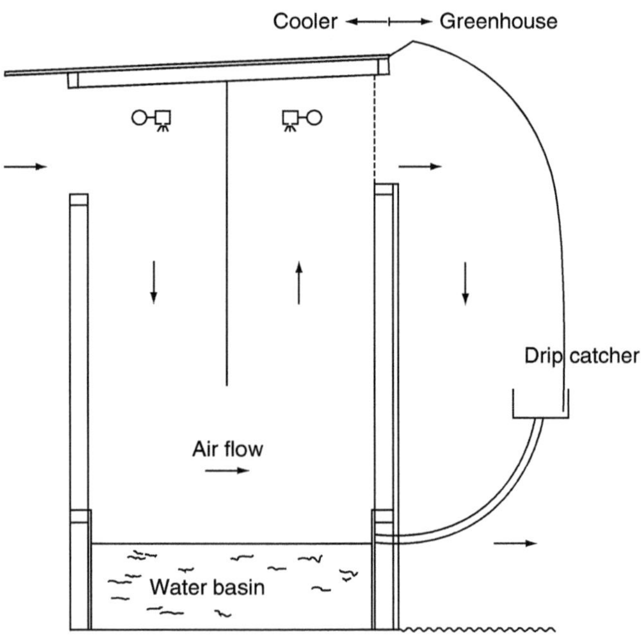

Fig. 11.27 First design of the spray cooler

to collect the water drops after the second chamber at the entrance to the greenhouse.

The nozzles were self-made of rigid PVC plastic material (Fig. 11.28). They have a cylindrical form, a centrifugal inlet at the top, and a central outlet of 3 mm

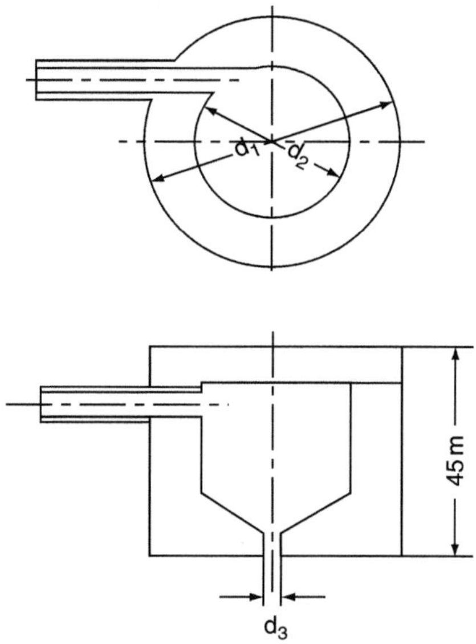

Fig. 11.28 Drawing of the nozzles, which are very simple and can be manufactured easily, $d_1 = 52$ mm, $d_2 = 30$ mm, $d_3 = 3$ mm

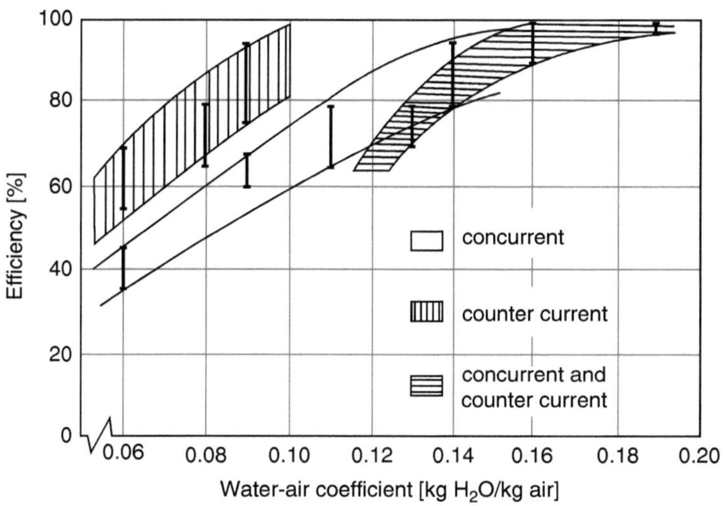

Fig. 11.29 The efficiency of the spray cooler versus the water–air coefficient. The water–air coefficient is the ratio of sprayed water to air flow

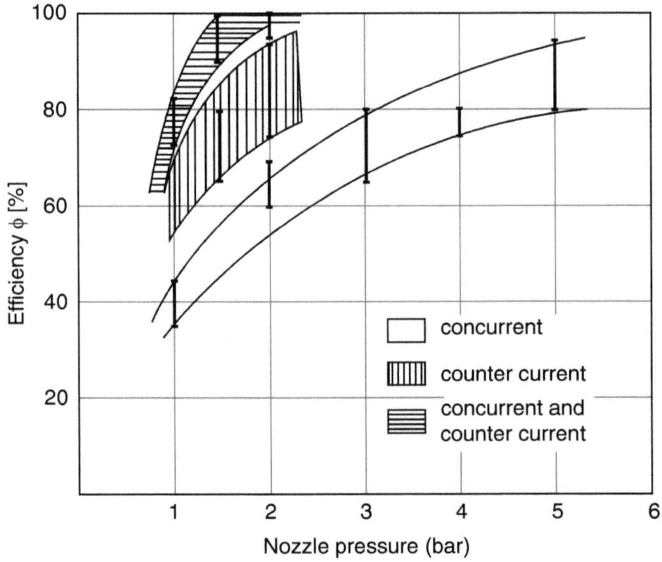

Fig. 11.30 Spray cooler efficiency versus the nozzle pressure

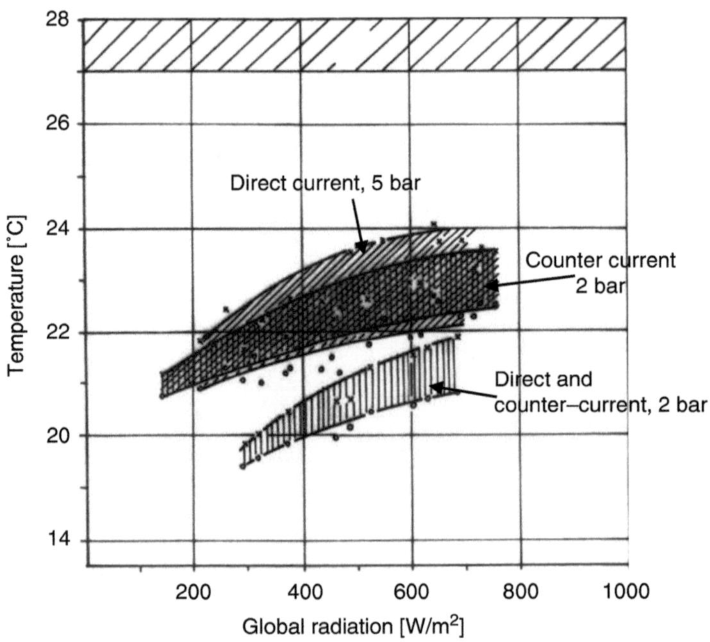

Fig. 11.31 Shows the temperatures for the different operation modes in the middle of the two spans of the greenhouse at an outside temperature of 27–28°C depending on outside global radiation

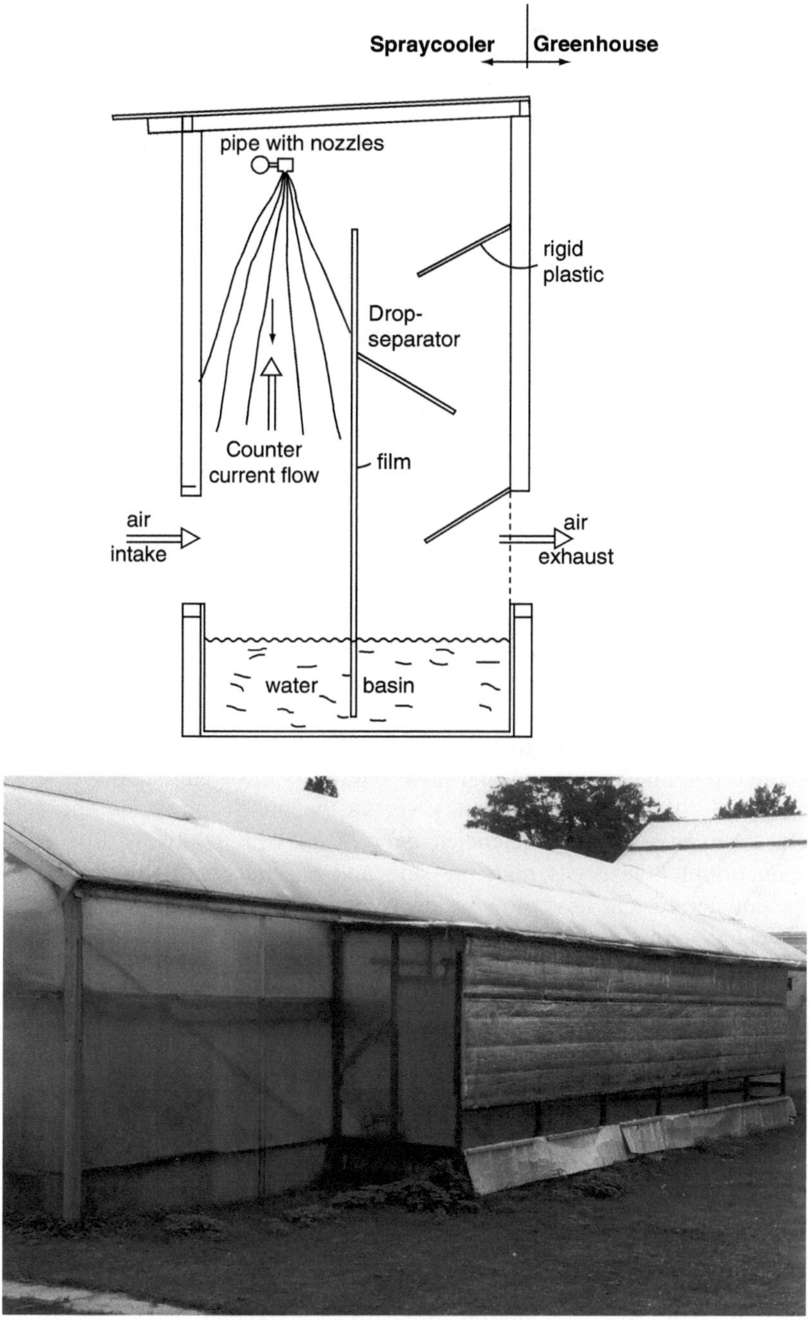

Fig. 11.32 New design of the spray cooler, type Hannover

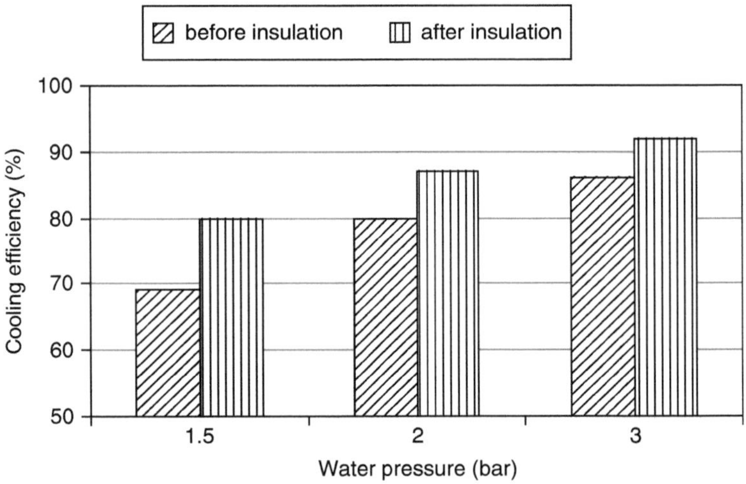

Fig. 11.33 Influence of water pressure and insulation on efficiency (Gombe 2007)

diameter at the bottom. The characteristics of the nozzles are low pressure loss, and resistance against corrosion and clogging by salt.

The cooling efficiency of the different operation modes was measured (Fig. 11.29).

The counter-current flow gives nearly the same results as the combined system with concurrent and counter-current flow, but needs less water supply.

Figure 11.30 shows the efficiency versus the nozzle pressure. The counter-current flow gives good efficiencies over 90%, with relatively low pressure. The concurrent flow needs high pressure and has lower efficiencies.

The counter-current flow provides nearly the same temperature decrease as the concurrent system (Fig. 11.31), but needs a lower supply of water and a lower working pressure for the nozzles. The counter-current operation mode has advantages in comparison to concurrent flow. The temperature in the greenhouse can be slightly decreased by the combined system, at a working pressure of 2×10^5 Pa, if more water is delivered and discharged.

The spray cooler has been changed (Fig. 11.32). In the first chamber is a counter-current spray cooler, and in the second chamber a very simple drop separator. Rigid plastic sheets are stretched in a longitudinal direction. The cooled air is turning around the sheets and the water drops are caught by centrifugal forces. No or very few drops enter the greenhouse.

The exchange coefficient is up to 80%. The temperature decreases by 5–6°C depending on outside humidity. The results are comparable with the fan and pad system. The spray cooling system is insensitive to clogging.

A new design of the spray cooler has been studied by Gombe (2007). The main objective was to improve the drop separation, and to evaluate nozzle characteristics and cooling efficiencies. The cooler worked in the counter-current

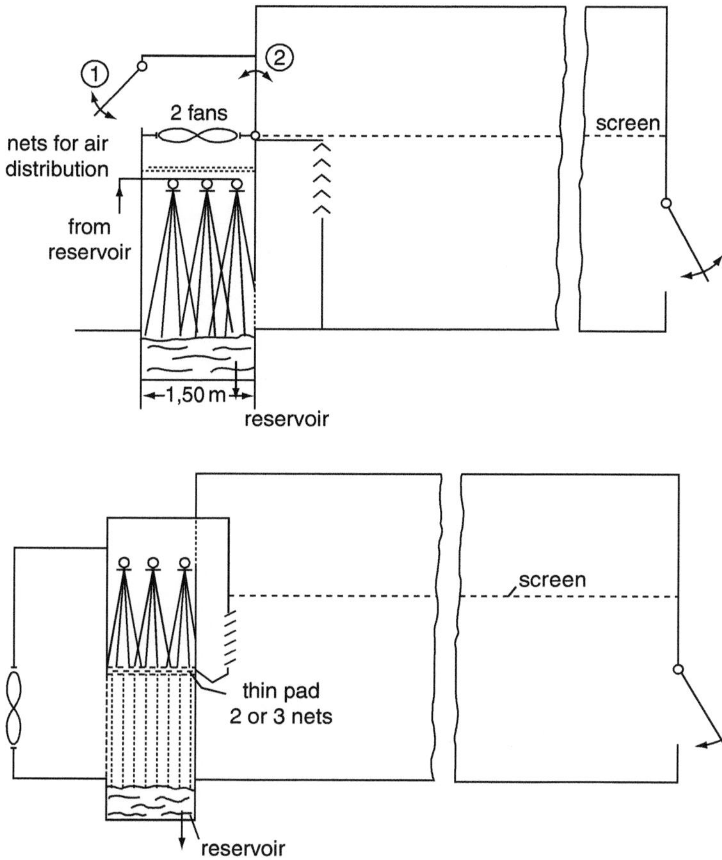

Fig. 11.34 Proposals of two positive pressure spray cooling systems with concurrent and counter current air flow respectively

Table 11.3 Results of spray cooling system in Kuwait (Jamal et al. 2002a, b)

	System A Counter-current flow	System B Concurrent flow
Cooling efficiency (%)	75	61
Mean melon yield (per greenhouse) (kg)	321	265
Cooling water consumption (l/m^2 d)	7.87	3.06

operation mode. The cooling chamber, fitted with three nozzle tubes, was 1.8 m wide and 2.8 m high. The improved drop separator made of corrugated rigid plastic sheets was 0.9 m wide.

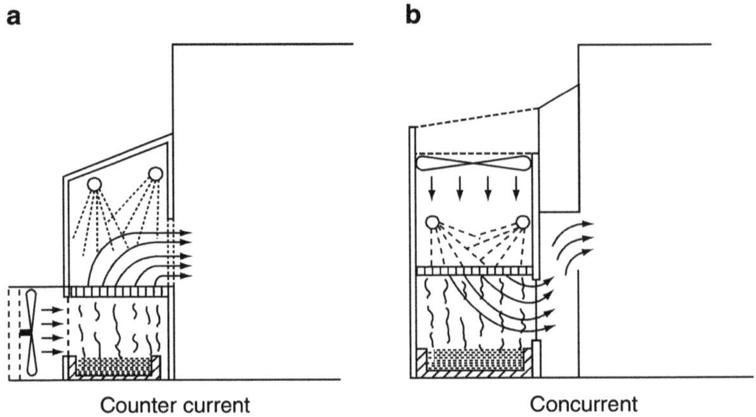

Fig. 11.35 Proposals of two positive pressure spray cooling systems with concurrent and counter-current air flow respectively

Fig. 11.36 Positive spray cooler in Kuwait 2002 (type A in Fig. 11.35), with inside view above water basin and nozzles

The nozzles should be installed as high as possible in the cooling chamber, separated from the air inlet. The higher the installation, the larger the possible contact time of the droplets in the air stream. Very small droplets cause better cooling efficiency, but large droplets can be separated better in a drop separator. The self-made nozzles in Fig. 11.28 proved to be a good compromise.

The insulation of the cooling unit is very important, and caused a significant improvement of the efficiency. The cooler was insulated by 50 mm-thick Styrofoam, lined with a reflective plastic film. Figure 11.33 shows the influence of water pressure and insulation on the cooling efficiency of the spray cooler.

Some positive and negative pressure spray coolers have been proposed for arid regions in a consultant report for the FAO (von Zabeltitz 1988b). Figure 11.34 shows two examples.

Some positive and negative pressure systems have been designed and investigated in Kuwait (Jamal et al. 2002a, b; Mulla 2006). The best results with regard to yield of sweet melon and overall profitability were achieved by positive pressure systems (Table 11.3). Figure 11.35 shows two positive pressure systems, A with counter current and B with concurrent air flow. Two PWC tubes with 15 nozzles, 75 cm distance on the tubes and 120 l/h water output were installed (Fig. 11.36).

Chapter 12
Heating

If growers want to produce a healthy crop, with good quality and high yield, they have to provide appropriate climatic conditions for crop production inside the greenhouse. Night temperatures and even day temperatures can sink below the biological optimum in a subtropical climate during winter. That means that greenhouses have to be heated for appropriate quality.

The advantage of heating is rapid growing and earliness, but heating is an economic problem because of high energy prices. Bailey (2006) reported: all types of heating systems must increase productivity and earliness sufficiently to justify the investment and running costs.

But heating has some more advantages that are often underestimated. The inside humidity and therefore the danger of disease infestation can be controlled much better by heating systems. The use of chemicals will be reduced, with positive consequences on the environment and health (Baille 2001).

The area of heated to unheated greenhouses can be a characteristic for the standard of greenhouse management in areas where the temperature can sink below the biological optimum for crop growth (Montero et al. 2009).

The choice of heating system has a significant influence on heat energy requirement (Tantau 1983; von Zabeltitz 1986a)

The greenhouses themselves have to be constructed according to the rules for energy conservation.

Heating systems have to be controlled very well. The whole crop can be destroyed if they break down in nights with frost, Fig. 12.1.

The following questions have to be answered for the design of appropriate heating systems for greenhouses:

1. Is heating necessary, based on crop requirement and climatic conditions?
2. How much heating energy is required for maintaining a design temperature on cold nights (heat requirement). How much heat energy or fuel will be required in the different months of the growing season (heat or fuel consumption)?
3. Which kind of energy sources are available?
4. What is the temperature level of the heating energy?

C. von Zabeltitz, *Integrated Greenhouse Systems for Mild Climates*,
DOI 10.1007/978-3-642-14582-7_12,

Fig. 12.1 Crop totally destroyed by breakdown of the heating system for one night

5. Which kind of heating system is appropriate for the available energy?
6. What is the expected expenditure for the use of energy?
7. What are the consequences for the greenhouse structure, grower, and crop cultivation?

To answer the first question, one has to compare the climatic conditions of a region with the requirements for crop growth (see Chaps. 2 and 3).

12.1 Heat Requirement

The heat requirement of a greenhouse can be calculated by the equation

$$q = \mathrm{u} \times (A_c/A_g) \times (t_{id} - t_{od})\ (\mathrm{W/m^2}) \tag{12.1}$$

$$u = u_t - u_a$$

A_c (m²) = surface of greenhouse cover.
A_g (m²) = greenhouse floor area.
t_{id} (°C) = design inside temperature depending on crop requirement.
t_{od} (°C) = design outside temperature.

The overall heat consumption coefficient u depends on the cladding material, the sealing of the greenhouse structure, the heating system, the irrigation system, the wind speed, the cloudiness and the rainfall. The u-value consists of two parts; the heat transfer coefficient for heat loss by heat transmission through the cladding material u_t, and the heat transfer coefficient for heat loss by air exchange through leakage u_a. The value u_a is about 10–30 % of the u-value. The tightness of the greenhouse has a significant influence on the heat loss.

The main influencing factors on the overall heat transfer coefficient u are the heating system and the cladding material of the greenhouse, including thermal screens for energy-saving.

Table 12.1 shows the u-values depending on the cladding material, and Table 12.2 depending on the heating system. The choice of the heating system has an influence on the possible energy saving.

The *design outside temperature* t_{od} can be taken as the mean minimum temperature of the coldest month (Müller 1996). This is for

- Almeria (Spain) 8°C
- Antalya (Turkey) 6.1°C
- Catania (Sicily) 7.7°C
- Gafsa (Tunisia) 4°C

The value of A_c/A_g depends on the greenhouse structure and becomes:

Tunnel greenhouse, 8 m width, 3.5 m height	$A_c/A_g = 1.5$
Multi-span greenhouse, 8 m width, 4 m ridge height, 2.5 m gutter height	$A_c/A_g = 1.33$

Table 12.1 Overall heat transfer coefficient u (W/m^2 K) for different cladding materials. Mean values of various measurements and calculations (mean wind speed 4 m/s and mixed heating system) (von Zabeltitz 1982, 1986; Tantau 1983; ANSI/ ASAE standard 2003; Meyer 1981, 1982)

Material	u (W/m^2 K)
Single glass	6.0–8.8
Double glass	4.2–5.2
Double acryl sheet (16 mm)	4.2–5.0
Single PE film	6.0–8.0
Double PE film	4.0–6.0
Thermal screen below single glass or film	3.2–4.8

Table 12.2 Overall heat transfer coefficient u (W/m^2 K) for different heating systems and single-layer greenhouse cladding (Tantau 1983, 1998)

Heating system	u (W/m^{2}K)	Relative value (%)
Tube heating at eaves height	8.2	100
Tube heating below table	7.4	90
Tube heating at side wall	8.1	99
Tube heating on soil, bench heating	6.7	82
Free discharge air heater, low fan speed	9.9	121
Free discharge air heater, middle fan speed	7.1	87
Free discharge air heater, high speed	8.0	97
Air heater with perforated plastic tube	7.0	85
Convector heating	7.8	95

The *heat requirement* for assumed design inside temperatures of 12°and 16°C in a multi-span greenhouse for Almeria, covered by single PE film ($u = 7$ W/m^{2}K) is

$$q_{(12)} = 7 \times 1.33 \times (12 - 8) = 37.2\mathrm{W/m}^2$$

$$q_{(16)} = 75\ \mathrm{W/m}^2$$

The corresponding heat requirements for Antalya, Catania, and Gafsa are:

Antalya:	$q_{(12)} = 55$	$q_{(16)} = 72$	W/m^2
Catania	$q_{(12)} = 4$	$q_{(16)} = 77$	W/m^2
Gafsa	$q_{(12)} = 75$	$q_{(16)} = 111$	W/m^2

A greenhouse span of 8 m width and 60 m length requires a heater with the following heat capacities for the design temperatures:

	Gafsa	Almeria	Antalya	Catania
12°C	36 kW	18 kW	26 kW	19 kW
16°C	53 kW	36 kW	35 kW	37 kW

12.2 Fuel Consumption

The yearly fuel consumption should be roughly calculated to get an impression about the economic situation if greenhouses are heated.

Normally, the calculation of heat consumption is based on hourly temperatures of all days when heating is necessary. Those data very often are not known.

The mean maximum t_{mmax} and mean minimum t_{mmin} temperatures can be obtained from different stations (Müller 1996). Hallaire (1950) developed a method for calculating the hourly temperatures and the mean day and mean night temperatures from the daily mean maximum and minimum temperatures (Hallaire 1950).

The mean hourly temperature of the day is

$$t_{\mathrm{h}} = t_{\mathrm{mind}} + f_{\mathrm{d}} \times A \tag{12.2}$$

The mean night temperature is

$$t_{\mathrm{mn}} = t_{\mathrm{mind}} + A\frac{\Sigma f_n}{24 - d_{\mathrm{l}}} \tag{12.3}$$

The mean day temperature is

$$tm_{\mathrm{d}} = t_{\mathrm{mind}} + A\frac{\Sigma f_{\mathrm{d}}}{d_{\mathrm{l}}} \tag{12.4}$$

Table 12.3 Coefficients for (12.2)–(12.4) (Hallaire 1950)

d_l	Σf_n	$\Sigma f_n/(24\text{-}d_l)$	Σf_d	$\Sigma f_d/d_l$
7	8.21	0.48	4.25	0.61
9	6.0	0.4	5.67	0.63
11	4.5	0.375	6.99	0.635
13	3.45	0.31	8.10	0.623
15	2.51	0.28	9.29	0.62
17	1.58	0.23	10.96	0.644

t_h = hourly temperature

$A = t_{maxd} - t_{mind}$ = difference of mean minimum and mean maximum day temperature

f_d and f_n = coefficients depending on day length d_l (Table 12.3).

Assuming that heating in mild subtropical and arid climates is necessary mainly during night hours, the mean fuel consumption for the winter months and summarised for the year can be estimated by the following equation, using the Hallaire method to calculate the mean night hours.

$$Q_{(month)} = u \times (A_c/A_g) \times (t_{id} - t_{st} - t_{mn}) \times n_n \times n_d (\text{Wh/m}^2\text{month}) \quad (12.5)$$

u (W/m^2 K) = overall heat consumption coefficient (Table12.1)

A_c/A_g = surface of greenhouse cover/greenhouse floor area

t_{id} (°C) = design inside temperature

t_{mn} (°C) = mean night temperature (12.3)

t_{st} (°C) = mean temperature increase at night by heat storage in the soil from daytime

n_n (–) = number of night hours

n_d = number of days per heated month.

The temperature increase by heat storage from day to night t_{st} can be assumed to be 1–2°C.

It can be shown that the calculation of mean fuel consumption with mean night temperature gives nearly the same value as the calculation just with night hours, which is more complicated.

To calculate the fuel consumption (l oil equivalent), the figures of (12.5) have to be divided by the heat capacity of the oil (about 10 kWh/l oil) and by the efficiency of the heater system (assumed to be 80%).

Examples.

Single film cover with u = 7 W/(m^2 K), A_c/A_g = 1.33; t_{id} = 16°C; t_{st} = 2°C.

The details for the calculation are shown in Annex 2.

The estimation of heat consumption by (12.5) gives the following results, which are roughly calculated values, because the heat consumption depends on various influencing factors as climate conditions inside and outside as well as greenhouse structure, surface, heating system, and soil storage capacity.

The yearly heat consumption Q_{year} as sum of Q_{month} of heated months for some Mediterranean locations results in

Almeria, Spain (36°50′N)	$Q_y = 35.6$ (kWh/m^2 year)
Antalya, Turkey (36°53′N)	$Q_y = 56.1$ (kWh/m^2 year)
Catania, Sicily, (37°30′N)	$Q_y = 46.4$ (kWh/m^2 year)

The fuel consumption (l oil equivalent/m^2 year) becomes $Q_{oil} = Q_{year}/10 \times 0.8)$

Almeria	$Q_{oil} = 4.4$ (l/m^2 year)
Antalya	$Q_{oil} = 7$ (l/m^2 year)
Catania	$Q_{oil} = 5.8$ (l/m^2 year)

Although all locations are on nearly the same latitude, the fuel consumption is very different.

The calculated values correspond very well with those calculated by a simulation program, HORTEX, developed by Rath (1992).

12.3 Heating Systems

One has to distinguish between the generation of the heat energy and the distribution of heat energy inside the greenhouse.

The following energy sources for generation of heat energy can be applied in horticulture:

- Combustion of fossil fuels, oil, gas, coal
- Combustion of biomass, wood, straw
- Combustion of biomass from waste of fruits such as stones and husks
- Geothermal energy
- Waste heat from industry
- Solar energy

Heat generation by combustion can be done in a:

- central warm water boiler
- decentralised warm water boiler
- directly fired air heater

The combustion of fuels or biomass results in high temperature energy for the heating systems. The geothermal energy, waste heat and solar energy are normally low-temperature energies below 60°C with special demands to the heating system.

There are still very simple, self-made heaters used in some areas (Fig. 12.2). They may keep the greenhouse frost-free, but they are not sufficient to produce a healthy crop and good quality.

Fig. 12.2 Simple but not sufficient heating systems. These heating systems are still used in several countries

Figure 12.3 shows the cross-section of a directly fired air heater. The combustion gas flows through a heat exchanger and leaves the heater through the chimney. The greenhouse air is blown through the heat exchanger by a fan, where it is warmed up.

Coming out of the heater, the warm air is distributed through perforated plastic film tubes, which lie between the rows of the crop.

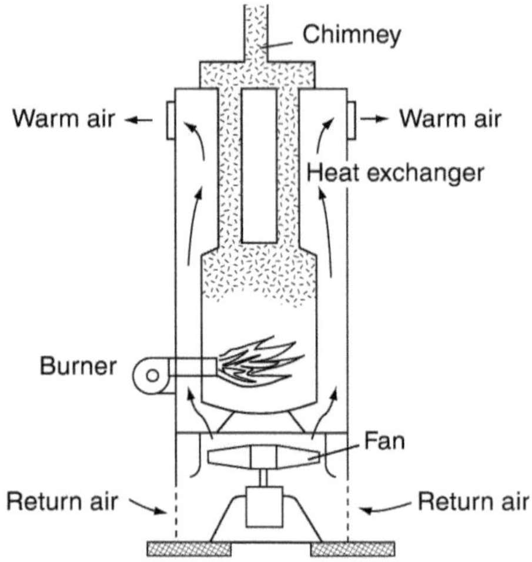

Fig. 12.3 Directly fired air heater

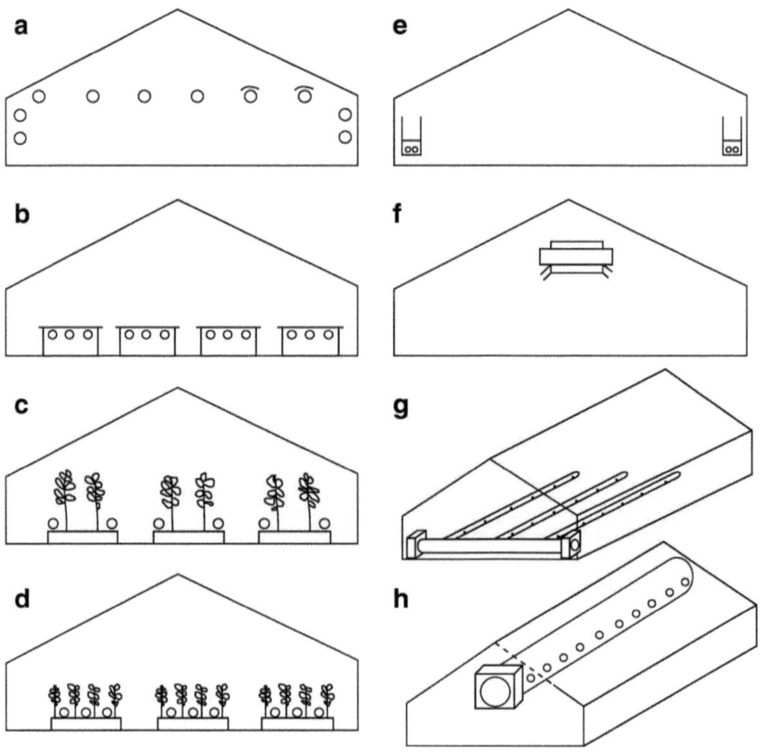

Fig. 12.4 Heating systems for greenhouses

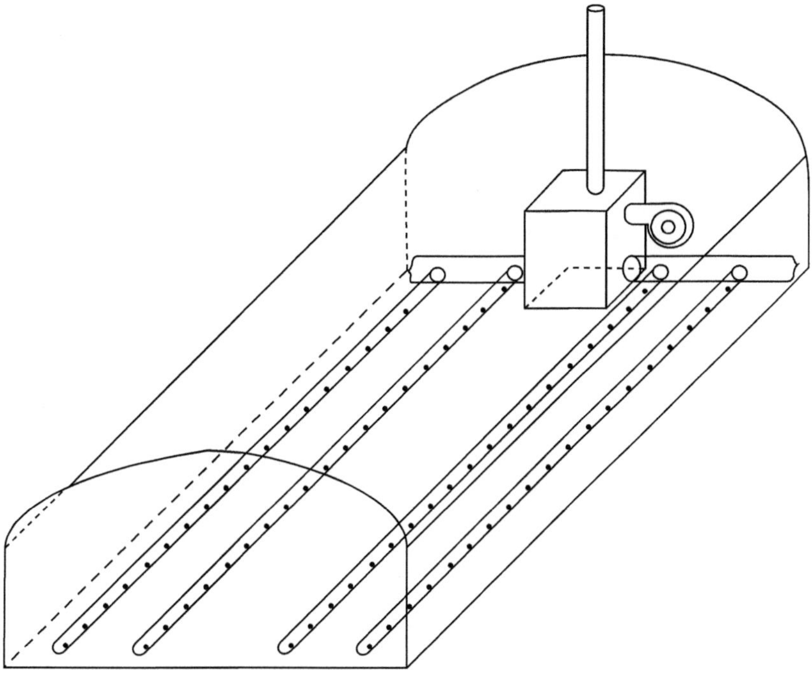

Fig. 12.5 Directly fired air heater and air distribution by perforated plastic tubes

The heat *distribution inside the greenhouse* is either by water- or air-heating systems.

If the heat energy comes from a warm-water boiler or from geothermal sources, it can be distributed inside the greenhouse by air heating systems or by warm-water heating systems.

Figure 12.4 shows various possible heating systems for greenhouses. The overall heat transfer coefficients differ between heating systems (Table 12.2).

Directly fired air heaters with air distribution through perforated plastic tubes are used very often in subtropical climates (Fig. 12.5). The plastic tubes lie between the rows of the crop.

The diameter of the tubes is about 30–60 cm. The discharge holes are located on opposite sites of the tube, about 30–45° above the horizontal if the tubes are positioned on the ground, and about 30–45° below the horizontal if the tube is hanging above the crop. The holes are typically spaced 0.3–1.0 m apart along the axis of the tube, depending on tube diameter and length. The total area of the holes should be in between 1.5 and 2 times the cross-sectional area of the tube.

One tube is generally sufficient for about a 9 m greenhouse width or less. More tubes are necessary for wider greenhouses (ANSI/ASAE Standard 2003).

How to do:

Fig. 12.6 Well-installed plastic tubes

How to do:

Fig. 12.7 The plastic tubes are installed too high above ground and plant level

This kind of air-heating system has advantages:

- The air will be distributed evenly across the greenhouse, and the temperature distribution can also be even.
- The air humidity between the plants will be reduced by forced air movement.

The consequence is a reduced disease infestation and necessity of fewer chemicals.

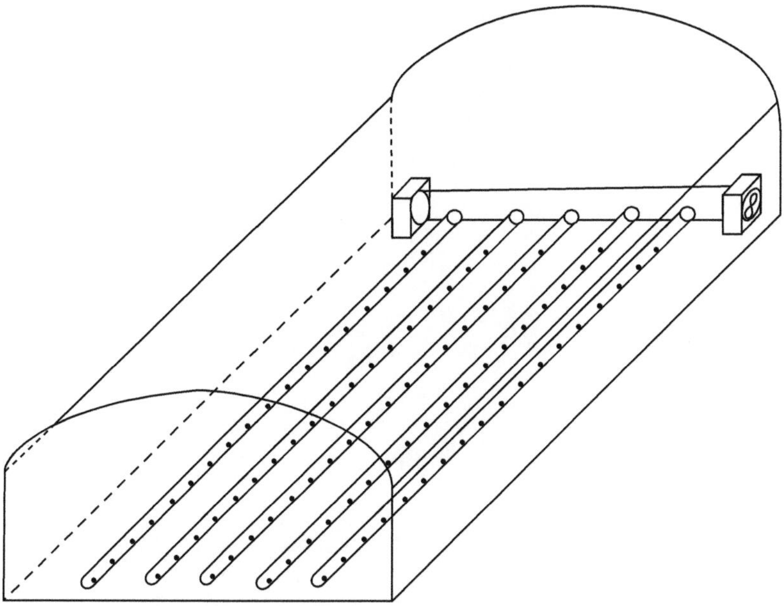

Fig. 12.8 Air heating by warm-water to air heat exchanger

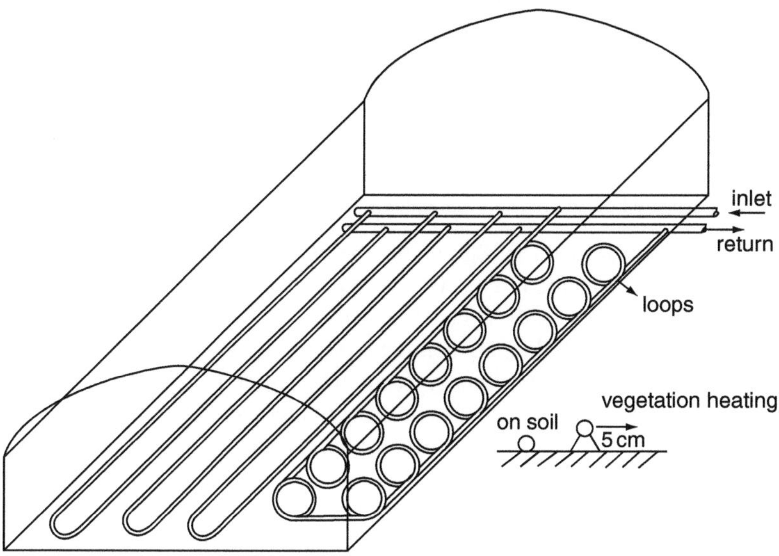

Fig. 12.9 Warm-water heating system by plastic tubes

It is very important that the perforated plastic tubes are installed as low as possible, next to the plant area, and that the tubes lie between the plant rows if possible, even if this is uncomfortable for the workers.

Fig. 12.10 Corrugated plastic tubes for greenhouse heating

If the tubes are installed too high above the plant area, the warm air rises up and does not heat the plant area. This is a waste of heat energy, and infestation by diseases can occur more easily (Figs. 12.6 and 12.7).

Figure 12.8 shows another air heating system. A water-to-air heat exchanger with a fan transfers the heat energy from the heater to the air, which will be distributed by perforated plastic tubes in the greenhouse.

Figures 12.9 and 12.10 shows a warm-water heating system. Flexible corrugated plastic tubes with a diameter of 20–25 mm lie on the ground and distribute the heat energy into the plant area. The plastic tubes are installed longitudinally or in form of loops. The installation and the number of tubes depend on the inlet temperature of the warm water. Tubes manufactured from polypropylene are suitable for temperatures up to 60°C. The tubes should lie directly on the ground if soil and air heating is wanted at the same time. The heat transfer only to air is better if there is a space between tube and soil of about 5 cm.

The heat transfer of tubes that lie on the ground floor is 0.67–1.6 W/(mK) per 1 m length of tube. The heat transfer is a little bit higher if there is free convection around the tube, which means if the whole tube is surrounded by air.

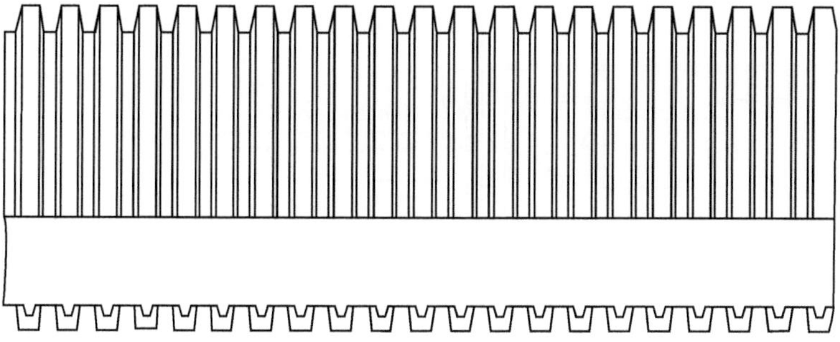

Fig. 12.11 Corrugated plastic tube for warm-water heating

Fig. 12.12 Cooling towers to cool geothermal water for irrigation in Tunisia

Figure 12.11 shows the cross-section of a corrugated PP tube that has an increased surface for better heat transfer and is very flexible.

12.4 Geothermal Energy for Greenhouse Heating

The use of geothermal energy for greenhouse heating is a very good solution, if the geothermal water is available not too deep in the subsoil, if the water temperature is suitable, and if the water is not too corrosive. This is the case for example in Tunisia and in Turkey.

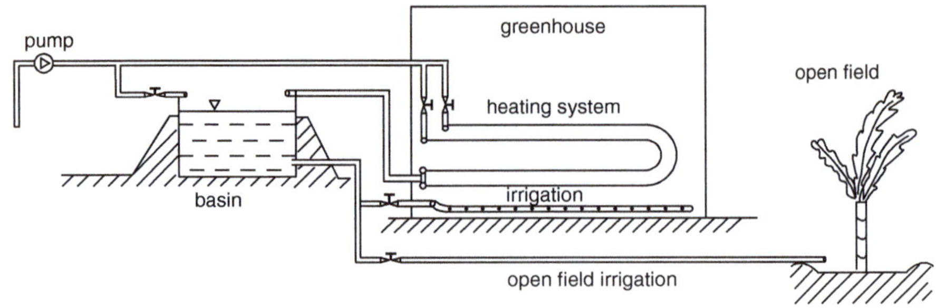

Fig. 12.13 Heating and irrigation with geothermal water in Tunisia

Fig. 12.14 Corrugated PP tubes for heating the greenhouse by geothermal water in Tunisia

Fig. 12.15 Corrugated PP tubes outside the greenhouse for cooling the geothermal water, when no heating is necessary, Tunisia

Geothermal water is used for irrigation in Tunisia, and has therefore to be cooled down. For this reason, huge cooling towers have been built for cooling the irrigation water at the start (Fig. 12.12).

Then growers started to use the geothermal water for irrigation and heating at the same time. Figure 12.13 shows a principle of such a system (von Elsner 1990). The geothermal water comes from the well and flows either into a basin which acts as a cooling pond, or through a heating system, if heating is necessary. After it has been cooled down to about 20°C, it will be used for irrigation in the greenhouse or open field. The salt content is 2–3 g/l. Corrugated PP tubes, 25 mm diameter, are used for heating, because the water is corrosive (Fig. 12.14). Cooling tubes outside the greenhouse are an alternative for cooling the irrigation water when no heating is necessary (Fig. 12.15).

The precondition for the irrigation system is a low outlet temperature of the water. This causes problems of temperature distribution in the greenhouse, if the inlet temperature is for example 60°C and the outlet has to be 20°C. The flow rate has to be very low, and the heat output of the tubes is very different. One possibility for minimising the problem is the installation of different numbers of tubes in the forward and return flow, for example ten forward and 14 return.

Another problem is the control of the heating system. In many cases, no electricity is available, and automatic control systems are too expensive. To adapt the heating system to outside and inside temperature, the heating system can be divided into two independent circles of 1/3 and 2/3 of the heat capacity. The adaptation can be made by switching the valves according to the heat capacity required.

To improve the control, simple thermostat valves with temperature sensors are used, when no electricity is available (Fig 12.16). These thermostat valves have to

Fig. 12.16 Simple control system by thermostat

Fig. 12.17 Heat exchanger for geothermal water in Turkey

be installed at the end of the return flow to avoid emptying of the heating system at times of standstill. This is a good, cheap and simple solution.

If the geothermal water is very corrosive as in Turkey, heat exchangers have to be installed, Fig. 12.17.

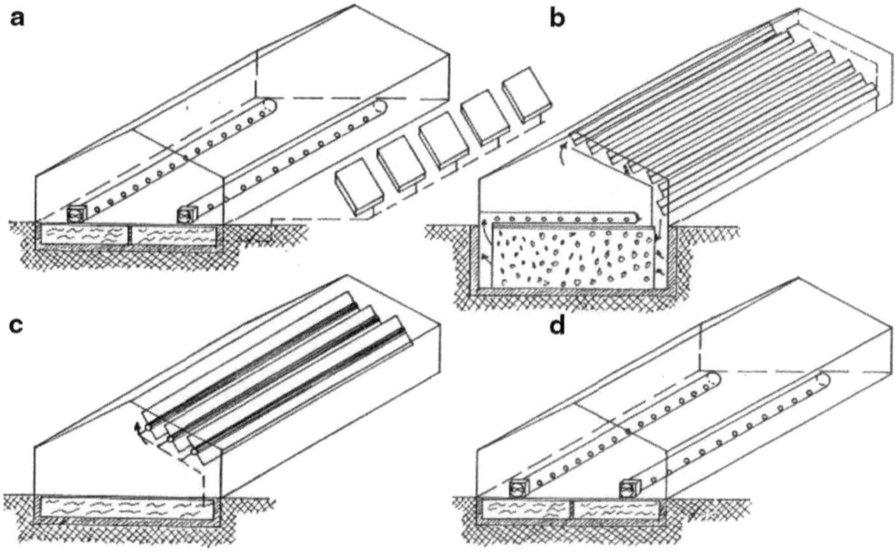

Fig. 12.18 Principles of solar heating systems for greenhouses. **a** Separate collectors near the greenhouse; water storage and heat distribution by water-to-air heat exchanger and perforated plastic tubes. **b** Solar air collector integrated into greenhouse structure; rock bed storage and warm air distribution by perforated plastic tubes. **c** Water collector integrated into the greenhouse structure. **d** Greenhouse as collector. Heat exchanger for energy collection and distribution

12.5 Solar Heating

Solar energy covers a part of the heating energy that is needed during the daytime. To use solar energy for heating during the night, two problems have to be solved (von Zabeltitz 1987, 1988a):

- The conversion of global radiation into thermal energy,
- The storage of the thermal energy for heating purposes during night time.

The *conversion of global energy* into thermal energy is based on the following principles (Fig 12.18):

1. Separate solar collectors (air or water collectors are placed outside the greenhouse and serve to heat the thermal storage.
2. Solar collectors are an integral part of the greenhouse, loading fluid or solid/air storages.
3. The greenhouse itself is a collector; a part of the global radiation that penetrates the greenhouse is being converted into thermal energy.

Normally short-term storage from day to night is used for the *storage of heat energy*. Long-term storage from summer to winter needs huge storage volumes. Storage materials are gravel, water solar ponds, soil and phase change material.

Gravel or rock storages act as store and heat exchanger simultaneously. With other storage materials, one needs extra heat exchangers.

With regard to heat storage, the following technical details have to be determined:

- The storage medium.
- The storage capacity. The maximum of storable energy in kWh/m^3 or kWh/kg.
- The loading and unloading energy per time unit.
- The efficiency. The relationship between useable energy out of the storage and the sum of input energy, including the energy that is necessary for loading and unloading (electricity).
- The storage position, i.e., where the storage is located.
- The storage configuration.

In order to design a solar heating system, the following questions must be answered:

1. What kind of solar system should be applied?
2. What is the amount of energy that will be converted from solar energy into thermal energy?
3. How much solar energy is available each day during the heating season, including hourly distribution?
4. How much energy is required to heat the greenhouse?
5. How much of the heat energy consumption can be covered by the thermal energy produced from solar energy?
6. What difference in temperature between inside and outside can be achieved by solar heating?
7. Is solar heating economically feasible?

The actual outside temperature as well as the global radiation have to be considered for the calculation of heat consumption. The effective hourly heat consumption q_h can be estimated by the following equation (Damrath 1982, 1983; von Zabeltitz 1988a).

$$q_h = (A_c/A_g) \times u \times (t_i - t_{oeff}) - q_o \times \tau \times \eta \, W/m_2$$

In this equation:

A_c/A_g (–) = relationship of greenhouse cover to floor area.

t_{oeff} (°C) = actual outside temperature.

t_i (°C) = inside temperature.

τ (–) = transmittance of greenhouse. $\tau = 0.6$–0.7 for single plastic film cover.

η (–) = conversion factor of global radiation energy to thermal energy inside the greenhouse. $\eta = 0.5$–0.7.

q_o (W/m^2) = outside global radiation.

u (W/m^2 K) = overall heat transfer coefficient.

The heat requirement has to be calculated for every hour of the day with corresponding values of temperature and global radiation. The daily sum of the hourly values results in the daily heat requirement. If the hourly value of q_h becomes

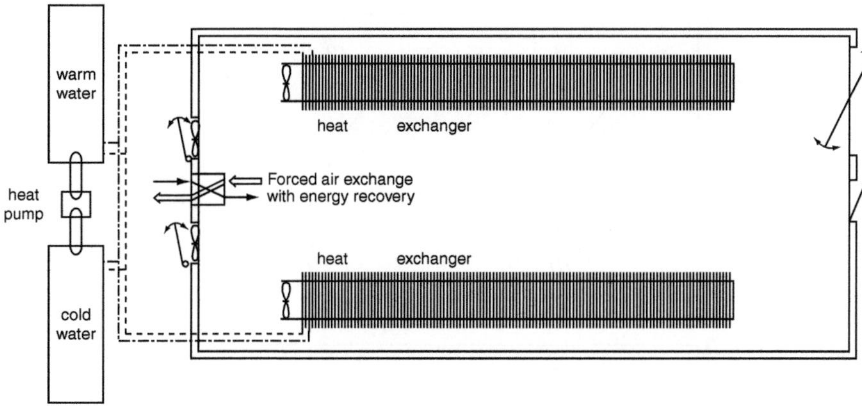

Fig. 12.19 Principle of solar greenhouse, Hannover, Germany

negative at daytime, the sum of these values results in the excess heat energy that can be theoretically stored out of the greenhouse for heating at night (Fig. 12.20).

Many systems using solar energy for greenhouse heating have been developed, both highly sophisticated and very simple ones (von Zabeltitz 1984, 1985, 1987, 1989). Some examples will be demonstrated, although some of them were not successful in practice. Maybe some new ideas can be created from this research and development.

Detailed experiments and calculations were carried out with the system "greenhouse as collector" in the Institute for Horticultural Engineering Hannover (Damrath 1982, 1983; von Zabeltitz 1984, 1986).

Figure 12.19 shows the principle of the solar greenhouse. Cold water of 2–6°C is pumped from cold water storage through an air–water heat exchanger in the greenhouse during the daytime. The greenhouse air that is warmed up by solar energy is thus cooled down, and energy is transferred to the water in the heat exchanger. The cold water storage will be heated up to 18–24°C in the daytime. A heat pump is installed between cold- and warm-water storage that increases the water temperature in the warm-water storage to a level suitable for heating at night. The greenhouse will be heated at night by the same heat exchanger. The heat pump can operate for 24 h independently from the climate control for heating and cooling, and therefore has relatively low power. The greenhouse itself remains closed, but has forced ventilation for excessively warm days, and an additional heat exchanger for fresh air exchange with heat recovery (Rüther 1989).

A theoretical simulation model has been developed with the results of experimental data for the design of solar systems (Damrath 1982).

The heat exchanger in the greenhouse is a water–air exchanger with connected perforated plastic film tubes for air distribution.

Figure 12.20 shows the calculated mean daily heat requirement and the energy that can be stored in the daytime depending on the season in Germany, Cyprus, and

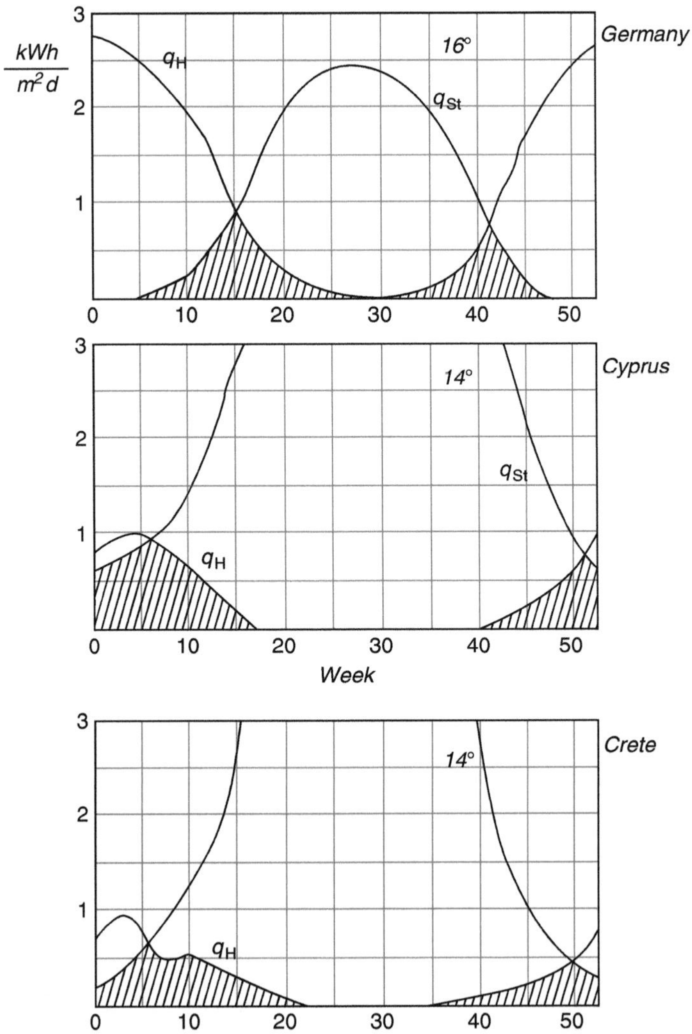

Fig. 12.20 Mean daily heat requirement q_{H} and storable energy q_{St} from the solar greenhouse

Crete. The dashed fields below the curves q_{H} and q_{St} show the amount of solar energy that can be used for heating with day-to-night storage. The contribution of solar energy amounts to 15–25% of the necessary heat energy in Germany, if the greenhouse is heated throughout the year to 16°C. Nearly 100% can be substituted in Mediterranean areas with 14°C inside temperature. The amounts in Fig. 12.20 are the maximum possible values with unlimited storage capacity. The design and layout of the heat exchanger, storage and heat pump have considerable influence on the gain of energy, and the components of the solar system influence each other. For short-term storage from day to night, a storage volume of 0.2 m^3/m^2 greenhouse

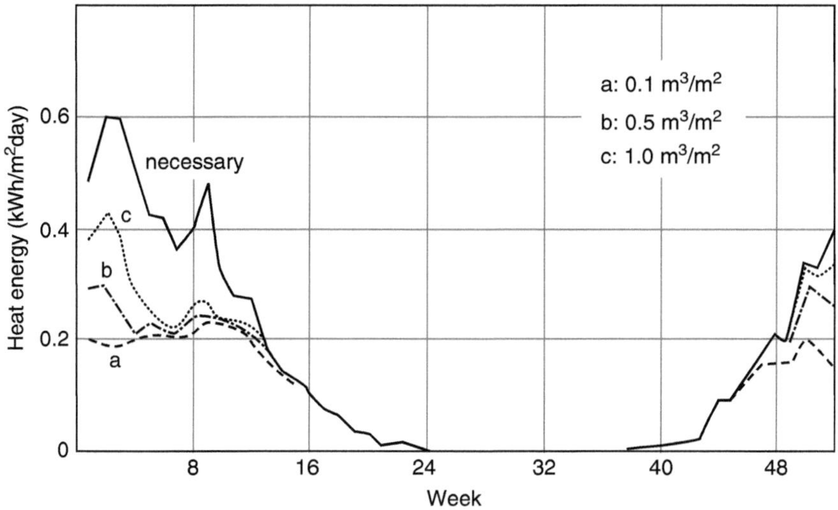

Fig. 12.21 Calculated heat energy and storable energy with different storage volumes for climate of Crete

floor area is a reasonable value. For long-term storage from summer to winter, 12–18 m^3/m^2 would be necessary. An energy balance has shown that significant amounts of electric energy are required for the heat pump and fans. Thus, the use of such a system may be questionable for temperate climates. The conditions can be better in subtropical climates.

Figure 12.21 shows a calculation for a greenhouse on the Island of Crete with the model of Damrath (1982) for an inside temperature of 12°C but without the use of a heat pump (Bredenbeck 1982). Heating at night took place with water temperatures that could be stored in the daytime. The curves a, b, c demonstrate the storable solar heat energy that could be stored out of the greenhouse with different storage volumes in comparison to the necessary heat energy for 12°C inside temperature. With 0.5 m^3/m^2 storage volume, 67% of the yearly heat energy can be covered (von Zabeltitz 1986).

A simplified "greenhouse as collector" system was installed in a commercial greenhouse under German climate conditions (Bredenbeck 1986, 1992). It was a rock bed storage below the greenhouse, Fig 12.22. During the daytime, the hot air was collected above a shading system and thermal screen, and at night was delivered below the thermal screen.

The calculated percentages of heat energy gained by solar energy and stored in the rock bed storage are given in Table 12.4 for climate conditions in Hannover and Munich, Germany (Bredenbeck 1992).

More solar heating systems, developed in several countries, are described by von Zabeltitz (1987), but only a few of them came into practical use by growers because of economic problems.

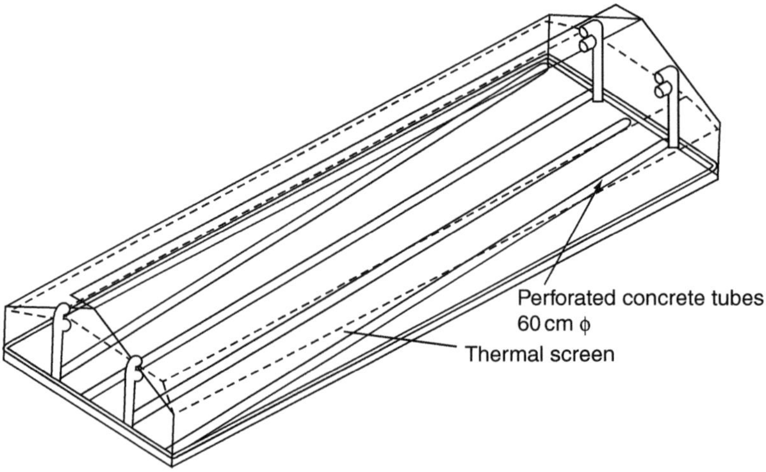

Fig. 12.22 Greenhouse as collector design in a commercial greenhouse

Table 12.4 Percentage of heat energy gained by solar energy in a greenhouse with rock bed storage thickness and air fluxes, inside temperature 18°C (Bredenbeck 1992)

Storage thickness (m)	Air flux (m^3/m^2 h)	Solar heat energy Hannover (%)	Solar heat energy Munich (%)
0.6	13	25	32
0.6	26	28	35
0.6	52	28	37
0.9	13	25	33
0.9	26	30	36
0.9	52	33	37

The *passive system*, using transparent, water-filled PE tubes, was installed by growers in practice. Figure 12.23 shows the principle of the system. Transparent PE plastic-film tubes are laid between the rows and filled with water. The tubes serve as collector, storage and heating system at the same time.

The layout is about 60–100 l water per m^2 floor area. The diameter is 30–35 cm. One can raise the temperature at night by 3–5°with these tubes. This is not really a heating system, but a good measure for improving the conditions in unheated greenhouses and for frost protection. The efficiency of transparent tubes is much better than that of black tubes (Figs. 12.23 and 12.24).

The *passive solar heating system* inside the greenhouse has a positive influence on:

- Air temperature at night
- Maximum air temperature during daytime
- Plant temperature
- Soil temperature

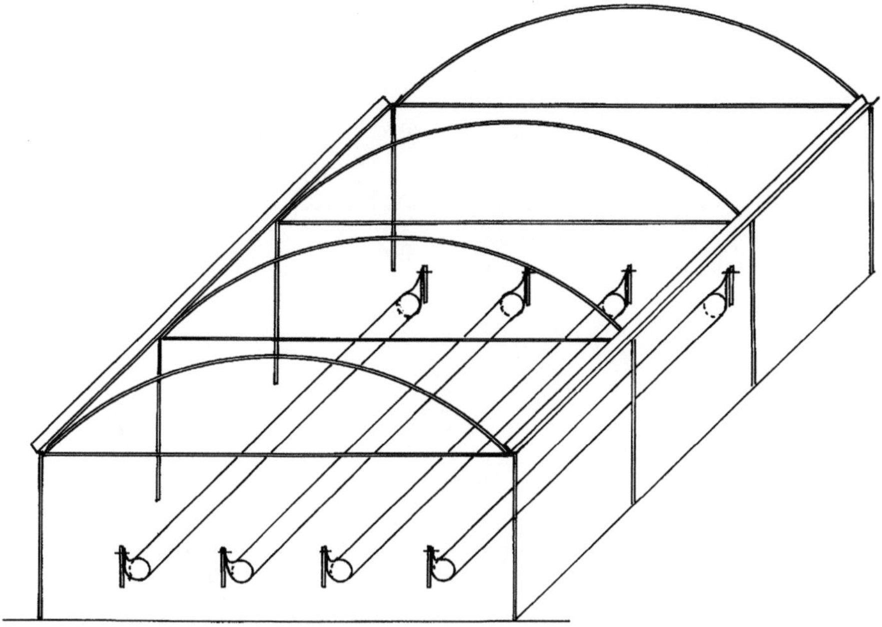

Fig 12.23 Principle of passive solar heating system using water-filled transparent plastic tubes

- Air humidity

The efficiency of the tubes depends on:

- The number of tubes or amount of water
- The tightness of the greenhouse structure
- The cladding material
- The height of the crop
- The outside weather conditions.

The greenhouse should have no leakage, and the cladding material should have high transmission for global radiation and low transmission for long-wave radiation.

Measurements and calculations have been carried out in Germany (Thomas 1994). Figure 12.25 shows the possible temperature difference inside to outside depending on the outside daily global radiation for a plastic-film greenhouse with double-inflated roof. The curves are degressive because of the influence of shading by plants and the ventilation necessary with increasing solar radiation. The inside temperature can be increased 3–4°C by the solar system, and 5°C by the solar system and additional thermal screen. The outside wind speed has a significant influence on the temperature difference.

Fig. 12.24 Some examples of solar sleeves in different greenhouses

Figure 12.26 shows the temperature difference depending on outside global radiation for different covering ratios of the solar system. If the water-filled plastic tubes cover 40% of the greenhouse floor area, the inside temperature can be increased by 4°C above outside temperature with a global radiation of 6 kWh/ m^2 K on the day before and a wind speed of 2.5 m/s (Fig. 12.27).

A joint experiment about the use of water-filled PE tubes was organised for the growing seasons 1986/87 and 1987/88 in 11 countries (von Zabeltitz 1989). Two typical examples may be mentioned from Greece and Tunisia:

Greece (M. Grafiadellis, G. Spanomitsios, K. Mattes).

Water-filled PE tubes, 32 cm diameter, on black PE film.

Minimum temperature increase from 1.6°C in unheated to 4.5°C in heated greenhouse.

Increase of plant temperature by 2–4°C

Reduction of relative humidity by 6–12%.

Main yield of tomato in g/plant in 1987 season:

	Unheated	Heated by water-filled tubes
Early yield	1,292	1,793
Total yield	6,313	7,793

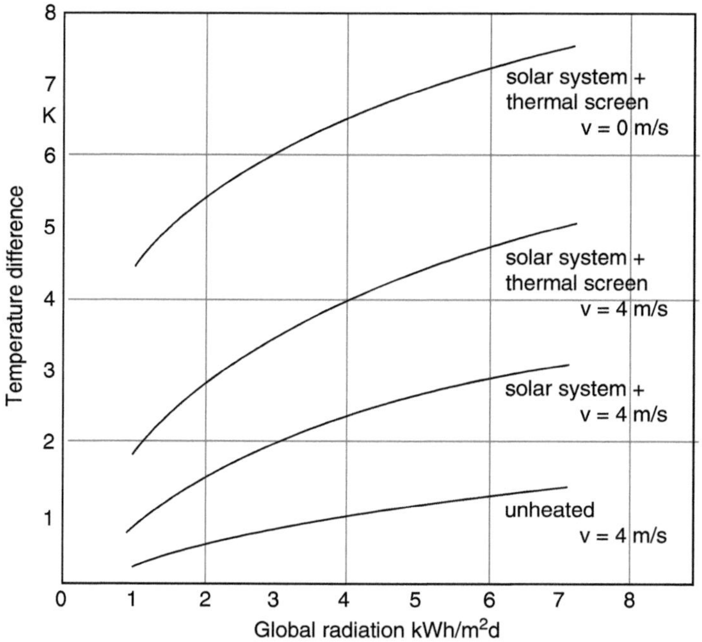

Fig. 12.25 Possible temperature difference with the passive solar heating system (Thomas 1994)

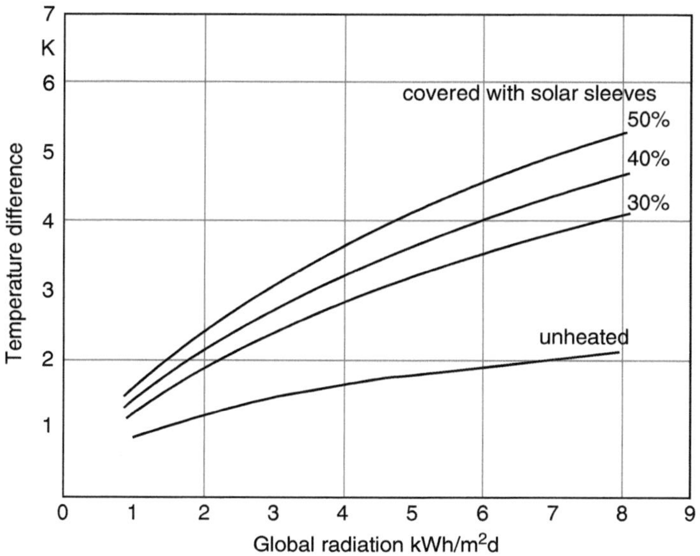

Fig. 12.26 Possible temperature difference for different covering ratios of the solar heating system depending on global radiation. Wind speed 2.5 m/s (Thomas 1994)

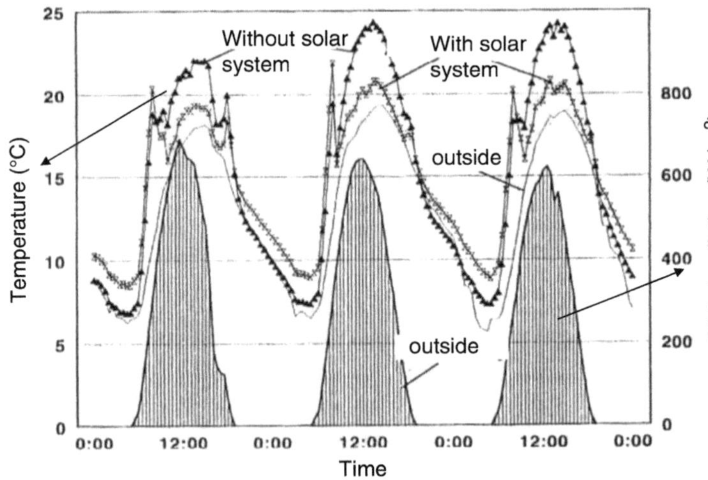

Fig. 12.27 Climate conditions in a greenhouse with a passive solar heating system for 3 days in April 1991. The solar system increases the inside temperature at night and decreases the temperature in the day time (Thomas 1994)

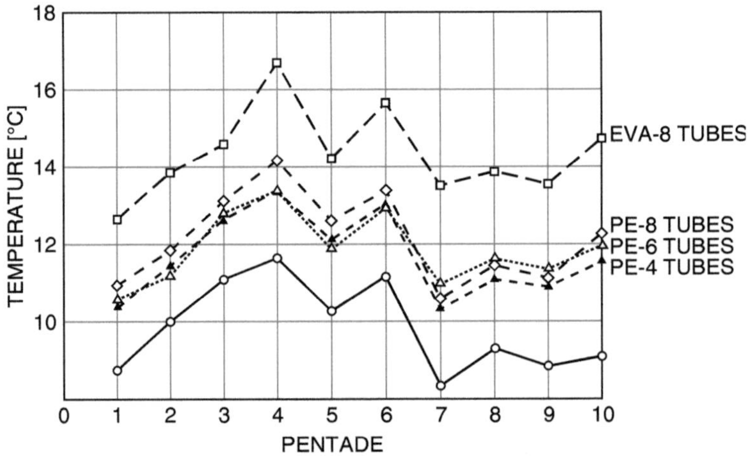

Fig. 12.28 Influence of the number of tubes and cladding material in a 9 m round-arched tunnel-type greenhouse on mean night temperatures in comparison to the unheated control greenhouse (growing season 1987/88). The minimum night temperature with water-filled tubes was about 2°C higher under EVA film than under PE film with eight tubes each. The relative humidity was about 85% under EVA, and thereby about 11–12% lower than under unheated PE film

Tunisia (A. Mougou and H. Verlodt)

One round-arched tunnel-type greenhouse covered by EVA film.

Two round-arched tunnel-type greenhouses covered by PE film.

Water-filled PE tubes, 32 cm diameter (Figs. 12.28–12.29).

Abou-Hadid et al. (1995) compared the effects of a warm-air heater and air distribution through perforated PE tubes with water-filled PE tubes, 40 cm

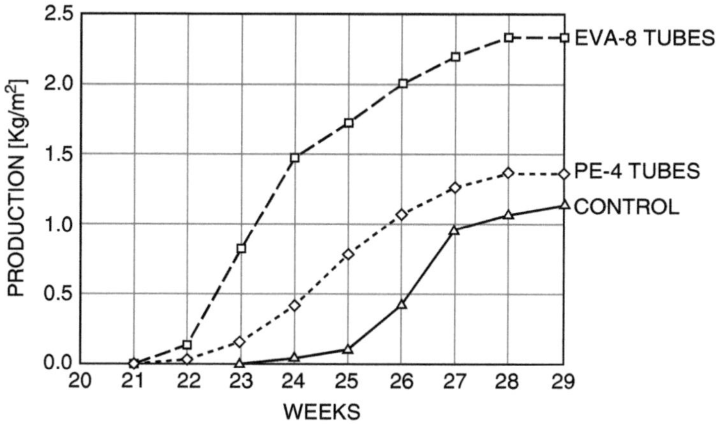

Fig. 12.29 Total yield of Gafsa muskmelons was significantly higher under EVA film than under unheated control

diameter, 80 l/m^2 coverage rate, and an unheated control greenhouse for the vegetable production in Egypt.

Results for Sweet Pepper.

The air heater was able to maintain the 15°C set point temperature, while the water-filled PE tube heating increased the temperature by 2.5–4.4°C above outside temperature. The highest temperature increase could be observed in the early season, when the pepper plants did not shade the tubes and when outside temperature dropped below 4–5°C. The total yield of sweet pepper with the different heating systems from January to May was:

Warm-water heater	10.6–10.94 kg/m^2
Water-filled PE tubes	5.3–6.3 kg/m^2
Unheated control	3.5–4.4 kg/m^2

The total yield with water-filled PE tubes was significantly higher than in the unheated control.

Results with French beans in double-layer plastic-film greenhouses.

The highest total yield and lowest level of malformed fruits could be obtained with warm-air heating, followed by water-filled PE tubes and unheated control.

The water-filled PE tubes can be an inexpensive method for improving the climate conditions during the early few months after planting, when cultivation starts with outside temperature below 5°C.

Cats and birds like the warm-water-filled plastic tubes for resting on. This brings a danger of producing leakage and as a result draining off the water. The danger is reduced when insect screens are installed.

Chapter 13
Crop Water Requirement and Water Use Efficiency

13.1 Crop Water Requirement

Irrigation water is a scarce factor in many regions. Therefore, crop water requirement has to be calculated, and irrigation systems have to be designed carefully. Knowledge of the evapotranspiration inside the greenhouse is important for successful plant growth, calculation of irrigation water consumption, and possible and economical rainwater collection and storage (see Chap. 14).

Evapotranspiration can be calculated by the FAO–Penman–Monteith method that has been developed for open field conditions (Allen et al. 1998). The Penman–Monteith equation for a reference evapotranspiration ET0, derived from an energy balance equation for an evaporating surface of a well-irrigated grass reference crop is:

$$\mathrm{ET0} = \frac{0.408 \times \Delta(q_{\mathrm{RN}} - q_{\mathrm{RG}}) + \gamma \frac{900}{T_{\mathrm{mean}} + 273} \times v \times (e_{\mathrm{S}} - e_{\mathrm{A}})}{\Delta + \gamma(1 + 0.34v)} \tag{13.1}$$

where:

$\Delta = f\,(T_{\mathrm{mean}})$: The slope of vapour pressure curve (kPa/°C), given in a Table (Annex 2, Allen et al. 1998).

$\gamma = f$ (altitude z): Psychometric constant (kPa/°C), given in a Table (Annex 2. Allen et al. 1998).

q_{RN} (MJ/m^2 day): Net radiation at crop surface.

q_{RG} (MJ/m^2 day): Soil heat flux density is very small, and can normally be neglected.

v (m/s): Air velocity.

$e_{\mathrm{S}} = f(T_{\mathrm{mean}})$ (kPa): Saturation vapour pressure, given in a Table (Annex 2, Allen et al. 1998).

$e_{\mathrm{A}} = f(T_{\mathrm{mean}})$ (kPa): Actual vapour pressure.

C. von Zabeltitz, *Integrated Greenhouse Systems for Mild Climates*,
DOI 10.1007/978-3-642-14582-7_13,

13.1.1 This Penman–Monteith Method Can Also Be Applied for Greenhouse Conditions, if the Parameters Are Adapted to Greenhouse Climate Conditions

The climate data for outside conditions can be taken from adequate references, for example Müller 1996 or climate data tools of FAO aquastat (www.fao.org/nr/water/aquastat/gis/index3.stm).

Given data and parameters are:

Mean max temperature, mean min temperature, mean relative humidity, number of sunshine hours, global radiation, mean wind velocity.

The equivalents for the radiation are

1 kWh = 3.61 MJ
1 MJ = 0.277 kWh
1 mm/day = l/m^2 day = 0.408 MJ/m^2 day

If the global radiation is not given, it can be calculated by a method given by Allen et al. (1998).

The inside temperature in unheated greenhouses is normally higher than the outside temperature, and the incoming global radiation is reduced. These factors have to be taken into consideration when estimating the evapotranspiration in greenhouses. The inside temperature in well-ventilated greenhouses during daytime can be assumed about 3–5° above outside temperature (see Chap. 9). The mean inside temperature during the night in unheated greenhouses is about 2°C above outside temperature, due to the storage effect of the soil (Thomas 1994; von Zabeltitz 1986a; Rath 1994).

The mean minimum and mean maximum temperatures for the calculation of the evapotranspiration inside the greenhouse can be assumed to be:

$$T_{\text{gmax}} = T_{\text{max}} + 4$$

$$T_{\text{gmin}} = T_{\text{min}} + 2$$

Mean inside temperature:

$$T_{\text{gmean}} = (T_{\text{gmax}} + T_{\text{gmin}})/2$$

The outside relative humidity decreases during daytime due to the increasing outside temperature. The relative humidity inside the greenhouse remains at a relatively high level due to the continuous evapotranspiration from crop and soil even if the greenhouse is ventilated during daytime. The mean relative humidity in the daytime inside a ventilated greenhouse can be assumed to be:

$RH_{\text{mean}} = 75\text{–}80\%$

Table 13.1 Crop coefficients k_C for various open field grown crops (Allen et al. 1998)

Crop	k_{Cini}	k_{Cmid}	k_{Cend}
Small vegetables (broccoli, cabbage, lettuce, onion)	0.7	1.05	0.95
Eggplant	0.6	1.05	0.9
Tomato	0.6	1.2	0.8
Cucumber	0.6	1.2	0.75
Watermelon	0.4	1.0	0.75

The incoming global radiation is reduced by the cladding material and construction components, and can be expressed by

$$q_{RSI} = \tau \times q_{RS} \tag{13.2}$$

where

$\tau = 0.6$–0.7 for single plastic-film covered greenhouses (see Chap. 6).

q_{RS} = outside global radiation.

The calculation of the reference evapotranspiration inside the greenhouse can be done with the help of a calculation sheet, given in Annex 3.

The actual crop evapotranspiration in the greenhouse is

$$\mathrm{AET_C} = k_C \times \mathrm{ET0} \quad (1/\mathrm{m}^2\,\mathrm{day} = \mathrm{mm/day}) \tag{13.3}$$

The crop coefficients k_C are given in tables for various crops in initial k_{Cini}, middle k_{Cmid}, and end k_{Cend} stage of cropping (Table 13.1, Allen et al. 1998).

The daily crop water requirement $\mathrm{CWR_d}$ can be calculated by

$$\mathrm{CWR_d} = \mathrm{AET_C}(1 + l_i) \times A_{Crop}/A_G \quad (\mathrm{mm/day}). \tag{13.4}$$

l_i = loss factor for irrigation.

$l_i = 0.03$–0.1 for drip irrigation systems.

A_{Crop}/A_G = crop-covered area to greenhouse floor area.

$A_{Crop}/A_G = 0.9$ for vegetables and cut flowers on ground beds.

The losses of different irrigation systems are given by De Pascale and Maggio (2005):

Drip irrigation	10–20%
Sprinkler irrigation	30–50%
Furrow irrigation	50–60%

Most of the greenhouses in warm climates will be irrigated by drip irrigation today. Modern irrigation systems can reduce the losses to 5–10%.

The monthly crop water requirement $\mathrm{CWR_m}$ is

$$\mathrm{CWR_m} = \mathrm{CWR_d} \times d_m \quad (\mathrm{mm/month}) \tag{13.5}$$

d_m = number of days in the month.

13.1.2 Example 1: Almeria (Spain)

The reference evaporation for an unheated greenhouse at Almeria (Spain) in May has been calculated by the adapted Penman–Monteith equation (see calculation sheet in Annex 3):

ET0 = 3.16 (mm/day) without considering the soil heat flux.

Fernandez et al. (2009) presented an equation for the calculation of the ET0 in unheated greenhouses:

$$\mathrm{ET0} = (0.288 + 0.0019 \times \mathrm{JD}) \times q_{\mathrm{RS}} \times \tau \quad (\mathrm{mm/day}) \tag{13.6}$$

JD = Julian days
q_{RS} (mm/day) = outside global radiation
τ = transmittance of the greenhouse
The calculation for Almeria in the middle of May results in:

ET0 = 3.61 (mm/day)

13.1.3 Example 2: Bangkok (Thailand)

Calculation of ET0 and CWR in April and May for Bangkok, Thailand.

ET0 (April) = 3.67 (mm/day)
ET0 (May) = 3.11 (mm/day)

The crop water requirement for a tomato crop is

CWR = ET0 × k_{C} × 1.05 × 0.9 (mm/day)

The crop coefficient for a high tomato crop in greenhouses can be higher than for tomatoes in open field, here k_{C} = 1.25 in mid-crop stage.

CWR = 3.67 × 1.25 × 1.05 × 0.9 = 4.34 (mm/day)

Harmanto et al. (2005) found 4.1–5.2 (mm/day) for the actual irrigation water of a tomato crop in the Bangkok climate.

13.1.4 Example 3: Antalya (Turkey) and Bangalore (India)

Figure 13.1: shows the calculated values for Antalya, Turkey, and Bangalore, India, Those figures are used for the calculation of rainwater collection (Chap. 14).

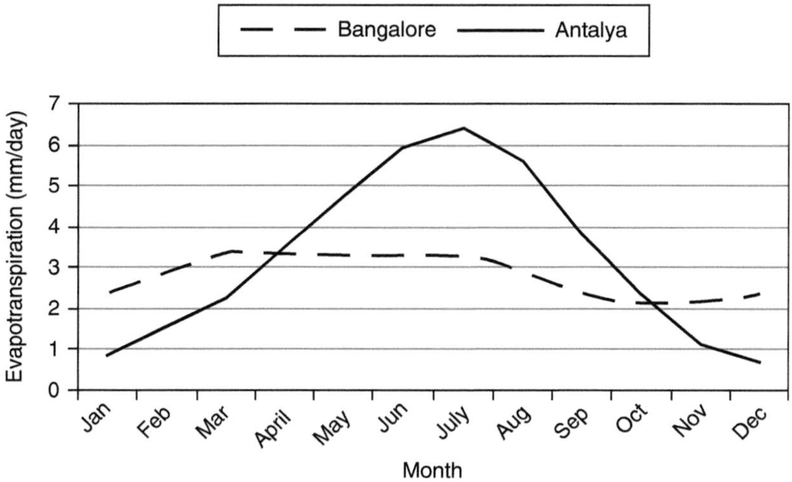

Fig. 13.1 Reference evapotranspiration ET0 for Antalya, Turkey, and Bangalore, India, calculated by the adapted Penman–Monteith equation

13.2 Water Use Efficiency

Water use efficiency is defined as ratio of yield to irrigation water requirement (De Pascale and Maggio 2005)

WUE = yield/irrigation water requirement (kg crop/m^3 irrigation water)

The irrigation water requirement IWR is CWR + eventual soil leaching requirement.

The WUE in greenhouses normally is much higher than in open field production because of

- Reduced evapotranspiration (less radiation, higher humidity).
- Increased crop yield by production techniques, climate control and pest control.
- Advanced irrigation techniques (drip irrigation, reuse of drainage water).

The mean WUE of some crops in Mediterranean countries is given by Pardossi et al. (2004) in Table 13.2:

Mean values for some crops in Mediterranean countries (Cyprus, Egypt, Greece, Israel and Spain) in comparison to the Netherlands were also given by Pardossi et al. (2004)

	WUE Mediterranean countries (kg/m^3)	WUE Netherlands (kg/m^3)
Tomato	21.8	58.2
Cucumber	14	28
Sweet pepper	30.3	77

Table 13.2 Water use efficiency WUE of tomato crops under different climate conditions and using different growing systems (Pardossi et al. 2004)

Growing conditions	Country	WUE (kg/m^3)
Open field soil culture	Israel	17
	France	14
Unheated plastic-film greenhouse		
Soil culture	Spain	25
Soil culture	France	24
Soil culture	Israel	33
Open substrate culture	Italy	23
Closed substrate culture	Italy	47
Climate-controlled greenhouse		
Open soil-less culture	France	39
Open soil-less culture	Netherlands	45
Closed soil-less culture	Netherlands	60

Table 13.3 Mean cooling efficiency in the pad depending on air flow rate (Sabeh 2007)

Air flow rate (m^3/s)	Pad efficiency (%)
4.5	83.2
9.4	80.6
13.0	77.4
16.7	73.5

Sabeh (2007) quantified and compared the water use efficiency of a fan and pad cooling system and a fog cooling system in a round-arched single-span greenhouse, 9.8 m by 28 m, 3.4 m gutter height, 6.3 m ridge height, with roof and side wall ventilation.

The fan and pad cooling system consisted of an 8.5 m by 1.2 m cellulose pad, 1.3 m above ground level, at the northern gable. Three exhaust fans producing different ventilation rates were installed at the southern gable.

The high-pressure fog cooling system operated at a pressure of 8,960 kPa (89.6 bar) and produced droplets less than 50 μm in diameter. A central overhead fog line was installed 3.1 m above the floor.

Table 13.3 shows the mean cooling efficiency for the pad with different air flows through the pad. Increasing ventilation rate decreases the cooling efficiency, because the higher air velocity reduces the contact time of the air with the water surface in the pad. The saturation of air by water vapour is lower.

The water use efficiency for the fan and pad system and a tomato crop with a total yield of 0.14 kg/m^2 day is given in Table 13.4.

The total WUE decreased with increasing ventilation rate because the fan and pad system uses more water for evaporation at higher ventilation rates. Increasing ventilation rate reduced the air temperature gradient between pad and fan from 8.6°C at 4.5 m^3/s to 4.0°C at 16.7 m^3/s, but the smaller temperature gradients were accompanied by lower relative humidity levels.

Table 13.4 Water use efficiency WUE_{fp} of a fan and pad cooling system, WUE_{oirr}, of an open irrigation system for tomato crop, and resulting WUE_{tot} for both systems together. Tomato yield 0.14 kg/m^2 day

Air flow (m^3/s)	WUE_{fp}	WUE_{oirr}	WUE_{tot}
4.5	44	31	18
9.4	22	31	13
13.0	17	31	11
16.7	14	31	10

Table 13.5 The water use efficiency WUE_{fog} for the high-pressure fog system, the WUE_{oirr} of an open irrigation system for a tomato crop, and the resulting WUE_{tot} for both systems

Air flow (m^3/s)	WUE_{fp}	WUE_{oirr}	WUE_{tot}
3.0	18	31	11
4.5	19	31	12
9.4	15	31	10

Table 13.5 shows the water use efficiencies for the high-pressure fog cooling system in the same greenhouse and under the same climate conditions. The inside temperature could be held at the control set point. The water use efficiency was highest with a ventilation rate of 4.5 m^3/s, and lowest with the highest ventilation rate. The central overhead nozzle line produced uniform greenhouse climate conditions.

Chapter 14
Rainwater Collection and Storage

Many regions have a high water surplus for the irrigation of plants during rainy seasons and a deficit of water in dry seasons. When growing plants in greenhouses, the rainwater running off the roofs of greenhouses can be collected and used for irrigation. If salty water is available, this can be mixed with rain water and then used as irrigation water. It is necessary to build greenhouses with sufficiently large gutters and storages for the collection of rain water. The crop water requirement has to be known in order to calculate the storage for rainwater and for irrigation systems.

The crop water requirement CWR can be calculated by use of the reference evapotranspiration ET0 (mm/day) according to FAO–Penman–Monteith with adapted parameters for unheated greenhouses (see Chap. 13).

The actual evapotranspiration AET of the crop inside the greenhouse is

$$\mathrm{AET_C} = \mathrm{ET0} \times k_C \quad (\mathrm{mm/day} = \mathrm{l/m^2 day}). \tag{14.1}$$

The daily crop water requirement is.

$$\mathrm{CWR_d} = \mathrm{AET}(1 + l_\mathrm{I}) \times A_\mathrm{crop}/A_\mathrm{G} \quad (\mathrm{mm/day}), \tag{14.2}$$

where

$l_\mathrm{I} = 0.03$–0.1 loss factor for the drip irrigation system.
$A_\mathrm{crop}/A_\mathrm{G} = 0.9$ for vegetables and flowers on ground beds.

The monthly crop water requirement CWR is

$$\mathrm{CWR_m} = \mathrm{CWR_d} \times d_\mathrm{m} \quad (\mathrm{mm/month}), \tag{14.3}$$

d_m = days in the month

C. von Zabeltitz, *Integrated Greenhouse Systems for Mild Climates*,
DOI 10.1007/978-3-642-14582-7_14, © Springer-Verlag Berlin Heidelberg 2011

14.1 Calculation of the Storage Volume

The following points have to be considered for the design of the storage basin:

- The area of greenhouses and the vacant area available for a storage basin.
- The distribution of precipitation and the amount of rainfall.
- The crop water requirement.
- Whether the storage basin is only to be used for storing rainwater or for mixing rain and salty water.

Normally, daily frequencies of precipitation should be taken into consideration for the calculation of the storage volume. Those values are unknown in most cases. Therefore, monthly precipitation can be used to estimate the storage volume in a first approximation.

The monthly collected amount of precipitation is.

$$\mathrm{CV_m} = \mathrm{Pre} \times f_\mathrm{C} \quad (\mathrm{l/m^2 month}), \tag{14.4}$$

where

Pre ($\mathrm{l/m^2 month}$) = mean monthly precipitation
$f_\mathrm{C} = 0.9$: Collecting factor for greenhouse roofs.

The collecting factor is the ratio of possible amount of collected rain water to the precipitation.

If rain water is to be used for irrigation, the monthly storable precipitation is:

$$\mathrm{STP_m} = \mathrm{CV_m} - \mathrm{CWR_m} - \mathrm{EV_{pond}} \quad (\mathrm{l/m^2 month}). \tag{14.5}$$

The evaporation of the storage basin surface $\mathrm{EV_{pond}}$ can be neglected if the basin is covered by a swimming plastic cover, for example.

If $\mathrm{STP_m}$ is positive, the storage will be filled, if $\mathrm{STP_m}$ is negative, the storage will be emptied.

The yearly storable precipitation is:

$$\mathrm{STP_y} = \sum \text{positive } \mathrm{STP_m} \quad (\mathrm{l/m^2 month}). \tag{14.6}$$

The yearly deficit is:

$$\mathrm{Def_y} = \sum \text{negative } \mathrm{STP_m} \quad (\mathrm{l/m^2 month}). \tag{14.7}$$

The yearly storage balance is:

$$\mathrm{STB_y} = \mathrm{STP_y} - \mathrm{Def_y} \quad (\mathrm{l/m^2 month}). \tag{14.8}$$

The following cases have to be distinguished:

1. If $STB_y > 0$ or $STP_y > Def_y$, the storable precipitation is sufficient for irrigation throughout the year. The storage volume becomes:

$$VST = Def_y \quad (l/m^2).$$

One can enlarge the volume, if the monthly variation of precipitation is high. The storage volume is:

$$VST = Def_y(1 + V_C),$$

V_C = coefficient of variation for precipitation (see Chap. 2).

2. If $STB_y < 0$ or $STP_y < Def_y$, the storable precipitation is not sufficient for irrigation. two cases have to be distinguished:
   (a) If the storable precipitation is higher than the maximum monthly collected precipitation CV_{mmax}
   If $STP_y > CV_{mmax}$, then

$$VST = STP_y$$

or

$$VST = STP_y(1 + V_C) \quad (lm^2)$$

   (b) If $STP_y < CV_{mmax}$, then

$$VST = CV_{mmax}$$

or

$$VST = CV_{mmax}(1 + V_C)$$

14.2 Example 1: Storage Volume for Climate Conditions in Bangalore (India)

Table 14.1: Calculation of the storage volume for collecting rainwater for irrigation in Bangalore (India). Precipitation Pre (Müller 1996); Monthly storable precipitation STP_m with (14.5) and $EV_{pond} = 0$; ET0 see Fig. 13.1;

$CWR_m = ET0 \times k_C(1 + l_i) \times A_{Cr}/A_G \times d_m$

For tomato, mean $k_C = 1.1$; $(1 + l_i) = 1.05$; $A_{Cr}/A_G = 0.9$

$k_C(1 + l_i) \times A_{Cr}/A_G = 1.04$;

$CWR_m = 1.04 \times ET0 \times d_m$

From Table 14.1 can be seen:

$STP_y = \Sigma pos\ STP_m = 164.9\ l/m^2$ year
$Def_y = \Sigma neg\ STP_m = 371.2\ l/m^2$ year

Table 14.1 Data for the calculation of the storage volume in Bangalore (India)

Month	Pre l/m^2 month	CV l/m^2 month	d_m	ET0 l/m^2 day	CWR_m l/m^2 month	STP_m l/m^2 month	STP_m accumulated
Jan	5	4.5	31	2.38	76.7	−72.2	+23.1
Feb	8	7.2	28	2.8	81.4	−74.2	−51.1
Mar	10	9	31	3.35	108	−99	−150.1
Apr	41	37	30	3.3	103	−66	−216.1
May	107	96.3	31	3.3	106.3	−10	−226.1
Jun	74	66.6	30	3.3	103	−36.4	−262.5
Jul	99	89.1	31	3.26	105.1	−16	−278.5
Aug	127	114.3	31	2.9	93.5	+20.8	+ 20.8
Sep	170	153	30	2.34	73	+80	+100.8
Oct	150	135	31	2.2	70.9	+64.1	+165.2
Nov	69	62.1	30	2.13	66.5	−4.4	+160.8
Dec	10	9	31	2.3	74.2	−65.2	+95.6

Case 2(a): Storage volume per m^2 greenhouse area $VST = STP_y = 165(l/m^2)$
The storage is empty from February until July.

Looking at the accumulated sum of STP_m, starting in August, the crop water requirement can be covered for 6 months from August to January by rainwater with a storage volume of 0.165 m^3 per m^2 greenhouse area.

14.3 Example 2: Storage Volume for Climate Conditions in Antalya (Turkey)

Table 14.2: Calculation of the storage volume for the collection of precipitation in Antalya (Turkey). Precipitation Pre (Müller 1996); STP_m with (14.5) and $EV_{pond} = 0$; ET0 from Fig. 13.1. $CWR_m = 1.04 \times ET0 \times d_m$ (see Example 1)

$STP_y = \Sigma pos\ STP_m = 608.2\ l/(m^2\ year)$
$Def_y = \Sigma neg\ STP_m = 1{,}651.4\ l/(m^2\ year)$
$STP_y > CV_{mmax}$:
Case 2(a): Storage volume $VST = STP_y = 608\ l/m^2$

The storage is empty from July until October. The crop water requirement can be covered for 8 months from November to June by rainwater with a storage volume of 0.61 m^3/m^2 greenhouse area.

14.4 Design of Rainwater Storage Basins

Different types of storage basins can be built, if enough space is available near the greenhouse:

- Simple basins, dug in the soil, if the soil at the bottom of the basin is sufficiently watertight.

Table 14.2 Data for the calculation of the storage volume in Antalya (Turkey)

Month	Pre l/m^2 month	CV l/m^2 month	d_m	ET0 l/m^2 day	CWR$_m$ l/m^2 month	STP$_m$ l/m^2 month	STP$_m$ accumulated
Jan	259	233	31	0.86	27.7	+205.3	+495.8
Feb	175	157.5	28	1.55	45.1	+112.4	+608.2
Mar	79	71	31	2.24	72.2	−1.2	+607
Apr	38	34	30	3.5	109.2	−75.2	+531.8
May	33	30	31	4.75	153.1	−283.1	+248.7
Jun	13	12	30	5.9	184.1	−172.1	+76.6
Jul	3	2.7	31	6.4	206.3	−203.6	−127
Aug	3	2.7	31	5.6	180.5	−177.8	−304.8
Sep	15	13.5	30	3.82	119.2	−105.7	−410.5
Oct	53	48	31	2.25	72.5	−24.5	−435
Nov	119	107	30	1.11	34.6	+72.4	+72.4
Dec	267	240	31	0.68	21.9	+218.1	+290.5

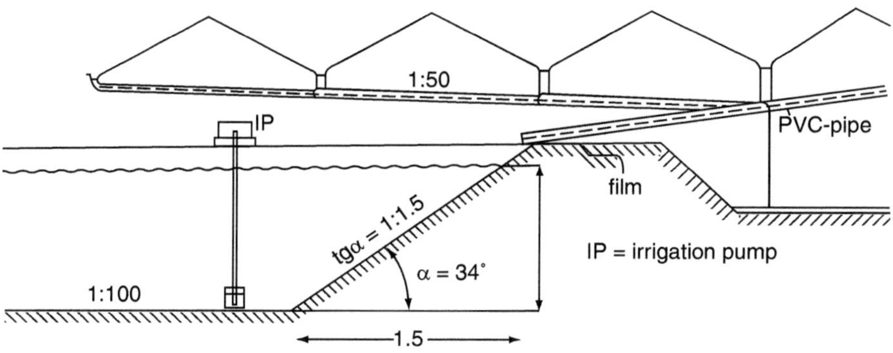

Fig. 14.1 Rainwater basin and rainwater run off from greenhouse gutters

- Earth basin lined with plastic film.
- Concrete basins; durable, need less maintenance, but are very expensive.

All storage basins should be covered at the surface by swimming plastic film, for example, to avoid too high evaporation.

Figure 14.1 shows the arrangement of a plastic-film water basin with water tubes from the gutters to the basin (von Zabeltitz and Baudoin 1999). To collect heavier rainfall, the gutters and tubes leading to the storage basin must have an adequate diameter. The tubes leading the water to the basin should have a slope of about 1:50–1:100. The following diameters are recommended:

Greenhouse floor area (m^2)	Tube diameter (mm)
<400	100
400–700	125
700–1,200	150

If the level of the storage basin is deep enough, the rainwater can be led via open gutter lined with plastic film.

The basin has to be situated at the deepest point of the site. If this is not the case and the greenhouses are placed deeper than the basin, the rainwater can be conducted into the basin by a siphon system (Fig. 14.2). Watertight tubes are installed

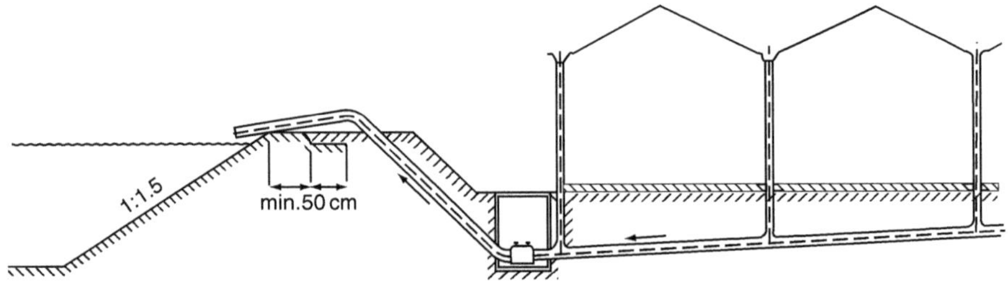

Fig. 14.2 Rainwater conducted to the basin by a siphon system

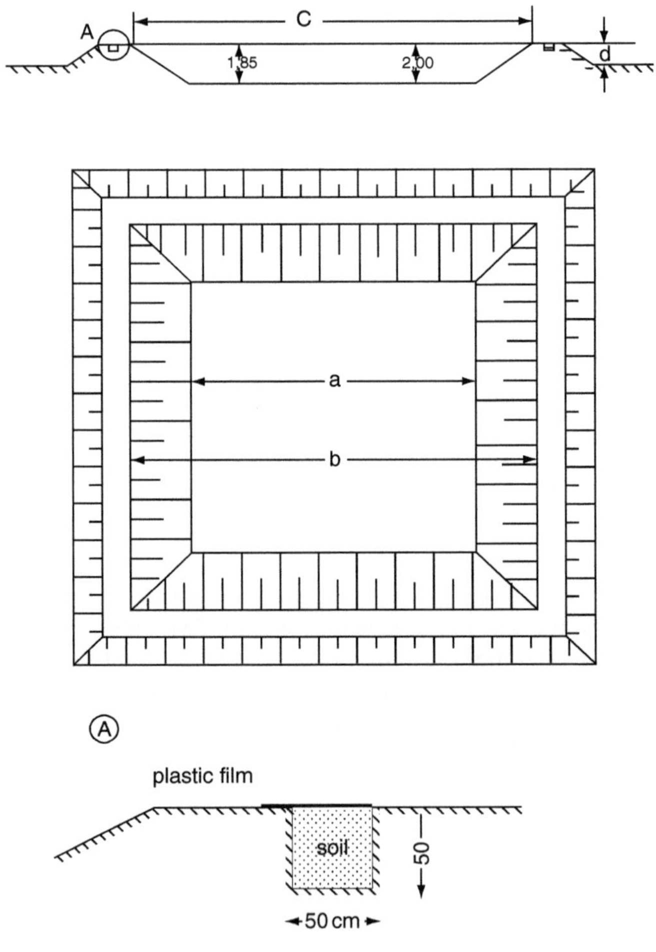

Fig. 14.3 Measurements of a rainwater basin

sloping downward to a deepest point near the basin, and from there into the basin. The gutters have to be above the water level of the basin. When rainwater in the tubes rises above the water of the basin, positive pressure develops, and the water flows from the tube into the basin.

When digging out the basin, the soil is thrown up around the basin as an embankment (Fig. 14.3).The angle of the embankment is about 34°, or it has a ratio of 1:1.5. Thus it is within the range of frictional angles of most types of soil.

In the case of a 2 m-deep square basin, the measurements a, b, c and d have the following values (m) for different quantities of water:

Water quantity(m^3)	a(m)	b(m)	c(m)	d(m)
200	7.25	13.03	12.8	0.75
400	11.7	17.5	17.25	09
600	15.05	20.8	20.6	1.0

Chapter 15
Desalination of Salty Water and Closed-System Greenhouses

The availability of sufficient irrigation water is a big problem in many arid and remote areas. The irrigation water for crop production in greenhouses will be provided by seawater desalination plants in some countries, which use oil for the necessary desalination energy. In Kuwait, for example, trucks transport irrigation water from sea water desalination plants to the greenhouse holdings. The water from desalination plants is relatively expensive, and is used mainly for human consumers.

Brackish or salty water is available in many arid regions. If greenhouse systems for reduced irrigation water requirement (closed-system greenhouses) and cost-effective solar water desalination systems can be developed, vegetable production can be extended to those arid regions.

A closed-system greenhouse with integrated solar water desalination was developed and evaluated between 1979 and 1984 at the University of Hannover (Strauch 1985a, b). An improved structure was designed and built in1987 as a pilot plant at the University of Adana, Turkey (Meyer et al. 1989; Baytorun et al. 1989).

Different solar water desalination systems were tested in parallel experiments in Hannover (Klasik et al. 1989).

15.1 Solar Desalination Systems

Salt water is heated up by solar energy in a basin or at an wetted absorber surface; the water evaporates and condensates on a transparent colder material that should be opaque to long-wave radiation (Figs. 15.1–15.4). The condensate runs down on the sloped surface into a gutter. The aim is to use as much solar energy as possible, and to design a simple and cost-effective construction with low maintenance and running costs. Only commercially available construction components, if possible greenhouse components, should be used for the design. A completely sealed system and good insulation to the soil are prerequisites for high efficiency. Losses are due

C. von Zabeltitz, *Integrated Greenhouse Systems for Mild Climates*,
DOI 10.1007/978-3-642-14582-7_15,

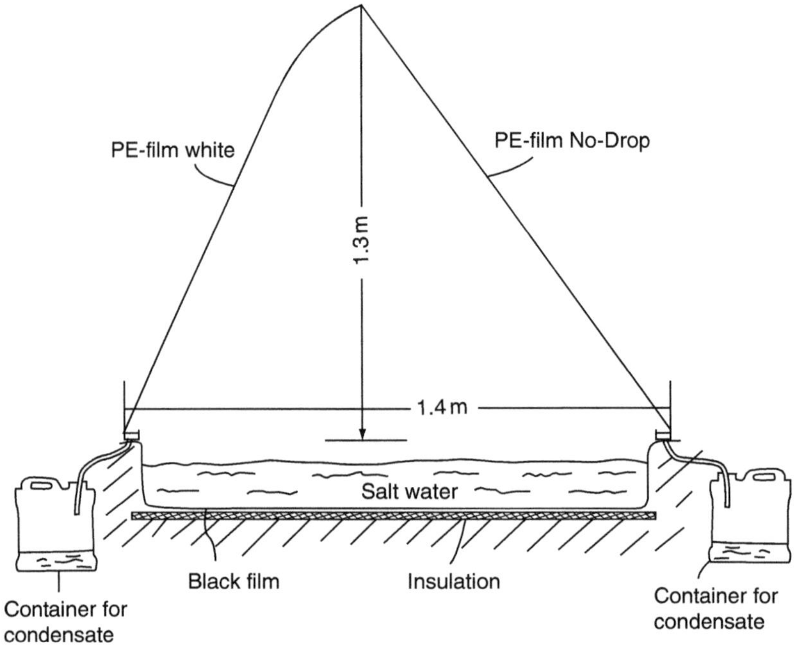

Fig. 15.1 The experimental type 1 consists of a plastic-film structure with a No- Drop film at the south side and a white PE film at the north side. The saltwater basin is covered with a black film and insulated from the soil at the bottom

to reflection at the covering material and water surface, as well as due to leakage of the structure.

The productivity of desalinated water [l/(m^2 day)] of the four systems depending on solar radiation is shown in Fig. 15.5. The active system type 3 had the highest productivity. The additional absorber inside the water basin of type 4 did not improve productivity. It is more suitable if the solar radiation gets through the whole water layer and is absorbed on the black film at the bottom. A part of the solar radiation will be absorbed directly by the water already.

The passive system 2 had a relatively good productivity. Type 1 with the plastic-film cladding had the worst productivity, and is not suitable for desalination systems, because the plastic film flutters through wind influence, and the condensed drops fall back into the saltwater basin.

Desalination systems do not work only during daytime, but a significant amount of water will be evaporated and condensed at night, because the water warms up during the day, evaporates at night and condenses on the colder cladding material. Figure 15.6 shows the condensed water production for day and night over a period of 48 days from end of May to beginning of July in Germany. The night-time production is higher than in daytime in the active systems.

Figure 15.7 shows the course of the temperatures for a 1-day period. The maximum water temperature of type 3 gets up to 50°C, while the temperature of

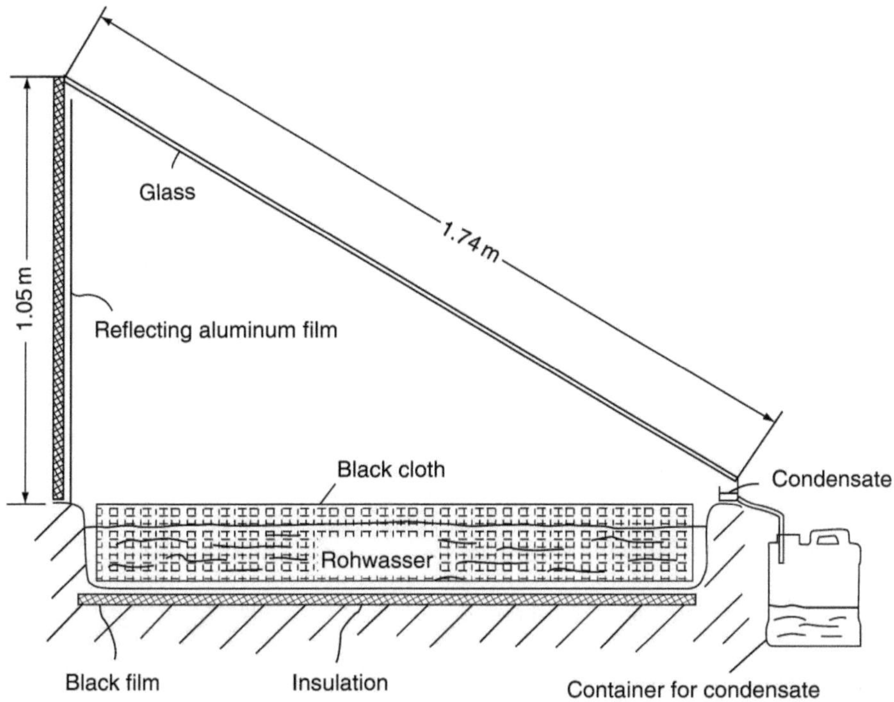

Fig. 15.2 The experimental type 2 consists of a vertical back wall covered by an aluminium film for reflection of sun radiation, and an inclined glass roof. Black absorbent cloths were positioned vertically and parallel at 10 cm distance in the basin for absorption of solar radiation. One part of the black cloths is positioned above the water surface. Water can rise up by capillarity and keep the cloths permanently wet. In thatway, it is possible to significantly enlarge the evaporation surface. Type 1 and type 2 are passive systems without any water pump

type 4 is lower than 45°C. The water temperatures correspond to the total productivity of the systems.

In addition, another passive system was designed and investigated, consisting of a water basin and two glass panes, 1.74 m long, on the south and north roof (Figs. 15.8 and 15.9). The productivity was 3 l/m^2 day for a global radiation of 5.9 kWh/m^2 day. Thus, the productivity was higher than in the other passive systems.

The following requirements should be fulfilled for simple solar desalination systems:

- The air volume in the system should be as small as possible.
- The construction components should be outside and not inside the system.
- The system must be absolutely sealed.
- The salt water should reach high temperatures.
- An air circulation inside can improve the productivity.

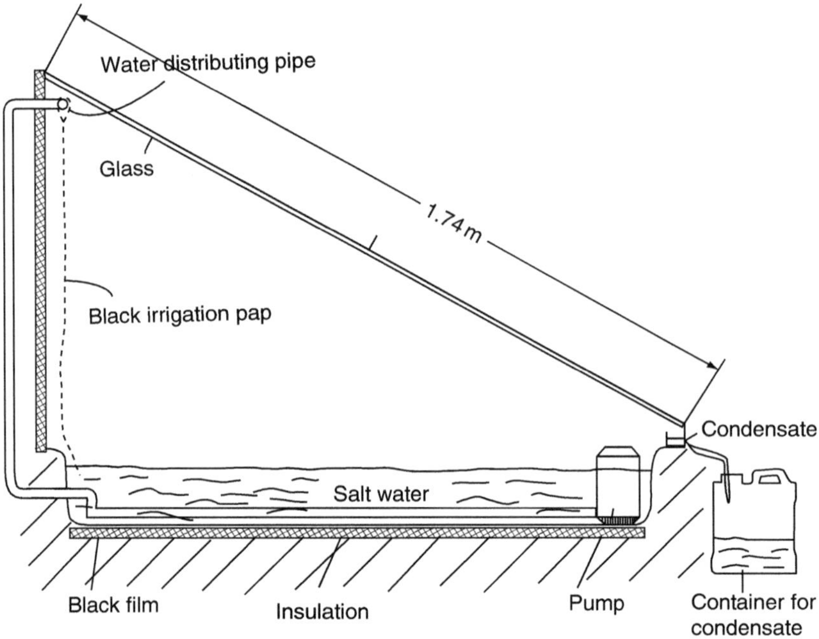

Fig. 15.3 Experimental type 3. The salt water is pumped and distributed over a black irrigation cloth at the back wall of the active system

15.2 Closed-System Greenhouse

Based on the experience with the simple solar water desalination systems a closed-system greenhouse with integrated solar water desalination was developed and evaluated first in a small greenhouse (Strauch 1985a, b), and then in a pilot plant designed in Germany, and erected and evaluated at the University of Adana, Turkey (Baytorun et al. 1989; Meyer et al. 1989). The greenhouse was designed to fulfil the following demands:

- Plant production in arid regions in a controlled environment with protection from wind, dust and low air humidity.
- Inside air temperatures which do not exceed suitable conditions for plants, even at high outside radiation and temperature.
- Reduction of water use by decreasing transpiration rate, and reduction of humidity losses through air exchange by making the greenhouse water-and airtight.
- Minimizing energy and water consumption by not using artificial cooling and heating.
- Recollection of condensed water from interior greenhouse surfaces.

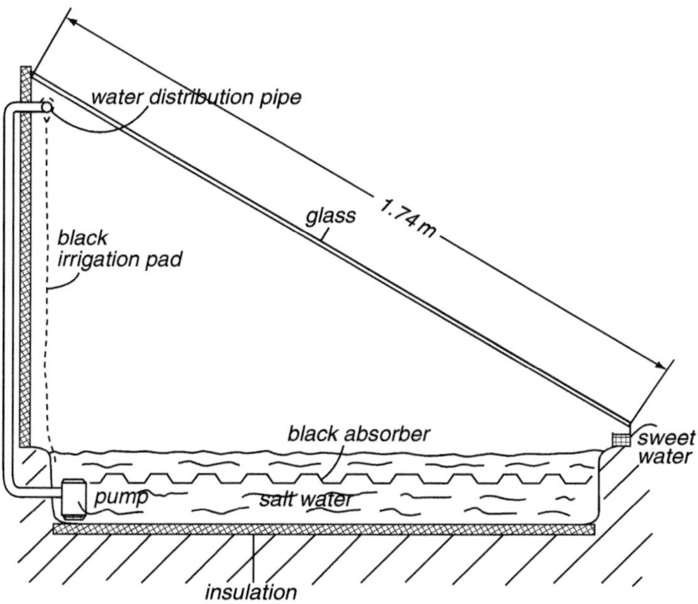

Fig. 15.4 Experimental type 4. An additional absorber sheet of corrugated aluminium covered by black shading material was positioned some cm below the water surface, to test the efficiency of an additional absorber inside the water basin

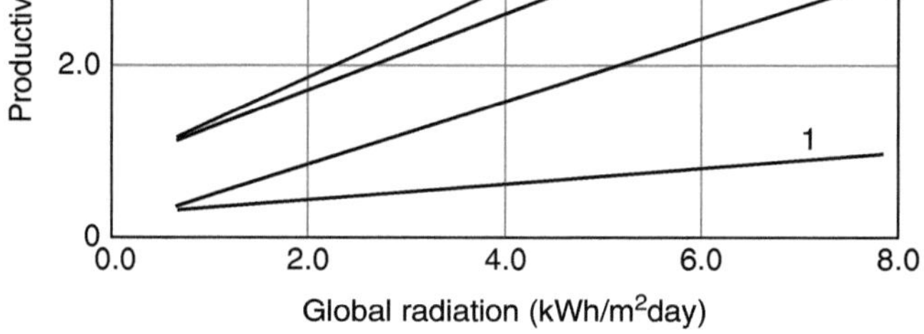

Fig. 15.5 Productivity of desalinated water (l/m^2 day) of the four desalination systems

- Independent water supply by solar desalination.
- Independent energy supply by solar cells.
- Construction with commercial greenhouse elements to reduce investment costs.

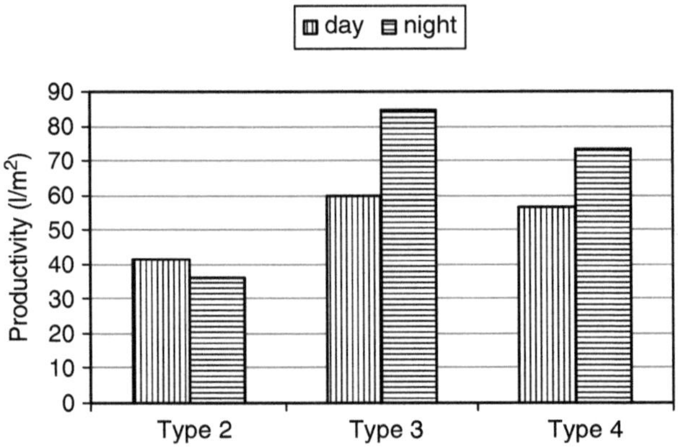

Fig. 15.6 Desalination productivity during day and night

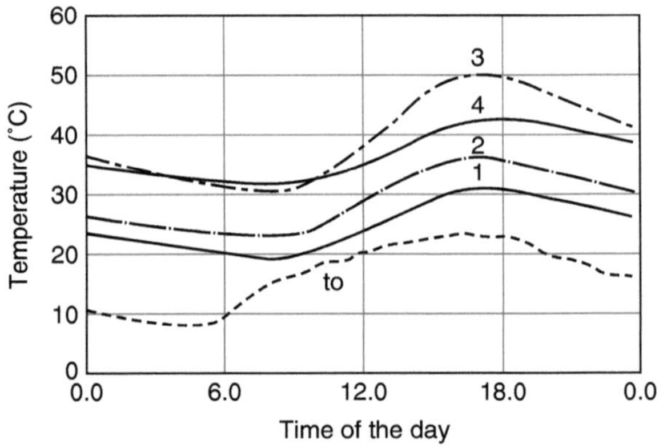

Fig. 15.7 Water temperatures for types 1–4 and outside temperature t_o

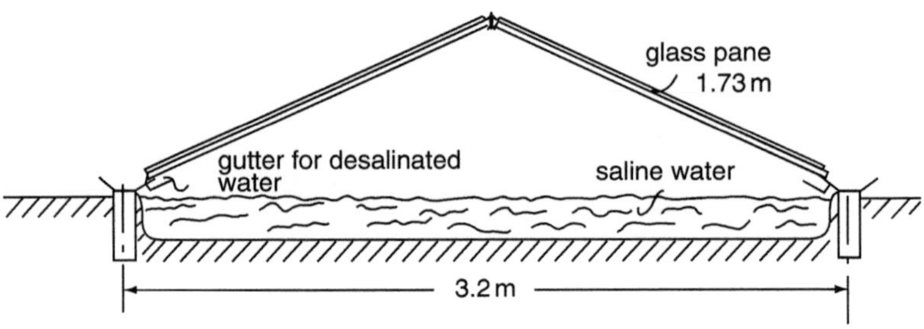

Fig. 15.8 Another passive desalination system

Figure 15.10 shows the cross-section of the greenhouse with a water desalination system at the southern side wall. Figure 15.11 shows the north and south view of the greenhouse. The special characteristics of the greenhouse were:

Fig. 15.9 Passive solar desalination system with two glass panes

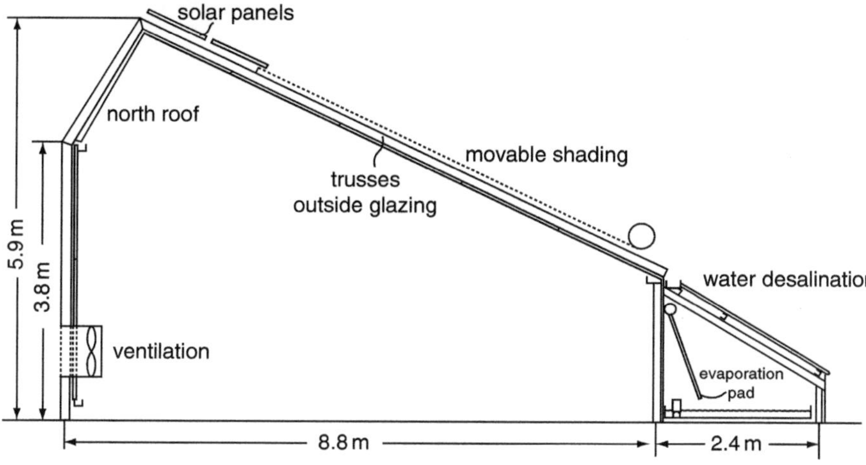

Fig. 15.10 Cross-section of the closed-system greenhouse, type Hannover

- The shape of the northern roof was designed to reflect the main part of the global radiation during high position of the sun.
- The southern roof was covered by normal glass. A special glass that absorbed a high amount of the incoming near infrared radiation to reduce the heat generation inside brought no advantage. Because of its own high temperature, the inside temperature was increased. Better would be a near infrared reflecting glass, but this was not available some years ago.

Fig. 15.11 Closed-system greenhouse type Hannover

- A movable lamellar outside shading system could be closed to reduce the incoming outside solar radiation at the southern roof.
- All trusses were designed outside the cladding material to prevent them warming up and heating the inside atmosphere. This construction is very recommendable for arid regions.
- A security or emergency low-capacity forced ventilation was installed at the northern side wall that could be used to circulate inside air only or to ventilate with outside air. The air exchange capacity of the fan was 6 (1/h).
- Solar panels were able to provide the necessary electricity power.
- Side wall and roofs were equipped with condensed water collection gutters to recollect evapotranspirated and condensed water at the inside surface.

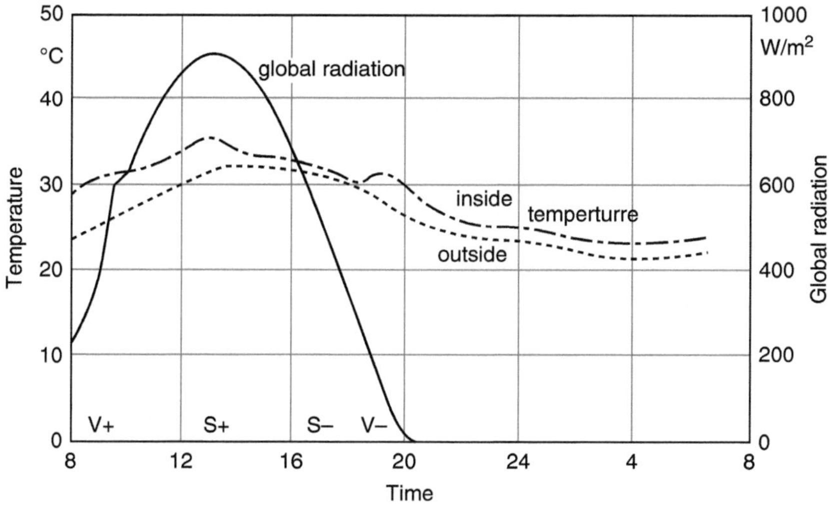

Fig. 15.12 Temperature inside and outside the closed system greenhouse. V+, V– means ventilation on and off. *S*+ and *S*– mean shading open and closed

The greenhouse was specially sealed, and could be kept closed as long as possible to reduce the requirement for desalinated irrigation water.

- A desalination unit was located at the southern side wall of the greenhouse. Because it was difficult to seal the system sufficiently at the site during mounting, it was recommended to separate greenhouse system and simple desalination system.

A tomato crop was grown in the greenhouse. Figure 15.12 shows an example of temperature course inside and outside with high global radiation and with climate control measures: ventilation, V+ and V– on and off, as well as shading, S+ and S– open and closed. Measurements demonstrate the possibility of keeping the inside temperature near the outside temperature.

The inside temperature could be kept near the outside temperature thanks to the outside shape, outside shading, outside trusses and low ventilation rate. The incoming solar radiation was mainly converted to latent heat by the evapotranspiration, and by preventing the conversion to sensible heat in the construction components which are located outside.

The CO_2 concentration decreased down to 150 Vpm in the closed atmosphere, which limited growth, but the security fan was able to keep the CO_2 concentration at a sufficient level. An CO_2 enrichment would be preferable. Too high humidity can be prevented by ventilation, but forced ventilation should be restricted to avoid too much humidity loss. The temperatures at night remained 1–3°C above outside temperature.

These measurements showed that plants can be grown without artificial cooling during a winter period in arid regions.

The mean irrigation water requirement for the middle period of tomato crop growth was 0.51 l/day plant or 1.17 l/m^2 day from April to beginning of June. That was less than the requirement in normal greenhouses.

The desalination productivity of the integrated desalination was about 3 l/m^{2}day, and thus less than expected in comparison to the experimental results of the separate desalination systems (see Sect. 15.1). Reasons were leakage in the structure and the internally mounted construction components in the desalination system in contrast to the greenhouse structure.

From the reduced irrigation water requirement in the closed system and the yield of desalinated water, one can expect a realistic area relation of desalination system to greenhouse floor area of 23–37% (Strauch 1985b). The clear desalination water can be mixed with salty water for irrigation.

The recollection of condensed water for irrigation depends on the tightness of the greenhouse structure. The amount of water recollected by condensation inside was about 20% of the irrigation water in the completely closed and sealed system, but would be reduced by the climate control measures of ventilation.

Summarising the results, the design of recollection systems inside and the corresponding water recollection was not recommendable for an economical use in practice.

Summarising all results, the following proposals for improvements were made (Meyer et al. 1989):

1. The desalination system should be designed as a separated passive system (Fig. 15.8).
2. The greenhouse system for arid regions could be a more simply designed one (Fig. 15.13) with the following characteristics:

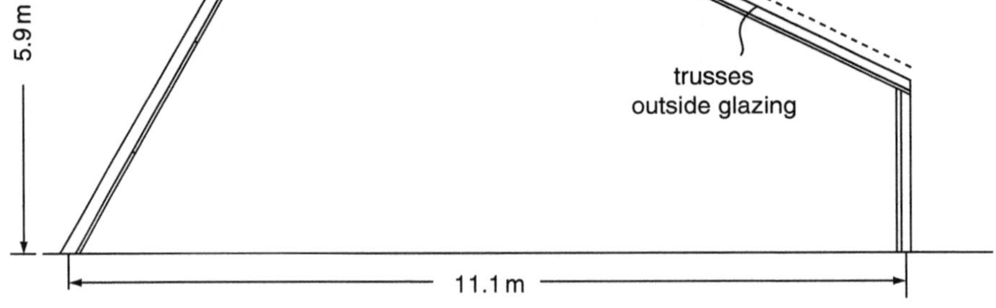

Fig. 15.13 New proposal for a closed-system greenhouse in arid regions

- Through north roof, with an inclination such that the main part of solar radiation around midday will be reflected.
- South roof cladding with normal glass, or near infrared reflecting material which is available today.
- Movable reflecting outside shading.
- All trusses outside the cladding material.

Such a greenhouse can help to solve some of the problems of crop growth in arid regions.

Chapter 16
CO_2 Enrichment

The CO_2 concentration inside a greenhouse can drop significantly below outside level when a dense crop is growing, even if the greenhouse is well-ventilated. The concentration can drop to less than 200 vpm during winter in mild climate regions. As the CO_2 concentration limits the photosynthesis of most vegetable species, the productivity decreases. The optimal CO_2 concentration for growth and yield seems to be 700–900 vpm (De Pascale and Maggio 2005, 2008). The CO_2 concentration should be kept to at least the outside level, but CO_2 enrichment is not a current practice in mild climates up to now.

The production loss due to CO_2 depletion may be higher than the production loss due to a reduced temperature through ventilation (Stanghellini et al. 2008). The enrichment of greenhouse air with CO_2 leads to better plant growth, shorter cropping times, and higher quality. Therefore a combined control of ventilation and CO_2 enrichment with low-cost CO_2 sources may result in improved and economically viable methods for crop growth in greenhouses.

Production losses in greenhouses are influenced by two main factors:

- Sufficient ventilation to avoid CO_2 depletion.
- Maintaining a higher temperature by heating on sunny, chilly days in spite of CO_2 depletion.

It is necessary to assess running and installation costs for CO_2 enrichment or heating to find out the optimal strategy for climate control. The compensation of CO_2 depletion by increased ventilation or even by CO_2 enrichment seems to be cheaper than compensation of production loss by heating. A good management strategy can be to ventilate as much or as little as necessary for temperature and humidity control, and to control CO_2 concentration inside the greenhouse up to outside level when ventilation is being used, and to higher levels when no or little ventilation is required for temperature control (Stanghellini et al. 2008, 2009).

C. von Zabeltitz, *Integrated Greenhouse Systems for Mild Climates*,
DOI 10.1007/978-3-642-14582-7_16,

The CO_2 concentration can be raised as follows (von Zabeltitz 1999):

- Technical carbon dioxide from bottles or tanks.
- Exhaust gases from gas burner for CO_2 enrichment with simultaneous heat production.
- Exhaust gases from directly fired air heater with gas burner.
- Straw between the plant rows, enriched with fertiliser and wetted. CO_2 is released during decomposition, but the amount of CO_2 cannot be controlled.

Exhaust gases from oil and coal heaters must not be used because of the content of sulphur dioxide.

Exhaust gases from gas burners can be led directly into the greenhouse (Fig. 16.1). Gas will be burned, and the CO_2 is blown with the circulating air into the greenhouse. Special control systems are necessary, and care must be taken that no carbon monoxide is formed. It is practical to mix the exhaust gas with fresh air. The water vapour and heat production, as well as the maximum allowed concentration, have to be taken into consideration.

CO_2 from an air heater (Fig. 16.2) can only be produced while it is in operation and when heating is needed, which is normally necessary only at night in mild climate regions. Some of the combustion gases are tapped off and used for CO_2 enrichment.

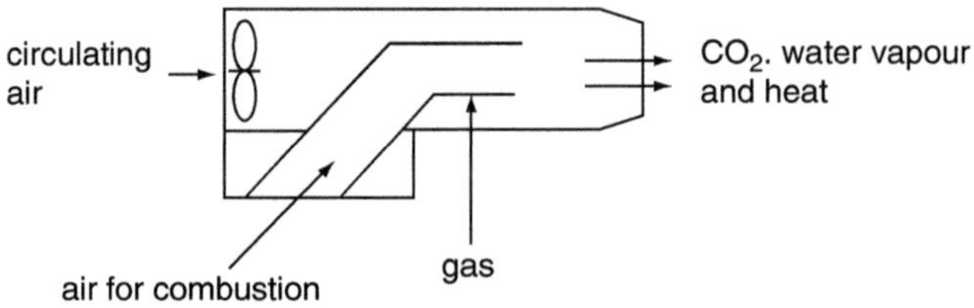

Fig. 16.1 Gas burner for CO_2 enrichment

Fig. 16.2 Directly fired gas burner for heating with CO_2 discharge for enrichment

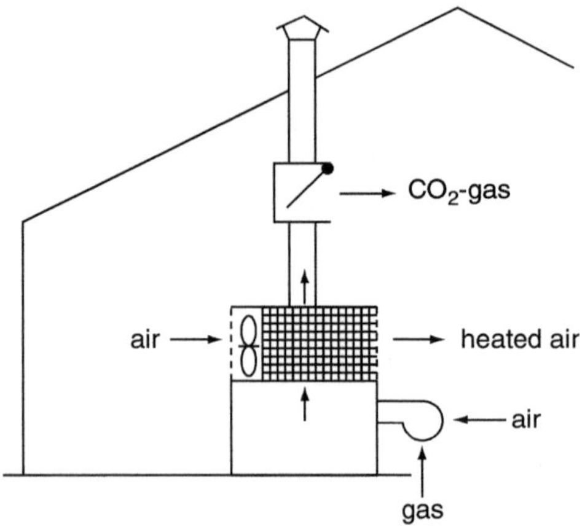

Annexes

Annex A

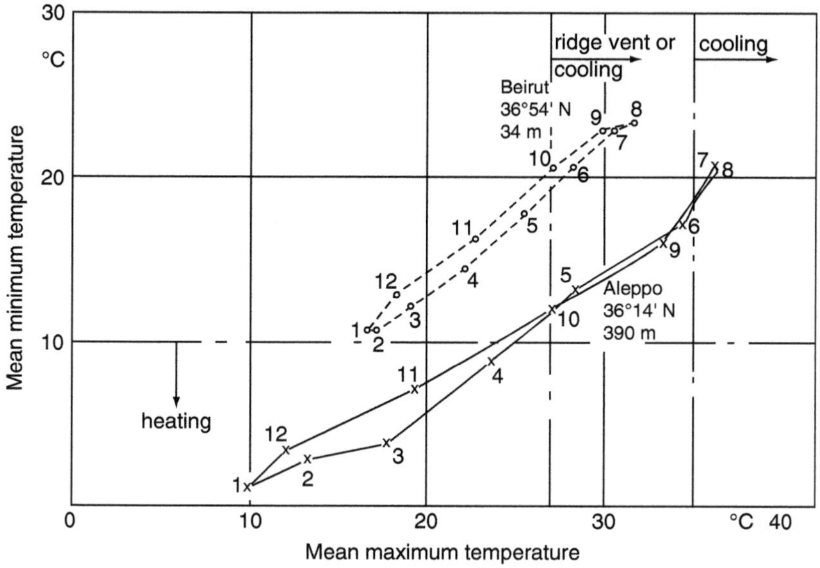

Fig. A.1 Minimum and maximum temperatures for Beirut (Mediterranean) and Aleppo (arid Syria)

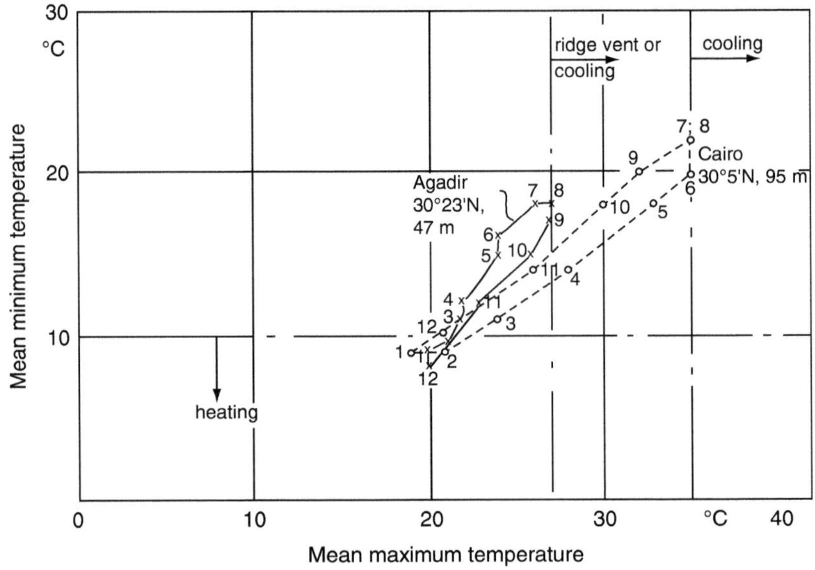

Fig. A.2 Mean and maximum temperatures for Agadir (Morocco) and Cairo (Egypt)

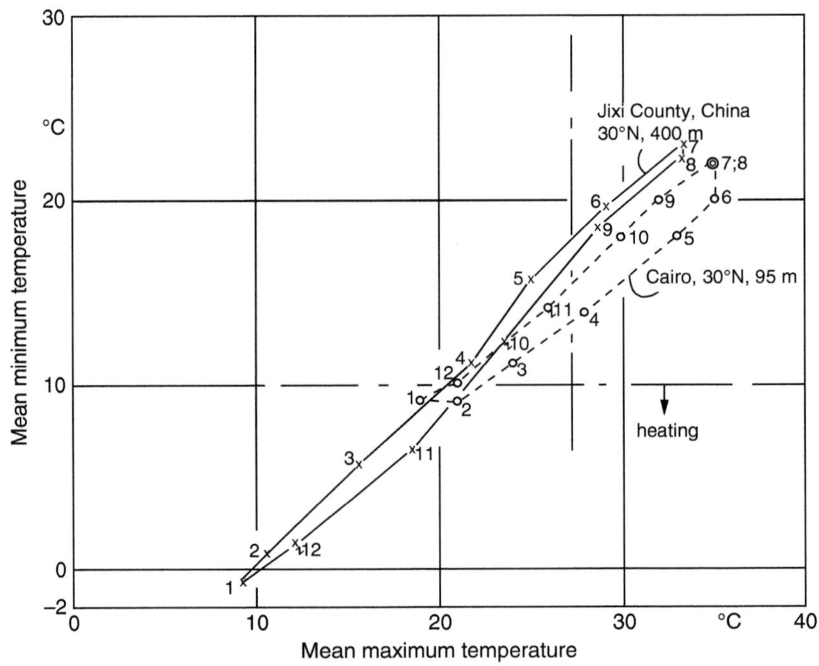

Fig. A.3 Mean temperatures for central China and Cairo at the same latitude

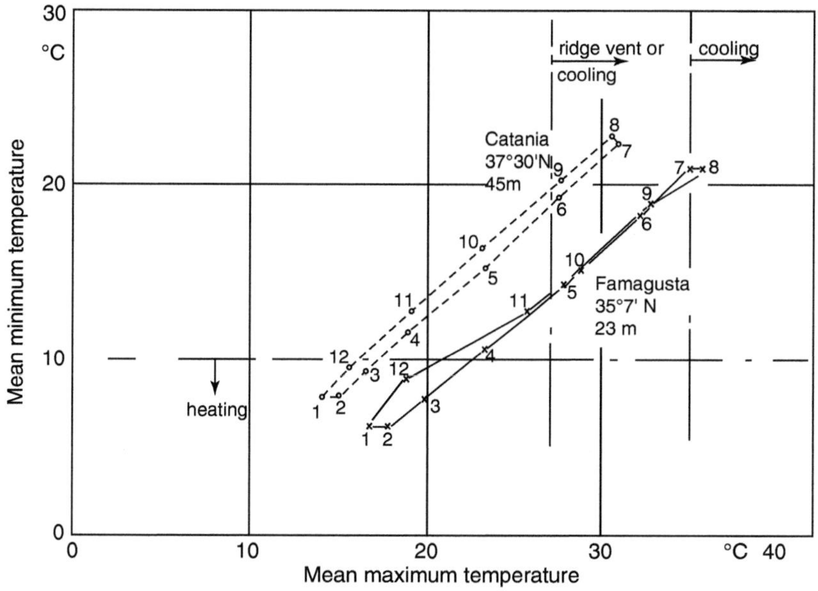

Fig. A.4 Mean temperatures for Catania, Sicily, and Famagusta, Cyprus

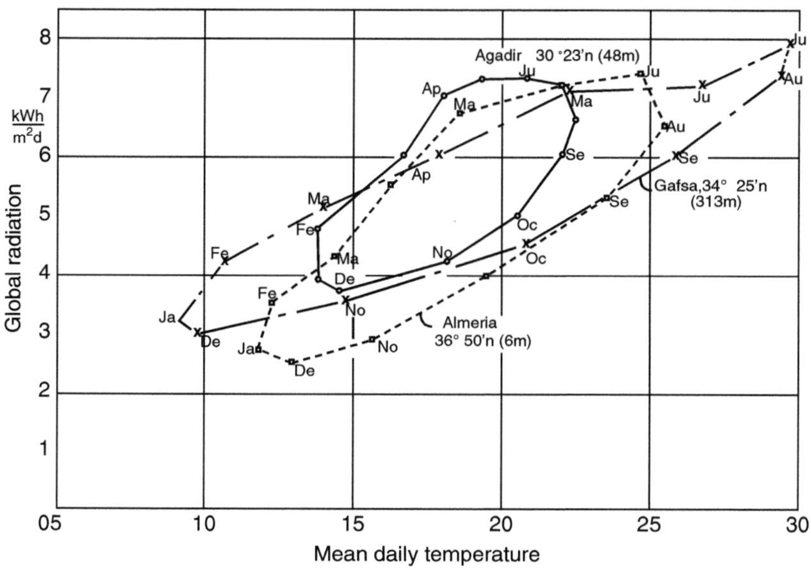

Fig. A.5 Means of daily global radiation for Agadir (Morocco), Gafsa (Tunisia) and Almeria (Spain)

Annex B

Calculation of fuel consumption in Sect. 12.2.

1. Almeria, Spain (36°50′N).

   $t_{id} = 16°C$, $t_{st} = 2°C$, $u = 7$ W/(m^2 K), $A_c/A_g = 1.33$, see 12.5.
   Mean night temperature (°C)

$$t_{mn} = t_{mmin} + A\frac{\Sigma f_n}{24 - d_l} \tag{12.3}$$

   t_{mmin} according to Müller (1996)
   $A = t_{maxd}–t_{mind}$ (Müller 1996)
   $\Sigma f_n/(24\text{-}d_l)$ Table12.3 (Hallaire 1950),
   d_l = mean daylight hours = f(latitude, month) according to Allen et al. (1998).

	t_{mmin}	A	d_l	$\Sigma f_n/(24\text{-}d_l)$	t_{mn}
December	9.2	7.4	9.5	0.375	12
January	8	7.6	9.7	0.375	10.8
February	8.5	7.6	10.7	0.345	11.1
March	10.5	7.3	11.7	0.33	12.9

   Fuel consumption $t_{id} = 16°C$.
   Q_{month} (kWh/m^2 month) = according to 12.5

	n_d	$n_n = 24\text{-}d_l$	Δt	Q_{month}
December	31	14.5	2	8.36
January	31	14.3	3.2	13.2
February	28	13.3	2.9	10.1
March	31	12.3	1.1	3.9

   Yearly fuel consumption $Q_y = \Sigma Q_{month} = 35.6$ (kWh/m^2 year)

2. Antalya, Turkey (36°53′N).

   The same assumptions.
   Mean night temperature

	t_{mmin}	A	d_l	$\Sigma f_n/(24\text{-}d_l)$	t_{mn}
December	7.8	8.9	9.5	0.375	11.1
January	6.1	8.9	9.7	0.373	9.4
February	6.7	8.9	10.6	0.345	9.8
March	7.8	10	11.7	0.33	11.1

   Fuel consumption Q_{month} (kWh/m^2 month), $t_{id} = 16°C$.

	n_d	$n_n = 24\text{-}d_l$	Δt	Q_{month}
December	31	14.5	2.9	12.1
January	31	14.3	4.6	18.98
February	28	13.3	4.2	14.7
March	31	12.3	2.9	10.3

   Yearly fuel consumption $Q_y = \Sigma Q_{month} = 56.1$ (kWh/m^2 year)

3. Catania, Sicily (37°30′N).

The same assumptions.
Mean night temperature

	t_{mmin}	A	d_l	$\Sigma f_n/(24\text{-}d_l)$	t_{mn}
December	9.4	6.3	9.5	0.375	11.8
January	7.7	6.4	9.7	0.375	10.1
February	7.9	7.2	10.6	0.345	10.4
March	9.3	7.3	11.7	0.33	11.7
April	11.6	7.4	13	0.31	13.9

Fuel consumption Q_{month} (kWh/m^2 month), $t_{id} = 16$°C.

	n_d	$n_n = 24\text{-}d_l$	Δt	Q_{month}
December	31	14.5	2.2	9.2
January	31	14.3	3.9	16.1
February	28	13.3	3.6	12.6
March	31	12.3	2.3	8.2
April	30	11	0.1	0.3

Yearly fuel consumption $Q_y = \Sigma Q_{month} = 46.4$ (kWh/m^2 year).

Annex C

Adapted calculation sheet for the reference evaporation ET0 in unheated greenhouses, using the Penman–Monteith equation (Allen et al. 1998) and example for Almeria in middle of May (36°50′N, 7 m altitude).

$$\mathrm{ET0} = \frac{0.408 \times \Delta(q_{RN} - q_{RG}) + \gamma \frac{900}{T_{mean}+273} \times v \times (e_S - e_A)}{\Delta + \gamma(1 + 0.34v)} \quad 1/(\mathrm{m^2 day})$$

Given parameters (Müller 1996)	
Mean max temperature T_{max} (°C)	22
Mean min temperature T_{min} (°C)	14.9
Altitude z (m)	7
Mean daily global radiation q_{RS} kWh/m^2 (1 kWh = 3.61 MJ)	6.7
$q_{RS} = 6.7 \times 3.61 = 24.2$ (MJ/m^2 day)	
Latitude	36°50′N
Actual duration of sunshine hours (h/day)	9.9
Adapted parameters for unheated greenhouses	
Mean max inside temperature $T_{gmax} = T_{max} + 4$(°C)	26
Mean min inside temperature $T_{gmin} = T_{min} + 2$(°C)	16.9
Mean inside temperature $T_{gmean} = (T_{gmax} + T_{gmin}/2)$	21.45
Inside global radiation $q_{RSi} = \tau \times q_{RS}$ MJ/(m^2 day)	16.9
For single plastic film $\tau = 0.7$	
Inside relative humidity $R_H = 75$–80 (%)	80

(*continued*)

Air velocity $v = 0.3$ m/s	0.3
Slope of vapour pressure $\Delta = f(T_{mean})$, (kPa/°C), Table 2.4, Annex 2 (Allen et al. 1998)	0.157
Psychrometrical constant $\gamma = f(z)$, (kPa/°C), Table 2.2, Annex 2, (Allen et al. 1998)	0.067
Vapour pressure deficit e_S–e_A	
Saturation vapour pressure $e_S = (e_{STgmin} + e_{STgmax})/2$ (kPa), Table 2.3, Annex 2, (Allen et al. 1998)	$(1.938 + 3.36)/2 = 2.65$
Actual vapour pressure $e_A = e_S \times R_H/100$ (kPa)	2.12
Radiation	
Net radiation $q_{RN} = q_{RNS}–q_{RNL}$ (MJ/m^2 day)	
Net solar radiation $q_{RNS} = 0.77 \times q_{RSI}$	13.01
$q_{RSI} = \tau \times q_{RS}$ (MJ/m^{2}day)	
Long-wave radiation: $q_{RNL} = (\sigma \times T_{gmax}{}^4 + \sigma \times T_{gmin}{}^4/2) \times (0.34–0.14\sqrt{e_A}) \times (1.35 q_{RSi}/q_{Ro}–0.35)$	
σT^4, Table 2.8, Annex 2 (Allen et al. 1998)	
$(\sigma \times T_{gmax}{}^4 + \sigma \times T_{gmin}{}^4/2)$	$(39.27 + 34.75)/2 = 37.01$
Clear sky radiation $q_{RO} = 0.75 q_{RA}$ (near sea level) or $q_{RO} = (0.75 + 2 \times 10^{-5} \times z) q_{RA}$	
Extraterrestrial radiation $q_{RA} = f$(latitude) (MJ/m^2 day) Table 2.6, annex 2, (Allen et al. 1998)	40
q_{RO} (MJ/m^2 day)	30
q_{RSi}/q_{RO} (MJ/m^{2}day)	$16.9/30 = 0.56$
q_{RNL} (MJ/m^{2}day)	2.1
Net radiation q_{RN} (MJ/m^{2}day)	10.91
Soil heat flux q_{RG} can be neglected	
ET0 (L/m^2 day)	3.16

References

Abou-Hadid AF, Beltagy AS, Mohamedien SA, Saleh MM, Medany MA (1995) Options for simple greenhouse heating systems. Acta Hortic 399:87–93

Allen G, Pereira LS, Raes D, Smith M (1998) Crop evapotranspiration. Guidelines for computing crop water requirements. FAO irrigation and drainage paper No. 56. FAO, Rome

Al-Massoum A, Hamar I, Ahmed MH (1998) A similitude model for testing greenhouse evaporative cooling pads under the hot-arid conditions of the U.A.R. Acta Hortic 456:329–337

Al-Mulla YA (2006) Cooling greenhouses in the Arabian Peninsula. Acta Hortic 719:495–505

ANSI/ASAE EP 406.4 (2003) Standard: heating, ventilation and cooling greenhouses. American Society of Agricultural Engineers, MI, USA

Antignus Y (2000) Manipulation of wave length dependent behaviour of insects. Virus Res 71:213–220

Antignus Y, Mor N, Ben Joseph R, Lapidat M, Cohen SH (1996) UV- absorbing plastic sheets protect crops from insect pests and from virus diseases vectored by insects. Environ Entomol 25(5):919–924

Antignus Y, Lapidat M, Hadar D, Messika Y, Cohen SH (1998) UV-absorbing screens serve as optical barriers to protect crops from virus and insect pests. J Econ Entomol 91(6):1401–1403

Antignus Y, Nestle D, Cohen S, Lapidat M (2001) Ultraviolet-deficient greenhouse environment affects whitefly attraction and flight behaviour. J Environ Entomol 30(2):394–399

Antignus Y, Ben-Yakir D (2004) Ultraviolet-absorbing barriers, an efficient integrated pest management tool to protect greenhouses from insects and virus diseases. In: Horowitz AR, Isaaya I (eds) Insect Pest Management: Field and Protected Crops. Springer, Berlin, Heidelberg, New York

Anton A, Montero JI (2003) Life cycle assessment: a tool to evaluate and improve the environmental impact of Mediterranean greenhouses. Acta Hortic 614:35–40

Anton A, Montero JI, Munoz P (2005) Use of high light transmission greenhouse covers of the geranium crop. Plasticulture 6(124):74–79

Anton R, Montero JI, Munoz P, Perez-Parra J, Baeza E (2006) Environmental and economical evaluation of greenhouse cooling systems in southern Spain. Acta Hortic 719:211–214

Arbel A, Yekutieli O, Barak M (1999) Performance of a fog system for cooling greenhouses. J Agric Eng Res 72:129–136

ASAE Engineering Practice: ASAE EP 406 (1981) Agricultural Engineers Yearbook. American Society of Agricultural Engineers, MI, USA

ASAE EP460 (2001) Commercial Greenhouse Design and Layout. American Society of Agricultural Engineers, MI, USA

Baeza EJ (2007) Optimizacion del discu de las sistemas de ventilacion en invernaderos tipo parral. Thesis doctoral, Escuela Politecnica Superior, Departimento de Ingenieria Rural, Universidad de Almeria

Baeza EJ, Perez-Parra J, Montero LI (2005) Effect of ventilation size on natural ventilation in Parral-type greenhouses by means of CFD simulation. Acta Hortic 691:465–472

Baeza EJ, Perez-Parra JJ, Lopez JC, Montero JI (2006) CFD study of the natural ventilation performance of a Parral-type greenhouse with different numbers of span and roof vent configurations. Acta Hortic 719:333–340

Baeza E, Perez-Parra JJ, Lopez JC, Gazquez JC, Montero JI (2008a) Numerical analysis of buoyancy driven natural ventilation in multi span type greenhouses. Acta Hortic 797:111–116

Baeza EJ, Perez-Parra JJ, Lopez JC, Montero JI (2008b) CFD simulation of natural ventilation of a Parral greenhouse with a baffle device below the greenhouse vents. Acta Hortic 801: 885–892

Bailey BJ (1998) Principals of environmental control, Chapter 3. In: von Zabeltitz C (ed) Energy conservation and renewable energies for greenhouse heating. CNRE guideline no. 2. FAO Regional office for Europe

Bailey BJ (2000a) Constraints, limitations and achievements in greenhouse natural ventilation. Acta Hortic 534:21–30

Bailey BJ (2000b) Wind driven ventilation in a large greenhouse. Acta Hortic 534:309–317

Bailey BJ (2006) Natural and mechanical greenhouse climate control. Acta Hortic 710:43–53

Bailey BJ, Montero JJ, Peres Parra J, Robertson AP, Baeza E, Kamaruddin R (2003) Airflow resistance of greenhouse ventilators with and without screen. Biosystems Eng 86(2): 217–229

Bailey BJ, Robertson AP, Lockwood AG (2004) The influence of wind direction on greenhouse ventilation. Acta Hortic 633:197–203

Baille A (1999a) Energy cycle, Chapter 11. In: Enoch HZ, Stanhill G (eds) Ecosystems of the World. Elsevier, Amsterdam, pp 265–286

Baille A (1999b) Greenhouse structure and equipment for improving crop production in mild winter climate. Acta Hortic 491:37–47

Baille A (2001) Trends in greenhouse technology for improved climate contropl in mild winter climate. Acta Hortic 559:161–170

Bakker JC, Bot CPA, Challa H, van de Braak NJ (eds) (1995) Greenhouse climate control, an integrated approach. Wageningen Pers, Wageningen

Baptista FJ, Bailey BJ, Randall JM, Meneses JF (1999) Greenhouse ventilation rate: theory and measurement with tracer gas techniques. J Agric Eng Res 72:363–374

Bartzanas T, Boulard T, Kittas C (2004) Effect of vent arrangement on windward ventilation of a tunnel greenhouse. Biosystems Eng 88(4):479–490

Bartzanas T, Fidaros D, Katsoulas N, Kittas C, Boulard T (2009) Experimental results and simulation analysis of climate in a greenhouse with insect screens. Greensys 2009, Quebec, Canada

Baudoin W, Grafiadellis M, Jiminez R, La Malfa G, Martinez-Garcia PF, Garnaud JC, Montero AA, Nisen A, Verlodt H, de Villele O, von Zabeltitz C (1991) Protected cultivation in the Mediterranean climate. FAO plant production and protection paper no. 90

Baytorun N (1986) Bestimmung des Luftwechsels bei gelüfteten Gewächshäusern (Air exchange in ventilated greenhouses), Gartenbautechnische Information. Institute for Horticultural Engineering Hannover no. 27

Baytorun N, von Zabeltitz C (1987) Die Wirkung bautechnischer Einflussgrößen auf den Luftwechsel gelüfteter Gewächshäuser. Gartenbauwissenschaft (Hortic Sci) 52(5):233–239

Baytorun N, Dumke C, Meyer J (1989) Closed system greenhouse with integrated solar desalination for arid regions. Gartenbauwissenschaft 54(2):62–65

Ben-Yakir D, Teitel M, Chen M, Barak M (2008a) Optimizing ventilation of protected crops while minimizing invasion by white fly and thrips. Acta Hortic 797:217–222

Ben-Yakir D, Hadar MD, Offir Y, Chen M, Tregermann M (2008b) Protecting crops from pests using Opt-Net screens and ChromatiNet shading nets. Acta Hortic 770:205–212

Bessler B (2000) Licht – Es kommt nicht nur auf die Menge an. Das Taspo Magazin 1:20–22

Bethke JA (1994) Considering installing screening? This is what you need to know. Greenhouse Manager 13(1):34–37

Bethke JA, Paine TD (1991) Screen hole size barriers for excluding of insect pests of glasshouse crops. J Entomol Sci 26(1):169–177
Bethke JA, Redak A, Paine TD (1994) Screens deny specific pests entry to greenhouses. Calif Agric 48(3):37–40
Bot GPA (1983) Greenhouse climate: from physical process to dynamic model. Ph.D. thesis, Agricultural University, Wageningen, Netherlands, p 240
Boulard T (2006) Greenhouse natural ventilation modelling: a survey of the different approaches. Acta Hortic 719:29–40
Boulard T, Baille A (1995) Modelling of air exchange rate in a greenhouse equipped with continuous roof vents. J Agric Eng Res 61:37–48
Boulard T, Draoui B (1995) Natural ventilation of a greenhouse with continuous roof vents: measurements and data analysis. J Agric Eng Res 61:27–36
Bournet PE, Ould Khaoua SA (2006) Predicted effects of roof vent combinations on the climate distribution in a glass house radiation and convection heat transfer. Acta Hortic 801:925–932
Bournet PE, Ould Khaoua SA, Boulard T (2007) Numerical prediction of the effect of vent arrangements on the ventilation and energy transfer in a multi span glasshouse using a bi-bond radiation model. Biosystems Eng 98:224–234
Bournet PE, Ould Khaoua SA (2008) Predicted effects of roof vent combinations on the climate distribution in a glass house radiative and convective heat transfer. Acta Hortic 801:925–932
Bredenbeck H (1982) Studie über die Nutzung der Solarenergie für die Gewächshausheizung in Griechenland. Gartenbautechnische Informationen Heft 16. Institut für Technik in Gartenbau und Landwirtschaft, Universität Hannover
Bredenbeck H (1985) Influence of different glazing materials on the light transmissivity of greenhouses. Acta Hortic 170:111–117
Bredenbeck H (1986) Energiesparendes Gewächshaussystem mit Solarenergienutzung. Gartenbautechnische Informationen Heft 26. Institut für Technik in Gartenbau und Landwirtschaft, Universität Hannover
Bredenbeck H (1992) Untersuchung angepasster Systeme für die solare Gewächshausheizung. Gartenbautechnische Informationen Heft 36. Institut für Technik in Gartenbau und Landwirtschaft, Universität Hannover
BUWAL (Bundesanstalt für Umwelt, Wald und Landwirtschaft) (1991) Ökobilanz von Packstoffen. C R Acad Sci, Bern, 1950–331:1533–1535
Castilla N (2002) Current situation and future prospects of protected cultivation in the Mediterranean region. Acta Hortic 582:135–147
Castilla N (2005) Invernaderos de plastico. Tecnologia y manejo. Ediciones Mundi-Prensa, Madrid, p 461
Castilla N, Lopez-Galvez J (1994) Vegetable responses in improved low-cost plastic greenhouses. J Hortic Sci 69(5):915–921
Castilla N, Hernandez J, Escobar I, Montero JI (1999) Alternative asymmetrical greenhouses for the Mediterranean area of Spain. Acta Hortic 486:83–86
Castilla N, Hernandez J, Escobar J, Anton A, Montero JJ (2001) Comparison of asymmetrical greenhouse types in the Mediterranean area of Spain. Acta Hortic 559:183–186
Castilla N, Hernandez J, Abou Hadid AF (2004) Strategic crop and greenhouse management in mild winter climate areas. Acta Hortic 633:183–196
Castilla N, Hernandez J (2005) The plastic industry in Spain. Chron Hortic 45(3):15–20
Castilla N, Hernandez J (2007) Greenhouse technological packages for high quality production. Acta Hortic 761:285–297
Castilla N, Montero JI (2008) Environmental control and crop production in Mediterranean greenhouses. Acta Hortic 797:25–36
Costa HS, Robb KL (1999) Effects of ultraviolet absorbing greenhouse plastic films at flight behaviour of Bemisia Argentifolia and Frankiniella Occidentalis. J Econ Entomol 92(3):557–562
Costa HS, Robb KL, Wilen CA (2002) Field trials measuring effects of ultraviolet absorbing greenhouse plastic film on insect populations. J Econ Entomol 95(1):113–120

Damrath J (1982) Solarenergienutzung im Gewächshaus. Eine energetische Darstellung des doppelt bedachten Gewächshauses mit solarunterstützter Heizung. Gartenbautechnische Informationen Heft 14. Institut für Technik in Gartenbau und Landwirtschaft, Universität Hannover

Damrath J (1983) Solarenergienutzung im Gewächshaus. Ergänzende Untersuchungen zur Solaranlage Typ Hannover. Gartenbautechnische Informationen Heft 15. Institut für Technik in Gartenbau und Landwirtschaft, Universität Hannover

De Jong T (1990) Natural ventilation of large multi span greenhouses. Ph.D. Thesis, Landbouwuniversiteit Wageningen, Netherlands

De Pascale ST, Maggio A (2005) Sustainable protected cultivation at a Mediterranean climate – perspectives and challenges. Acta Hortic 691:29–42

De Pascale ST, Maggio A (2008) Plant stress management in semiarid greenhouses. Acta Hortic 797:205–215

Demrati H, Boulard T, Bekkaoui A, Birden L (2001) Natural ventilation and microclimate performance of a large-scale banana greenhouse. J Agric Eng Res 80(3):261–271

Dierickx W (1998) Flow reduction of synthetic screens obtained with both a water and airflow apparatus. J Agric Eng Res 71:67–73

Dierickx W, Gabriels D, Cornils W (2001a) A wind tunnel study on wind speed reduction of technical textiles used as windscreen. Geotextiles Membr 19:59–73

Dierickx W, Gabriels D, Cornils W (2001b) Wind tunnel study on wind speed reduction through successive synthetic wind screens. J Agric Eng Res 79(1):117–123

Dierickx W, Gabriels D, Cornils W (2002) Wind tunnel studies on oblique windscreens. Biosystems Eng 82(1):87–95

Dierickx W, Cornils W, Gabriels D (2003) Wind tunnel studies on rough and smooth surface, turbulent approach flow and on inclined screens. Biosystems Eng 86(2):151–166

DIN EN ISO 14040 (1997-08) Umweltmanagement – Prinzipien und allgemeine Anforderungen. Beuth Verlag, Berlin

DIN EN ISO 14043 (2000-07) Umweltmanagement – Ökobilanzen Auswertung. Beuth Verlag, Berlin

Duckham AN, Hasefield GB (1971) Farming systems of the world. Chatto & Windus, London

EN 13031-1 (2001) European standard: greenhouses. Design and construction. Part 1: Commercial production greenhouse structures

Fähnrich I, Meyer J, von Zabeltitz C (1989) Influence of condensation on light transmission and heat transfer through greenhouse covering materials. Plasticulture 4(4):13–18

Fatnassi H, Boulard T, Bouirden L, Sappe G (2002) Ventilation performance of a large Canarian-type greenhouse equipped with insect proof nets. Biosystems Eng 82:97–105

Fatnassi H, Boulard T, Poncet C, Chave M (2006) Optimization of greenhouse insect screening with computational fluid dynamics. Biosystems Eng 93(3):301–312

Fatnassi H, Boulard T, Leyronas C, Barden M, Nicot P (2008) Greenhouse tunnel ventilation dependence on tomato crop height and leaf area index. Acta Hortic 801:1285–1292

Fernandez MD, Baeza E, Cespedes A, Perez-Parra J, Gazquez JC (2009) Validation of on-farm crop water requirements (pRho) model for horticultural crops in an unheated plastic greenhouse. Acta Hortic 807:295–300

Florez-Velasquez J, Montero JI, Baeza EJ (2009) Analysis of mechanical ventilation in a three span greenhouse using computational fluid dynamics (CFD). In: Proceedings Greensys, Quebec, Canada. Acta Hortic (in press)

Gazquez JC, Lopez JC, Baeza E, Sanchez-Guerrero MC, Modrano E, Lorenzo P (2006) Yield response of a sweet pepper crop to different methods of greenhouse cooling. Acta Hortic 719:507–713

Gazques JC, Perz-Parra JI, Baeza E, Suez M, Parra A (2008) Greenhouse cooling strategies for Mediterranean areas. Acta Hortic 801:425–431

Gazquez JC, Lopez JC, Baeza E, Perez-Parra J, Perez C, Meca DE, Acosta JA (2009) Screenhouses on the Mediterranean basin: pest incidence and productivity of a tomato crop. Greensys 2009, Quebec, Canada

Gbiorczyk K (2003) New test methods for evaluating the antifog effect of greenhouse films. Ph.D. Thesis, Gartenbautechnische Informationen, Institute for Horticultural engineering, University Hannover

Gbiorczyk K (2003) New test methods for evaluating the antifog effect of greenhouse films. Gartenbautechnische Information, Heft 58. Institute for Horticultural Engineering, University of Hannover

Gbiorczyk K, von Elsner B (2004) The effect of roof inclination on the condensation behaviour of plastic films used as greenhouse covering materials. Acta Hortic 633:127–136

Geoola F, Peiper UM (1994) Outdoor testing of the condensation characteristics of plastic film covering materials using a model greenhouse. J Agric Eng Res 57:167–172

Giacomelli G, Castilla N, Vanhenten E, Mears D, Sase S (2008) Innovation in greenhouse engineering. Acta Hortic 801:75–88

Gombe ZO (2007) Investigation and optimization of an evaporative spray cooling system for greenhouses. Master thesis, Department of Horticulture, Section of Biosystems and Horticultural Engineering, Leibniz University of Hannover

Gratraud J (1990) The influence of dirt laid on the greenhouse cover. Colloque de Annemasse, CIPA, Paris

Gunham T, Demir V, Yagcioglu AK (2007) Evaluation of the suitability of some local materials as cooling pads. Biosystems Eng 96(3):369–377

Hallaire M (1950) Les températures moyennes nocturnes, diurnes et nycthémérales exprimées en fonction du minimum et du maximum journaliers de température. C R Acad Sci 331:1533–1535

Harmanto M (2006) Evaluation of net greenhouses for tomato production in the tropics. Ph.D. Thesis, University of Hannover, Germany

Harmanto M, Salokhe VM, Babel MS, Tantau HJ (2005) Water requirement of drip-irrigated tomatoes grown in greenhouse in tropical environment. Agric Water Manage 71:225–242

Harmanto M, Tantau HJ, Salokhe VM (2006a) Optimization of ventilation opening area of naturally ventilated net greenhouse in humid tropical environment. Acta Hortic 717: 165–171

Harmanto M, Tantau HJ, Salokhe VM (2006b) Microclimate and air exchange rates in greenhouses covered with different nets in the humid tropics. Biosystems Eng 94(2):239–253

Heisler GM, Dewalle DR (1988) Effects of windbreak structure on wind flow. Agric Ecosyst Environ 22(23):41–69

Hemming S (2005) Auswirkungen von diffusem Licht im Gewächshaus (influence of diffuse light in the greenhouse). In: Annual meeting of the GKL German Society for plastics in Agriculture and Horticulture

Hemming S, Waaijenberg D (2001) Künftige Folien filtern Strahlung gezielt. Dtsch Gartenbau 55(37):34–36

Hemming S, Waaijenberg D, Campen JB, Bot GPA (2006a) Development of a greenhouse for tropical lowland in Indonesia. Acta Hortic 710:135–142

Hemming S, Kempkes F, van der Braak N, Deck T, Marissen N (2006b) Filtering natural light at the greenhouse covering – better greenhouse climate and higher production by filtering out NIR. Acta Hortic 711:411–416

Hemming S, van der Braak N, Dueck T, Jongschaap R, Maarisson N (2006c) Filtering natural light by using model simulations – more production and better plant quality by diffuse light? Acta Hortic 711:105–110

Hemming S, Kempkes F, van der Braak N, Dueck T, Marissen N (2006d) Greenhouse cooling by NIR reflection. Acta Hortic 719:97–105

Hita O, Romacho J, Soriano T, Morales MJ, Escobar J, Surez-Rey E, Hernandez J, Castilla N (2007) Comparison of two Mediterranean greenhouses “technological packages” with cherry tomatoes. Acta Hortic 747:309–313

Hoffmann S (1999a) Zur Wirkung von photoselektiven Bedachungsmaterialien auf Zierpflanzen. Ph.D. Thesis, Gartenbautechnische Informationen, Institut für Technik in Gartenbau und Landwirtschaft, Universität Hannover

Hoffmann S (1999b) Die Wirkung von UV Strahlung auf Blatt- und Blütenfarben von Zierpflanzen. Gartenbauwissenschaft 64(3):88–93

Hoffmann S (1999c) Die Wirkung von photoselektiven Bedachungsmaterialien auf Zierpflanzen. Gartenbauwissenschaft 64(3):100–105

Hoffmann S (1999d) Die Wirkung von photoselektiven Bedachungsmaterialien auf das Wachstum von Zierpflanzen. II Wirkung des UV Bereiches auf das Streckungswachstum. Gartenbauwissenschsft 64(4):183–189

Hoffmann S (1999e) Die Wirkung von photoselektiven Bedachungsmaterialien auf das Wachstum von Zierpflanzen. III Wirkung des Verhältnisses von Hellrot zu Dunkelrot auf das Streckenwachstum. Gartenbauwissenschaft 64(6):253–260

Hoffmann S, Waaijenberg D (2001) Tropical and subtropical greenhouses, a challenge for new plastic films. Acta Hortic 373:163–169

IS 14462 (1997) Indian standard. Recommendations for layout, design and construction of greenhouse structures

Ishigami Y, Tetsuko T, Goto E (2009) Analysis of the aerial environment of a tomato greenhouse equipped with different fog cooling systems. In: Proceedings of Greensys 2009, Quebec Canada. Acta Hortic (in press)

Ishii M, Sase S, Moriyama H, Kurata K, Ikeguchi A, Kubota C, Hahashi M, Sabeh N, Romero P, Giacomelli GA (2006) The effect of evaporation cooling in a naturally ventilated greenhouse on air and leaf temperature, relative humidity and water use in a semiarid climate. Acta Hortic 719:491–498

Ishii M, Sase S, Okushima L, Moriyama H, Hayashi M, Yamamoto Y (2008) Comparison of field measurement and CFD simulation for a naturally ventilated multi-span greenhouse in summer. Acta Hortic 801:941–948

ISO International standard 14040 (1998) Environmental management – Life cycle assessment – Principles and framework. ISO, Geneva

ISO international standard 14042 (1998) Environmental management – Life cycle assessment – Impact assessment. ISO, Geneva

Jackson IJ (1977) Climate, water and agriculture in the tropics. Longman, London

Jaffrin A, Morison A (1994) Role of structure, dirt and condensation on the light transmission of greenhouse covers. Plasticulture 101(1):33–44

Jamal M, Ramad N, Al-Lengawi J, Muraished N, Al-Herbi M, Fouad JA (2002a) Evaluation and comparison of different cooling systems in greenhouses. Expert consultation meeting on IPPM greenhouse ventilation and cooling in Kuwait. JCARDA, International Center for Agricultural Research in the Dry Areas, p 1–12

Jamal M, Al-Harbi M, Fouad J (2002b) Comparison of different new cooling systems in greenhouses of Kuwait. ICARDA-APRP Annual report 2001–2002, pp 50–54

Kacira M, Sase S, Okushima L (2004a) Effects of side vents and span numbers on wind induced natural ventilation of a gothic multi span greenhouse. Jpn Agric Res Q 38(4):227–233

Kacira M, Sase S, Okushima L (2004b) Optimization of vent configuration by evaluating the greenhouse and plant canopy ventilation rates under wind induced ventilation. Trans ASAE 47:2059–2067

Katsoulas N, Bartzanas T, Boulard T, Kittas C (2006) Effect of vent openings and insect screens on greenhouse ventilation. Biosystems Eng 93(4):427–436

Kempkes F, Stanghellini C, Hemming S, Dai J (2008) Cover materials excluding near infrared radiation: effect on greenhouse climate and plant processes. Acta Hortic 797:477–482

Kempkes F, Stanghellini C, Hemming S (2009) Cover materials excluding near infrared radiation: what is the best strategy in mild climates? Acta Hortic 807:67–72

Kenney WA (1987) A method for estimating windbreak porosity using digitized photographic silhouettes. Agric For Meteorol 39:91–94

Kim A, Giacomelli GA, Yoon J-Y, Sase S, Son J-E, Nam S-W, Lee I-B (2007) CFD modelling to improve the design of a fog system for cooling greenhouses. Jpn Agric Res Q 41(4):283–290

Kirsten W (1973) Die natürliche Einstrahlung in Gewächshäuser in Abhängigkeit von deren Form und Aufstellungsrichtung sowie der Beschaffenheit des Hüllstoffes (Natural light transmittance of greenhouses depending on shape and orientation as well as on cladding material). Ph.D. thesis, University of Hannover, Germany, p 90

Kittas C, Boulard T, Mermier M, Papadakis G (1996) Wind induced air exchange rates in a greenhouse tunnel with continuous side openings. J Agric Eng Res 65:37–49

Kittas C, Draoui B, Boulard T (1995) Quantification of the ventilation of a greenhouse with a roof opening. Agric For Meteorol 77:95–111

Klasik R, Müller G, von Zabeltitz C (1989) Ein Vergleich solarer Wasserentsalzungsanlagen. Gartenbauwissenschaft 54(1):13–15

Krug H, Liebig HP, Stützel H (2002) Gemüseproduktion. Ulmer Verlag, Stuttgart

Kumar P, Poehling HM (2006) UV-blocking plastic film and nets influence vectors and virus transmission on greenhouse tomatoes in the humid tropics. Environ Entomol 35(4):1069–1082

La Malfa G, Leonardi C (2001) Crop practices and techniques: trends and needs. Acta Hortic 559:31–42

Lee IB, Short TH (2000) Two dimensional numerical simulation of natural ventilation in a multi span greenhouse. Trans ASAE 43(3):745–753

Li S, Willits DH (2008) Comparing low pressure and high pressure fogging systems in naturally ventilated greenhouses. Biosystems Eng 101(1):69–77

Li S, Willits DH, Yunker CA (2006) Experimental study of a high pressure fogging system in naturally ventilated greenhouses. Acta Hortic 717:393–400

Lieftink DA, van Osten RM (1986) Geleidelijke verbetering van anticondensdruppelmaterialien. Vakblad voor de Bloemistrij 41:118–119

Loeffler AE, Gordon AM, Gillespie TJ (1992) Optical porosity and wind speed reduction by coniferous wind breaks in Southern Ontario. Agroforest Syst 17:119–133

Lopez-Marin J, Gonzalez A, Garcia Alonso Y, Espi E, Salmeron A, Fontecha A, Real AJ (2008) Use of cool plastic film for greenhouse covering in Southern Spain. Acta Hortic 801:181–186

Lüchow K, von Zabeltitz C (1992) Investigation of a spray cooling system in a plastic film greenhouse. J Agric Eng Res 52:1–10

Ludolph B, Bessler B (2002) UV-B durchlässiges Glas im Praxistest. Das Taspo Magaz 2:20–22

Manor G, Geva A, Secker J, Tabak S, Singer L (2005) Cleaning machine for dusty greenhouse soft roofs. Acta Hortic II(691):641–647

Medany MA, Hassanin MK, Farag AA (2009) Effect of black and white nets as alternative covers to sweet pepper production under greenhouses in Egypt. Acta Hortic 807:121–126

Meyer J (1981) Bewegliche Energieschirme. Gartenbautechnische Informationen, Heft 11. Institut für Technik in Gartenbau und Landwirtschaft, Universität Hannover

Meyer J (1982) Bewertung von beweglichen Energieschirmen im Hinblick auf den Energieverbrauch. Gartenbautechnische Informationen, Heft 12. Institut für Technik in Gartenbau und Landwirtschaft, Universität Hannover

Meyer J, Dumke C, Baytorun AN (1989) Entwicklung, Bau und Erprobung eines geschlossenen Gewächshausystems mit solarer Wasserentsalzung für Trockengebiete. Gartenbautechnische Informationen, Heft 32, Institut für Technik in Gartenbau und Landwirtschaft, Universität Hannover

Molina-Aiz FD, Valera DL, Lopez A (2009) Air flow at the openings of a naturally ventilated Almeria-type greenhouse with insect proof screens. In: Proceedings Greensys, Quebec, Canada. Acta Hortic (in press)

Möller J, Tanny J, Cohen S (2003) Micrometeorological characterisation in a greenhouse. Acta Hortic 614:445–451

Montero JI (2006) Evaporative cooling in greenhouses. Effect on microclimate, water use efficiency and plant response. Acta Hortic 719:373–383

Montero JI (2009) Greenhouse technology for the Mediterranean area: trends and needs. Advances in protected horticulture research. In: Proceedings of 2009 High-level International Forum on Protected Horticulture, China

Montero JI, Anton A, Munoz M (1999) Performance of an insect-proof screen as covering material in a greenhouse with improved opening surface. Agric Med 129:5–12

Montero JI, Hunt GR, Kamaruddin R, Anton A, Bailey BJ (2001) Effect of ventilator configuration on wind-driven ventilation in a crop protection structure for the tropics. J Agric Eng Res 80(1):99–107

Montero JI, Anton A (2003) Greenhouse characteristics and micro climate conditions. Acta Hortic 614:323–333

Montero JI, Anton A, Munoz P (2005) Tomato response to a high light transmission greenhouse film. Plasticulture 6(124):80–89

Montero JI, Stangellini C, Castilla N (2008a) Greenhouse technology for sustainable production in mild winter climate areas: trends and needs. Acta Hortic 807:33–44

Montero JI, Anton A, Mele M, Cid MC, Raya V (2008b) Suggestions to improve leeward ventilation of large span greenhouses. Acta Hortic 801:949–954

Montero JI, Stanghellini C, Castilla N (2009a) Technology for sustainable production in mild winter climate areas: Trends and needs. Acta Hortic 807:33–44

Montero JI, van Henten EJ, Son JE, Castilla N (2009) Greenhouse engineering: new technologies and approaches. Greensys 2009, Quebec Canada

Müller MJ (1996) Handbuch ausgewählter Klimastationen der Erde. Forschungsstelle Bodenerosion der Universität Trier, Mertesdorf, Germany

Munoz PJI, Montero AA, Giuffrida F (1999) Effect of insect screens and roof openings on greenhouse ventilation. J Agric Eng Res 73:171–178

Munoz P, Anton A, Nunez M, Paranjpe A, Arino J, Castell X, Montero JI, Rieradevall J (2008) Comparing the environmental impacts of greenhouse versus open field tomato production in the Mediterranean region. Acta Hortic 801:1591–1596

Mutwiwa UN (2004) Effects of UV absorbing plastic films on the behaviour of the greenhouse white fly. Master Thesis, Institute of Horticultural and Agricultural Engineering, University of Hannover

Mutwiwa UN, Borgemeister C, von Elsner B, Tantau HJ (2005) Effects of UV absorbing plastic films on greenhouse white fly. J Econ Entomol 98(4):1221–1228

Mutwiwa UN, Tantau HJ, von Elsner B, Salokhe PM (2007) Greenhouses in the humid tropics: problems and solutions. In: Proceedings of Solaris 3rd International Conference on Solar Radiation and Day Lighting

Mutwiwa UN, von Elsner B, Tantau HJ, Max JF (2008) Cooling naturally ventilated greenhouses in the tropics by near infrared reflection. Acta Hortic 801:259–265

NEN 3859 (1990) Greenhouse structural requirements. Greenhouse Standard, Netherlands

Nisen A (1969) L'éclairement naturel des serres, 1st edn. Duculot SA, Gembloux

Nisen A, Sirjacobs M, von Zabeltitz C (1984) Protected cultivation in Mediterranean climate, greenhouses in Egypt. FAO, Rome

Nisen OF (2002) Natural ventilation of a greenhouse with top screens. Biog Eng 81:443–452

Nisen A, Olympios C, Verlodt H, von Zabeltitz C (1990) Protected cultivation in the Mediterranean climate. Greenhouses for Cyprus. FAO, Rome

Ohyama K, Kozai T, Toida H (2008) Greenhouse cooling with continuous generation of upward-moving fog for reducing wetting of plant foliage and air temperature fluctuations: a case study. Acta Hortic 797:321–326

Ould Khaoua SA, Bournet PE, Migeon C, Boulard T, Chasseriaux G (2006) Analysis of greenhouse ventilation efficiency based on computational fluid dynamics (CFD). Biosystems Eng 95 (1):83–88

Papadakis G, Mermier M, Meneses J, Boulard T (1996) Measurement and analysis of air exchange rates in a greenhouse with continuous roof and side openings. J Agric Eng Res 63:219–228

Pardossi A, Tognoni F, Incrossi L (2004) Mediterranean greenhouse technology. Chron Hortic 44 (2):28–34

Pauer U (2002) Construction manual for greenhouses – 54 m^2 Kenya type. Ministry of Agriculture and Irrigation, Malawi. German Technical Cooperation, GTZ and German Development Service, DED

Pearson S, Wheldon AE, Hadley P (1995) Radiation transmission and fluorescence of nine greenhouse cladding materials. J Agric Eng Res 77(1):7–38

Perez-Parra J, Baeza E, Montero JI, Bailey BJ (2004) Natural ventilation of greenhouses. Biosystems Eng 87(3):355–360

Perez-Parra J, Montero JI, Baeza EJ, Lopez Hernandez JC (2006) Determination of global wind coefficients for the development of ventilation rate calculation models for a parral multi span greenhouse. Acta Hortic 710:143–150

Pollet IV (2002) Radiation transmission through dry and condensate covered greenhouse cladding materials. Ph.D. thesis, Faculty of Agricultural and Applied Biological Sciences, Ghent University, Belgium

Rapisarda C, Tropea GG, Gascone G, Mazzarella R (2006) UV-absorbing plastic films for the control of Bemisia tabacci and tomato Yellow Leaf Curl Disease (TYLCD) on protected cultivation in Sicily. International Symposium on Greenhouse Cooling. Acta Hortic 719:597–604

Rath TH (1992) Einsatz wissensbasierter Systeme zur Modellierung und Darstellung von gartenbautechnischem Fachwissen am Beispiel des hybriden Expertensystems HORTEX. Gartenbautechnische Informationen, Heft 34, Institut für Technik im Gartenbau der Universität Hannover

Rath TH (1994) Einfluss der Wärmespeicherung auf die Berechnung des Heizenergiebedarfes von Gewächshäusern mit Hilfe des k′-Modelles. Gartenbauwissenschaft 59:39–44

Romacho I, Hita O, Soriano T, Morales MI, Scobar I, Suarez-Rey EM, Hernandez J, Castilla N (2006) The growth and yield of cherry tomatoes in net covered greenhouses. Acta Hortic 719:529–534

Runkle S, Heins RD, Jaster P, Till Ch (2002) Environmental conditions under experimental near infrared reflecting greenhouse film. Acta Hortic 578:181–185

Russo G, Scarascia Mŭŭgnoza G (2005) LCA methodology applied to various typology of greenhouses. Acta Hortic 691:837–843

Rüther M (1989) Energieeinsparung durch Wärmerückgewinnung beim Zwangsluftwechsel in abgedichteten Gewächshäusern. Gartenbautechnische Informationen, Heft 31, Institut für Technik in Gartenbau und Landwirtschaft, Universität Hannover

Sabeh NC (2007) Evaluation and minimizing of water use by greenhouse evaporative cooling systems in a semi-arid climate. Dissertation, Department of Agricultural and Biosystems Engineering, University of Arizona

Sabeh NC, Giacomelli GA, Kubota C (2006) Water use for pad and fan cooling of a greenhouse in a semi-arid climate. Acta Hortic 719:409–416

Santos B, Rios D, Nasco R (2006) Climate conditions in tomato screenhouses in Tenerife (Canary Islands). Acta Hortic 719:215–221

Sase S (2006) Air movement and climate uniformity in ventilated greenhouses. Acta Hortic 719:313–323

Sase S (2008) Greenhouse energy and resource use – cooling in semi-arid greenhouses. Acta Hortic 797:299–306

Sase S, Christianson LL (1990) Screening greenhouses – some engineering considerations. ASAE paper No. NABEC 90 – 201, p 1–13

Sase S, Moriyama H, Kurata K, Sabeh N, Romero P, Giacomelli GA, Kubota C, Hayashi M (2006) Effect of natural ventilation on relative humidity and water use for fog cooling in a semiarid greenhouse. Acta Hortic 719:385–392

Sase S, Okushima L, Suzuki M (2008) Effect of insect screen configuration on natural ventilation in a single span glasshouse. Acta Hortic 801:237–243

Schultz W (1996a) Einfluss der spektralen Durchlässigkeit von Gewächshäusern auf die Pflanzenqualität. In: Neue Wege beim Kunststoffeinsatz im Landbau. KTBL Arbeitspapier 234, pp 18–27

Schultz W (1996b) Welche Folie für das Gewächshaus. Spektrale Durchlässigkeit und Einfluss auf Klima und Pflanzenqualität. Dtsch Gartenbau 10:618–624

Schultz W (1997) Verfahren zur Untersuchung und Bewertung des Kondensationsverhaltens von Gewächshausfolien. Gartenbautechnische Information, Heft 41. Institute for Horticultural Engineering, University of Hannover

Schüsseler P, von Zabeltitz C (2004) Umweltgerechte Techniken in der Pflanzenproduktion. Ulmer Verlag, Stuttgart

SETAG (1993) Guidelines for Life Cycle Assessment. A code of practice. SETAC, Brussels

Shahak Y, Gussakovsky EE (2004) Color nets: crop protection and light quality manipulation in one technology. Acta Hortic 659:143–151

Shahak Y, Gal E, Ben-Yakir D (2008) Photoselective shade netting integrated in greenhouse technology for improved performance of vegetable and ornamental crops. Acta Hortic 797:75–80

Shahak Y (2008) Photoselective netting for improved performance of horticultural crops. A review of ornamental and vegetable studies carried out in Israel. Acta Hortic 770:161–168

Shahak Y, Ratner K, Zur N, Offir Y, Matan E, Yehezkel H, Messika Y, Posalski I, Ben-Yakir D (2009) Photoselective netting: an emerging approach in protected agriculture. Acta Hortic 807:79–84

Short TH (2004) Greenhouse structures and plant growth control systems for energy efficient crop production. Acta Hortic 633:17–27

Soriano T, Montero JI, Sancez-Guerrero MC, Modrano E, Anton A, Hernandez J, Morales MJ, Castilla N (2004) A study of direct solar radiation transmission in asymmetrical multi span greenhouses using scale models and simulation models. Biosystems Eng 88(2):243–253

Soriano T, Hernandez J, Montero JI, Castilla N (2009) Solar radiation transmission in Mediterranean plastic greenhouses. Acta Hortic 807:73–78

Stanghellini C, Kempkes FLK, Knies P (2003) Enhancing environmental quality in agricultural systems. Acta Hortic 609:277–289

Stanghellini C, Incrossi L, Gasquez JC, Dimauro B (2008) Carbon dioxide concentration in Mediterranean greenhouses: how much lost production? Acta Hortic 801:1541–1549

Stanghellini C, Kempkes K, Incrossi L (2009) Carbon dioxide fertilization in Mediterranean greenhouses: when and how is it economical? Acta Hortic 807:135–140

Strauch KH (1985a) Geschlossene Gewächshaussysteme mit integrierter solarer Wasserentsalzungsanlage für aride Gebiete. Gartenbautechnische Informationen Heft 22, Institut für Technik in Gartenbau und Landwirtschaft, Universität Hannover

Strauch KH (1985b) A closed system greenhouse with integrated solar desalination for arid regions. Acta Hortic 170:29–36

Tanny J, Cohen SH, Teitel M (2003) Screenhouse microclimate and ventilation: an experimental study. Biosystems Eng 4(3):331–341

Tanny J, Teitel M, Barak M, Esquira Y, Amir R (2008) The effect of height on screenhouse microclimate. Acta Hortic 801:107–114

Tantau HJ (1983) Heizungsanlagen im Gartenbau. Verlag Eugen Ulmer, Stuttgart

Tantau HJ (1998) Heizungssysteme im Gewächshaus. In: Erdgas im Gartenbau. Ruhrgas AG, Essen

Teitel M (2006) The effect of screens on the micro climate of greenhouses and screen houses – a review. Acta Hortic 719:575–586

Teitel M (2007) The effect of screened openings on greenhouse microclimate. Agric Forest Meteorol 143:159–175

Teitel M, Haim Y, Tanny T, Seginer I (2008a) Flow through inclined screens: comparison between experiments and CFD results. Acta Hortic 797:139–144

Teitel M, Liron O, Haim Y, Seginer I (2008b) Flow through inclined and concertina-shape screens. Acta Hortic 801:99–105

Teitel M, Dvorkin D, Tanny J (2009) Flow patterns downstream screens of different porosities. Greensys June 2009, Quebec, Canada (in press)

Thomas J (1994) Untersuchungen zur Optimierung und Eignung wassergefüllter Folienschläuche als passive Kollektor- und Speicherelemente im Gewächshaus. Gartenbautechnische Informationen, Heft 38, Institut für Technik in Gartenbau und Landwirtschaft, Universität Hannover
Toida H, Kozai T, Ohyama K, Handarto H (2006) Enhancing fog evaporation using upward air stream to improve greenhouse cooling performance. Biosystems Eng 93(2):205–211
Torita H, Satov H (2007) Relationship between shelterbelt structure and mean wind reduction. Agric For Meteorol 145:186–194
Trewartha G, Lyle T, Horn H (1980) An Introduction to Climate. McGraw-Hill, New York
Van der Velden NJA, Janse J (2004) Sustainability of greenhouse fruit vegetables: Spain versus the Netherlands: development of a monitoring system. Acta Hortic 655:275–281
Verhaegh AP (1996) Efficientje van energi en Gewasbeschermingsmiddelen tomaten en rosen in kassen. Publikatie 4.142. Landbouw-Economish Instituut, Den Haag
Verlodt H (1999) Climate and vegetable management under greenhouse conditions. In: Greenhouse vegetable production and protection in Albania. Proceedings of FAO workshop in Tirana, 9–11 March 1999
Von Elsner B (1990) Gewächshausheizung mit Thermalwasser in Südtunesien. Dtsch Gartenbau 30:1943–1944
Von Elsner B (2005a) Vergleich von Gewächshausbedachungen (Comparison of greenhouse cladding material). Annual meeting of GKL (German Society for Plastics in Agriculture)
Von Elsner B (2005b) Interference pigments in photoselective shading. Acta Hortic 711:412–417
Von Elsner B, Briassoulis D, Waaijenberg D, Mistriotis A, von Zabeltitz C, Gratraud J, Russo G, Suay-Cortes R (2000a) Review of structural and functional characteristics of greenhouses in European Union countries. Part I: Design requirements. J Agric Eng Res 75:1–16
Von Elsner B, Briassoulis D, Waaijenberg D, Mistriotis A, von Zabeltitz C, Gratraud J, Russo G, Suay-Cortes R (2000b) Review of structural and functional characteristics of greenhouses in European Union countries. Part II: Typical designs. J Agric Eng Res 75:111–126
Von Zabeltitz C (ed) (1982) Energieeinsparung und alternative Energiequellen im Gartenbau. Verlag Eugen Ulmer, Stuttgart
Von Zabeltitz C (1984) Gewächshausheizung mit Sonnenenergie. Grundlagen der Landtechnik 34(4):158–162
Von Zabeltitz C (1985) New construction of a plastic film greenhouse. Acta Hortic 170:25–28
Von Zabeltitz C (1986a) Gewächshäuser, Planung und Bau. (Greenhouses, design and construction). Ulmer Verlag, Stuttgart
Von Zabeltitz C (1986b) Greenhouse heating with solar energy. Energ Agr 5:111–120
Von Zabeltitz C (ed) (1987) Greenhouse heating with solar energy. REUR Technical Series 1. FAO Regional Office for Europe, Rome, p 199
Von Zabeltitz C (ed) (1988a) Energy conservation and renewable energies for greenhouse heating. REUR Technical Series 3. FAO Regional Office for Europe, Rome
Von Zabeltitz C (1988b) Cooling and heating of greenhouses in Kuwait. Report of a consultancy mission to Kuwait. Food and Agriculture Organization of the United Nations, Rome
Von Zabeltitz C (1988c) Greenhouses for warmer climates. Plasticulture 80(4):39–50
Von Zabeltitz C (ed) (1989) Passive solar heating of greenhouses with water-filled polyethylene tubes. Gartenbautechnische Informationen, Heft 30, Institut für Technik in Gartenbau und Landwirtschaft, Universität Hannover
Von Zabeltitz C (1990a) Appropriate greenhouse constructions for mild climates. J Agric Trop Subtrop 91:65–77
Von Zabeltitz C (1990b) Experiences with a new plastic film greenhouse. Acta Hortic 281:75–82
Von Zabeltitz C (1994) Measurements of climate in different greenhouse types on Seychelles. Report to FAO Rome
Von Zabeltitz C (1996a) Protective shelters for Seychelles. In: Paper presented on the Workshop for the Improved Vegetable Production Project Seychelles, Seychelles, 22 May 1996
Von Zabeltitz C (1996b) Tropical shelter structures. Farmers handbook, vol 10. Ministry of Agriculture and Marine Resources, MAMR, Seychelles

Von Zabeltitz C (1997a) Die richtige Gewächshauslüftung wählen. Gartenbau 42:6–7

Von Zabeltitz C (1997) Umweltschonung und technischer Fortschritt – Widerspruch oder Notwendigkeit. Farewell lecture, University of Hannover, Gartenbautechnische Informationen

Von Zabeltitz C (1997c) Why greenhouses in the tropics. Plasticulture 113(1):21–26

Von Zabeltitz C (1998) Greenhouse constructions for protected cultivation in Malawi. Consultant report for GTZ, Germany

Von Zabeltitz C (1999) Greenhouse structures. In: Stanhill G, Zvi Enoch H (eds) Greenhouse ecosystems. Ecosystems of the world, vol 20. Elsevier, Amsterdam, pp 17–69

Von Zabeltitz C (2000) Greenhouses for India. In: Singh HP, Dadlani NK (eds) Commercial Floriculture. FAO/Ministry of Agriculture, Rome/New Delhi, pp 173–191

Von Zabeltitz C (2004) Umweltgerechte Techniken in der Pflanzenproduktion. Ulmer Verlag, Stuttgart

Von Zabeltitz C, Meyer J, Dickob D, Dierend A, Hölscher TH (1989) Planungshilfen für den Einsatz elektrischer Verbraucher im Unterglasgartenbau (Planning of electricity use in protected cultivation). Bericht Nr. 13. Arbeitsgemeinschaft Elektrizitätsanwendung in der Landwirtschaft, Essen

Von Zabeltitz C, Stubben D, Bischoff H (1992) Stoff- und Energiebilanz für Gewächshauskonstruktionen. Dtsch Gartenbau 38:2248–2250

Von Zabeltitz C, Lüchow K (1993) Investigation of a spray cooling system for greenhouses. In: Proceedings of Regional Consultative Meeting on Greenhouse Production in the Mediterranean Region, Agadir

Von Zabeltitz C, Baudoin W (1999) Greenhouses and shelter structures for tropical regions. FAO plant production and protection paper no. 154, p 122

Waaijenberg D (1990) Standard for film covered greenhouses. Acta Hortic 281:129–137

Waaijenberg D (1995) Chapter 4.2: Construction. In: Bakkaer JC, Bot GPA, Challer H, van de Braak NJ (eds) Greenhouse climate control. Wageningen Pers, Wageningen

Waaijenberg D (2006) Design, construction and maintenance of greenhouse structures. Acta Hortic 710:31–42

Waaijenberg D, Denkov D (1992) Reduction of wind damage to tunnel greenhouses by structural improvements. In: Proceedings of workshop on European greenhouse standards, Barcelona, Technical series 25, FAO, Rome, pp 48–56

Weimann G (1985) Light transmissivity of different film coverings on greenhouses. Acta Hortic 170:143–146

Weimann G (1986) Untersuchungen zu neuen Foliengewächshauskonzeptionen (Investigation of new plastic film greenhouse systems). Gartenbautechnische Informationen, Heft 24, Institut für Technik in Gartenbau und Landwirtschaft. Universität Hannover, p 152

Willits DH (2003) Cooling fan ventilated greenhouses. A modeling study. Biosystems Eng 84 (3):315–329

Willits DH (2006) Fan ventilated greenhouse cooling: some considerations for design. Acta Hortic 719:83–95

Wilson JD (1987) On the choice of a windbreak porosity profile. Bound-Lay Meteorol 38:37–49

Index

MIX
Papier aus verantwortungsvollen Quellen
Paper from responsible sources
FSC® C105338

If you have any concerns about our products,
you can contact us on
ProductSafety@springernature.com

In case Publisher is established outside the EU,
the EU authorized representative is:
Springer Nature Customer Service Center GmbH
Europaplatz 3, 69115 Heidelberg, Germany

Printed by Libri Plureos GmbH
in Hamburg, Germany